Practice Book

CONCEPTUAL Physics Fundamentals

Paul G. Hewitt
City College of San Francisco

San Francisco Boston New York
Cape Town Hong Kong London Madrid Mexico City
Montreal Munich Paris Singapore Sydney Tokyo Toronto

Publisher: Adam Black
Project Editor: Chandrika Madhavan
Production Supervisor: Mary O'Connell
Main Text Cover Designer: Yvo Riezebos Design
Supplement Cover Designer: 17th Street Studios
Cover Printer: Phoenix Color
Text Printer: Bind-Rite Graphics

ISBN-10: 0-321-53074-8
ISBN-13: 978-0-321-53074-5

2 3 4 5 6 7 8 9 10—BRG— 10 09 08
www.aw-bc.com/physics

Welcome to the
CONCEPTUAL PHYSICS FUNDAMENTALS PRACTICE BOOK

These practice pages supplement *Conceptual Physics Fundamentals.* Their purpose is as the name implies—practice—not testing. You'll find it is easier to learn physics by *doing* it—by practicing. AFTER you've worked through a page, check your responses with the reduced pages with answers beginning on page 131.

Pages 197 to 239 show answers to the odd-numbered exercises and solutions to the problems in the textbook.

Enjoy your physics!

Table of Contents

Name ______________________________ Date ____________

CONCEPTUAL Physics FUNDAMENTALS PRACTICE PAGE

Chapter 1 About Science

Making Hypotheses

The word science comes from Latin, meaning "to know." The word *hypothesis* comes from Greek, "under an idea." A hypothesis (an educated guess) often leads to new knowledge and may help to establish a theory.

Examples:

1. It is well known that objects generally expand when heated. An iron plate gets slightly bigger, for example, when placed in an oven. But what of a hole in the middle of the plate? One friend may say the size of the hole will increase, and another may say it will decrease.

a. What is your hypothesis about hole size, and if you are wrong, is there a test for finding out?

____

b. There are often several ways to test a hypothesis. For example, you can perform a physical experiment and witness the results yourself, or you can use the library or internet to find the reported results of other investigators. Which of these two methods do you favor, and why?

____

____

2. Before the time of the printing press, books were hand-copied by scribes, many of whom were monks in monasteries. There is the story of the scribe who was frustrated to find a smudge on an important page he was copying. The smudge blotted out part of the sentence that reported the number of teeth in the head of a donkey. The scribe was very upset and didn't know what to do. He consulted with other scribes to see if any of their books stated the number of teeth in the head of a donkey. After many hours of fruitless searching through the library, it was agreed that the best thing to do was to send a messenger by donkey to the next monastery and continue the search there. What would be your advice?

____

Making Distinctions

Many people don't seem to see the difference between a thing and the abuse of the thing. For example, a city council that bans skateboarding may not distinguish between skateboarding and reckless skateboarding. A person who advocates that a particular technology be banned may not distinguish between that technology and the abuses of that technology. There's a difference between a thing and the abuse of the thing.

On a separate sheet of paper, list other examples where use and abuse are often not distinguished. Compare your list with others in your class.

Chapter 1 About Science

Pinhole Formation

Look carefully on the round spots of light on the shady ground beneath trees. These are *sunballs*, which are images of the sun. They are cast by openings between leaves in the trees that act as pinholes. (Did you make a pinhole "camera" back in middle school?) Large sunballs, several centimeters in diameter or so, are cast by openings that are relatively high above the ground, while small ones are produced by closer "pinholes." The interesting point is that the ratio of the diameter of the sunball to its distance from the pinhole is the same ratio of the Sun's diameter to its distance from the pinhole. We know the Sun is approximately 150,000,000 km from the pin-hole, so careful measurements of of the ratio of diameter/distance for a sunball leads you to the diameter of the Sun. That's what this page is about. Instead of measuring sunballs under the shade of trees on a sunny day, make your own easier-to-measure sunball.

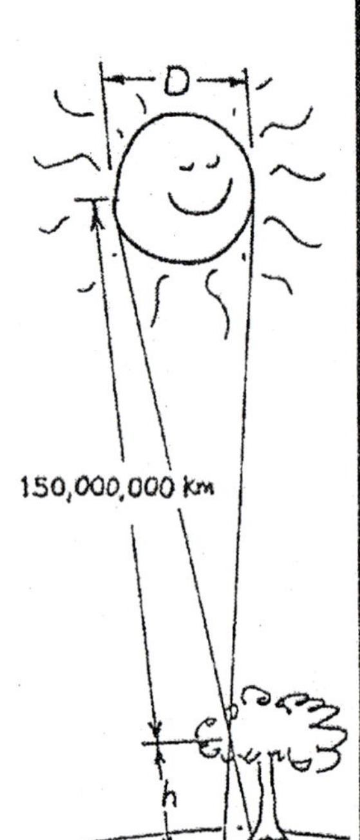

1. Poke a small hole in a piece of card. Perhaps an index card will do, and poke the hole with a sharp pencil or pen. Hold the card in the sunlight and note the circular image that is cast. This is an image of the Sun. Note that its size doesn't depend on the size of the hole in the card, but only on its distance. The image is a circle when cast on a surface perpendicular to the rays—otherwise it's "stretched out" as an ellipse.

2. Try holes of various shapes; say a square hole, or a triangular hole. What is the shape of the image when its distance from the card is large compared with the size of the hole? Does the shape of the pinhole make a difference?

____

____

3. Measure the diameter of a small coin. Then place the coin on a viewing area that is perpendicular to the Sun's rays. Position the card so the image of the sunball exactly covers the coin. Carefully measure the distance between the coin and the small hole in the card. Complete the following:

$$\frac{\text{Diameter of sunball}}{\text{Distance of pinhole}} = \underline{\qquad\qquad}$$

With this ratio, estimate the diameter of the Sun. Show your work on a separate piece of paper.

WHAT SHAPE DO SUNBALLS HAVE DURING A PARTIAL ECLIPSE OF THE SUN?

4. If you did this on a day when the Sun is partially eclipsed, what shape of image would you expect to see?

____

Name ______________________________ Date ____________

CONCEPTUAL Physics FUNDAMENTALS PRACTICE PAGE

Chapter 2 Atoms

Atoms and Atomic Nuclei

ATOMS ARE CLASSIFIED BY THEIR ATOMIC NUMBER, WHICH IS THE SAME AS THE NUMBER OF ____________ IN THE NUCLEUS.

TO CHANGE THE ATOMS OF ONE ELEMENT INTO THOSE OF ANOTHER, ____________ MUST BE ADDED OR SUBTRACTED!

Use the periodic table in your text to help you answer the following questions.

1. When the atomic nuclei of hydrogen and lithium are squashed together (nuclear fusion) the element that is produced is

   ____

2. When the atomic nuclei of a pair of lithium nuclei are fused, the element produced is

   ____

3. When the atomic nuclei of a pair of aluminum nuclei are fused, the element produced is

   ____

4. When the nucleus of a nitrogen atom absorbs a proton, the resulting element is

   ____

5. What element is produced when a gold nucleus gains a proton? ____________________

6. What element is produced when a gold nucleus loses a proton? ____________________

7. What element is produced when a uranium nucleus ejects an elementary particle composed of two protons and two neutrons?

   ____

8. If a uranium nucleus breaks into two pieces (nuclear fission) and one of the pieces is zirconium (atomic number 40), the other piece is the element

   ____

I LIKE THE WAY YOUR ATOMS ARE PUT TOGETHER!

SIGH

9. Which has more mass, a nitrogen molecule (N_2) or an oxygen molecule (O_2)?

   ____

10. Which has the greater number of atoms, a gram of helium or a gram of neon?

    ____

Chapter 2 Atoms
Subatomic Particles

Three fundamental particles of the atom are the ______________, ______________, and ______________. At the center of each atom lies the atomic ______________ which consists of ______________ and ______________. The atomic number refers to the number of ______________ in the nucleus. All atoms of the same element have the same number of ______________, hence, the same atomic number.

Isotopes are atoms that have the same number of ______________ but a different number of ______________. An isotope is identified by its atomic mass number, which is the total number of ______________ and ______________ in the nucleus. A carbon isotope that has 6 ______________ and ______________ is identified as carbon-12, where 12 is the atomic mass number. A carbon isotope having 6 ______________ and 8 ______________, on the other hand is carbon-14.

1. *Complete the following table*:

ISOTOPE	ELECTRONS	NUMBER OF PROTONS	NEUTRONS
Hydrogen-1	1		
Chlorine-36		17	
Nitrogen-14			7
Potassium-40	19		
Arsenic-75		33	
Gold-197			118

2. Which results in a more valuable product—*adding* or *subtracting* protons from gold nuclei?

____

3. Which has more mass, a lead atom or a uranium atom?

____

4. Which has a greater number of atoms, a gram of lead or a gram of uranium?

____

Of every 200 atoms in our bodies, 126 are hydrogen, 51 are oxygen, and just 19 are carbon. In addition to carbon we need iron to manufacture hemoglobin, cobalt for the creation of vitamin B-12, potassium and a little sodium for our nerves, and molybdenum, manganese, and vanadium to keep our enzymes purring. Ah, we'd be nothing without atoms!

Hewitt Drew it!

Name ______________________________ Date ____________

CONCEPTUAL Physics FUNDAMENTALS PRACTICE PAGE

Chapter 3 Equilibrium and Linear Motion

Static Equilibrium

1. Little Nellie Newton wishes to be a gymnast and hangs from a variety of positions as shown. Since she is not accelerating, the net force on her is zero. That is, $\Sigma F = 0$. This means the upward pull of the rope(s) equals the downward pull of gravity. She weighs 300 N. Show the scale reading(s) for each case.

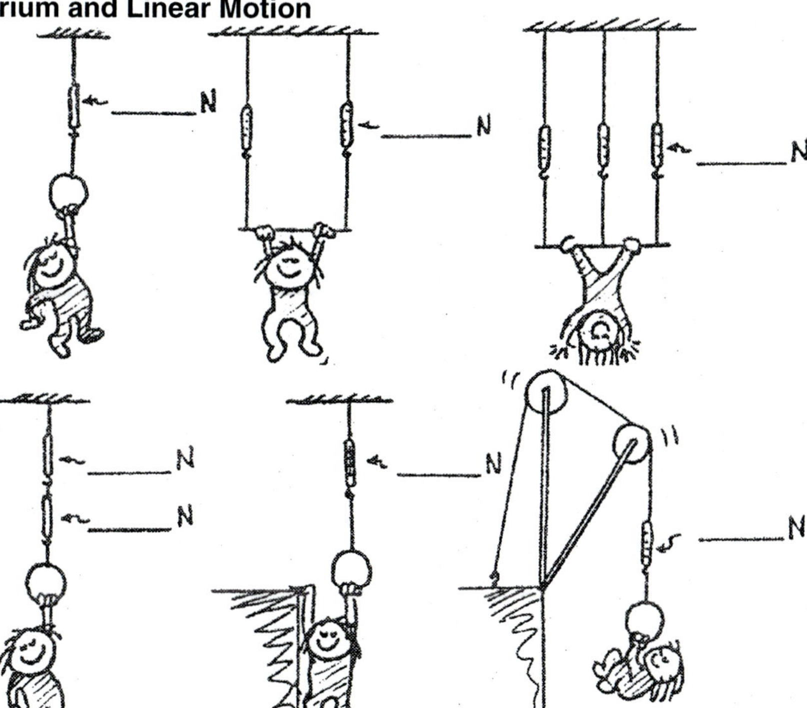

2. When Burl the painter stands in the exact middle of his staging, the left scale reads 600 N. Fill in the reading on the right scale. The total weight of Burl and staging must be

_________ N.

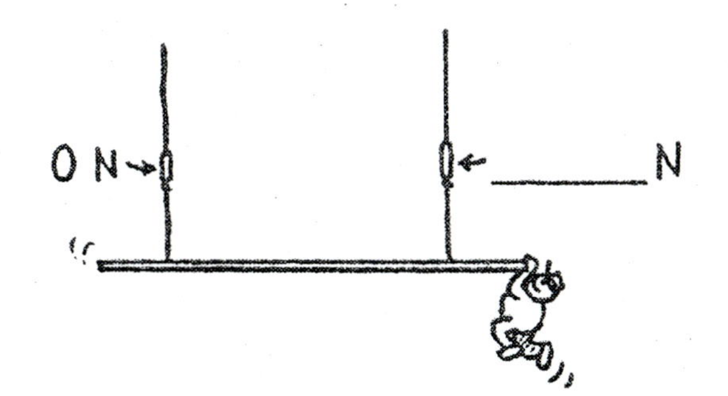

3. Burl stands farther from the left. Fill in the reading on the right scale.

4. In a silly mood, Burl dangles from the right end. Fill in the reading on right scale.

Chapter 3 Equilibrium and Linear Motion

The Equilibrium Rule: $\Sigma F = 0$

1. Manuel weighs 1000 N and stands in the middle of a board that weighs 200 N. The ends of the board rest on bathroom scales. (We can assume the weight of the board acts at its center.) Fill in the correct weight reading on each scale.

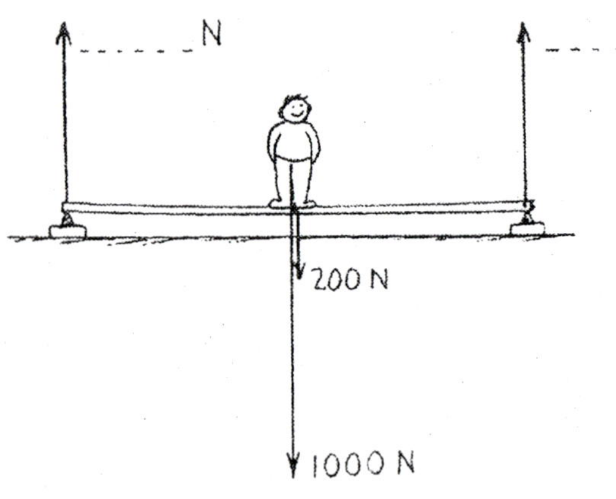

850 N

______ N

200 N

1000 N

2. When Manuel moves to the left as shown, the scale closest to him reads 850 N. Fill in the weight for the far scale.

3. A 12-ton truck is one-quarter the way across a bridge that weighs 20 tons. A 13-ton force supports the right side of the bridge as shown. How much support force is on the left side?

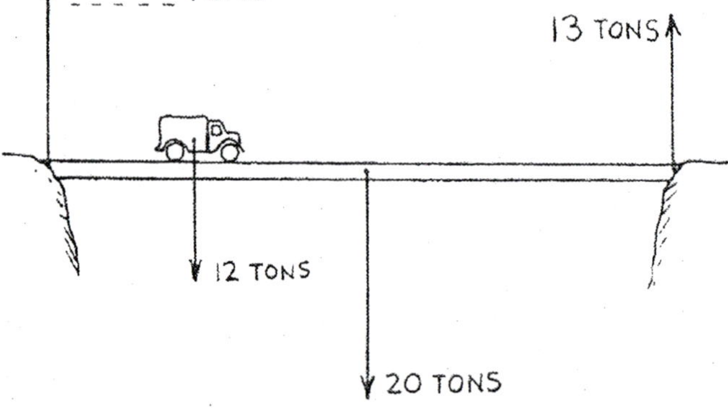

4. A 1000-N crate resting on a surface is connected to a 500-N block through a frictionless pulley as shown. Friction between the crate and surface is enough to keep the system at rest. The arrows show the forces that act on the crate and the block. Fill in the magnitude of each force.

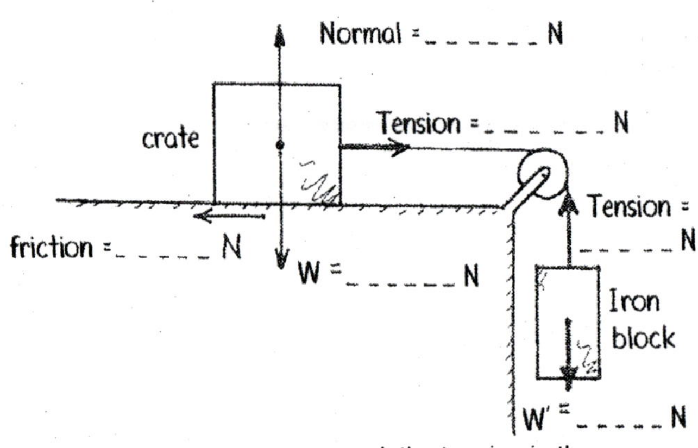

5. If the crate and block in the preceding question move at constant speed, the tension in the rope

   [is the same] [increases] [decreases].

   The sliding system is then in [static equilibrium] [dynamic equilibrium].

Name ______________________ Date __________

Chapter 3 Equilibrium and Linear Motion

Vectors and Equilibrium

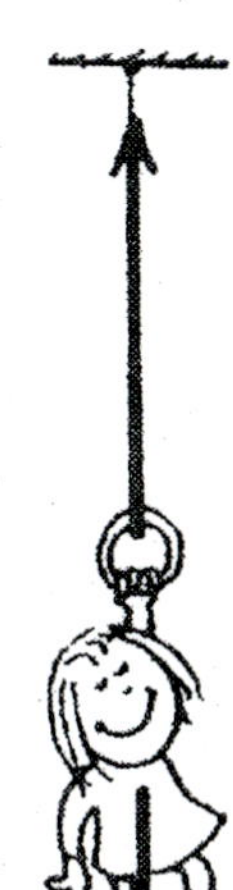

1. Nellie Newton dangles from a vertical rope in equilibrium: $\Sigma F = 0$. The tension in the rope (upward vector) has the same magnitude as the downward pull of gravity (downward vector).

2. Nellie is supported by two vertical ropes. Draw tension vectors to scale along the direction of each rope.

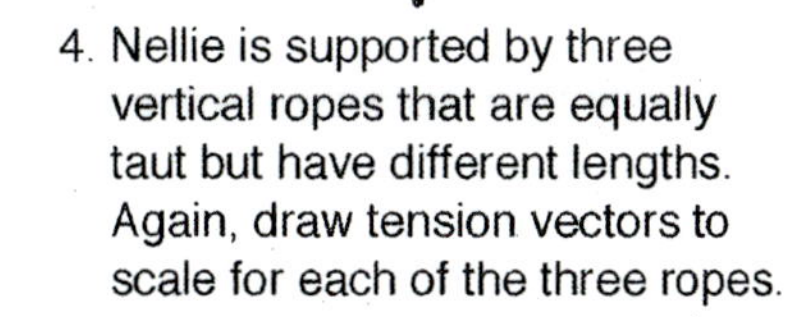

3. This time the vertical ropes have different lengths. Draw tension vectors to scale for each of the two ropes.

4. Nellie is supported by three vertical ropes that are equally taut but have different lengths. Again, draw tension vectors to scale for each of the three ropes.

Circle the correct answer:

5. We see that tension in a rope is [dependent on] [independent of] the length of the rope. So the length of a vector representing rope tension is [dependent on] [independent of] the length of the rope.

Rope tension does depend on the angle the rope makes with the vertical, as Practice Pages for Chapter 6 will show!

Hewitt Drew it!

Name ____________________ Date __________

CONCEPTUAL **Physics** FUNDAMENTALS PRACTICE PAGE

Chapter 3 Equilibrium and Linear Motion

Free Fall Speed

1. Aunt Minnie gives you \$10 per second for 4 seconds. How much money do you have after 4 seconds? __________

2. A ball dropped from rest picks up speed at 10 m/s per second. After it falls for 4 seconds, how fast is it going? __________

3. You have \$20, and Uncle Harry gives you \$10 each second for 3 seconds. How much money do you have after 3 seconds? __________

4. A ball is thrown straight down with an initial speed of 20 m/s. After 3 seconds, how fast is it going? __________

5. You have \$50, and you pay Aunt Minnie \$10/second. When will your money run out? __________

6. You shoot an arrow straight up at 50 m/s. When will it run out of speed? __________

7. So what will be the arrow's speed 5 seconds after you shoot it? __________

8. What will its speed be 6 seconds after you shoot it? __________

   Speed after 7 seconds? __________

Free Fall Distance

1. Speed is one thing; distance is another. How high is the arrow when you shoot up at 50 m/s when it runs out of speed? __________

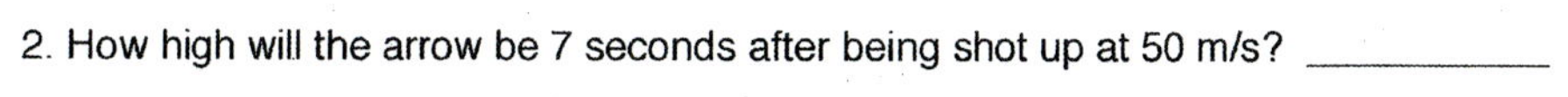

2. How high will the arrow be 7 seconds after being shot up at 50 m/s? __________

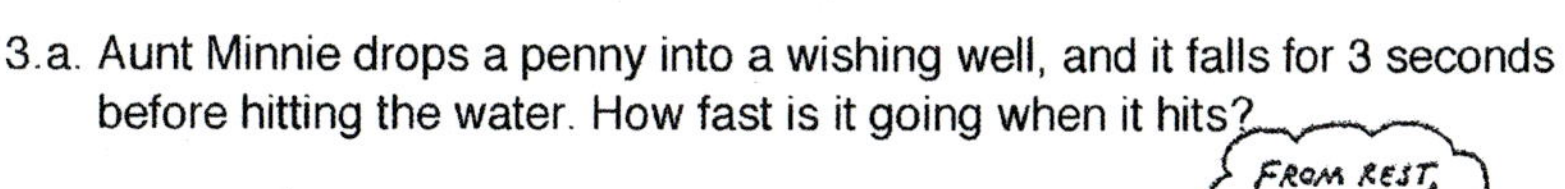

3. a. Aunt Minnie drops a penny into a wishing well, and it falls for 3 seconds before hitting the water. How fast is it going when it hits? __________

   b. What is the penny's average speed during its 3-second drop? __________

   c. How far down is the water surface? __________

FROM REST, $v = 10t$ $d = 5t^2$

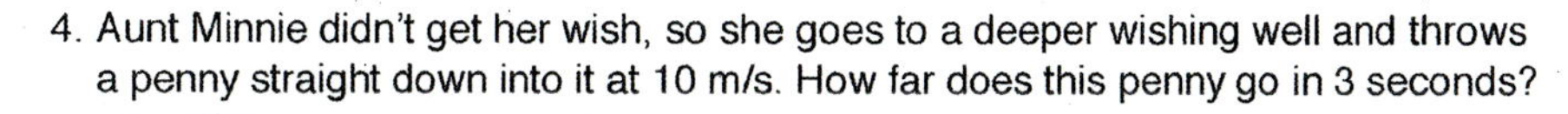

4. Aunt Minnie didn't get her wish, so she goes to a deeper wishing well and throws a penny straight down into it at 10 m/s. How far does this penny go in 3 seconds? __________

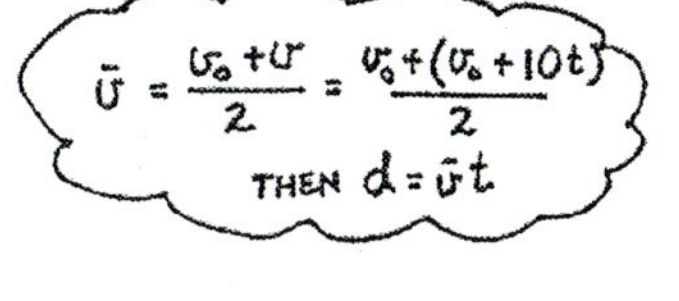

$\bar{v} = \frac{v_0 + v}{2} = \frac{v_0 + (v_0 + 10t)}{2}$
THEN $d = \bar{v}t$

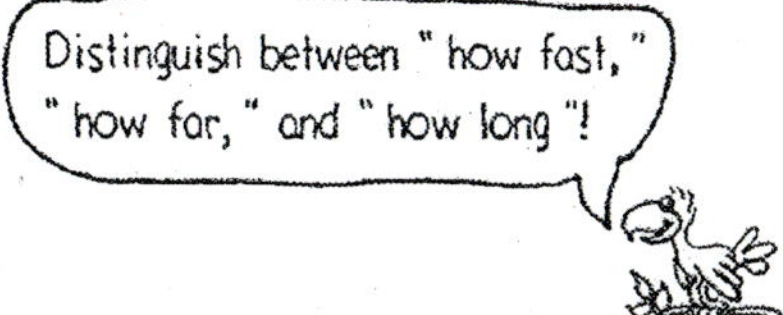

Distinguish between "how fast," "how far," and "how long"!

Chapter 3 Equilibrium and Linear Motion

Acceleration of Free Fall

A rock dropped from the top of a cliff picks up speed as it falls. Pretend that a speedometer and odometer are attached to the rock to indicate readings of speed and distance at 1-second intervals. Both speed and distance are zero at time = zero (see sketch). Note that after falling 1 second, the speed reading is 10 m/s and the distance fallen is 5 m. The readings of succeeding seconds of fall are not shown and are left for you to complete. So draw the position of the speedometer pointer and write in the correct odometer reading for each time. Use $g = 10\text{ m/s}^2$ and neglect air resistance.

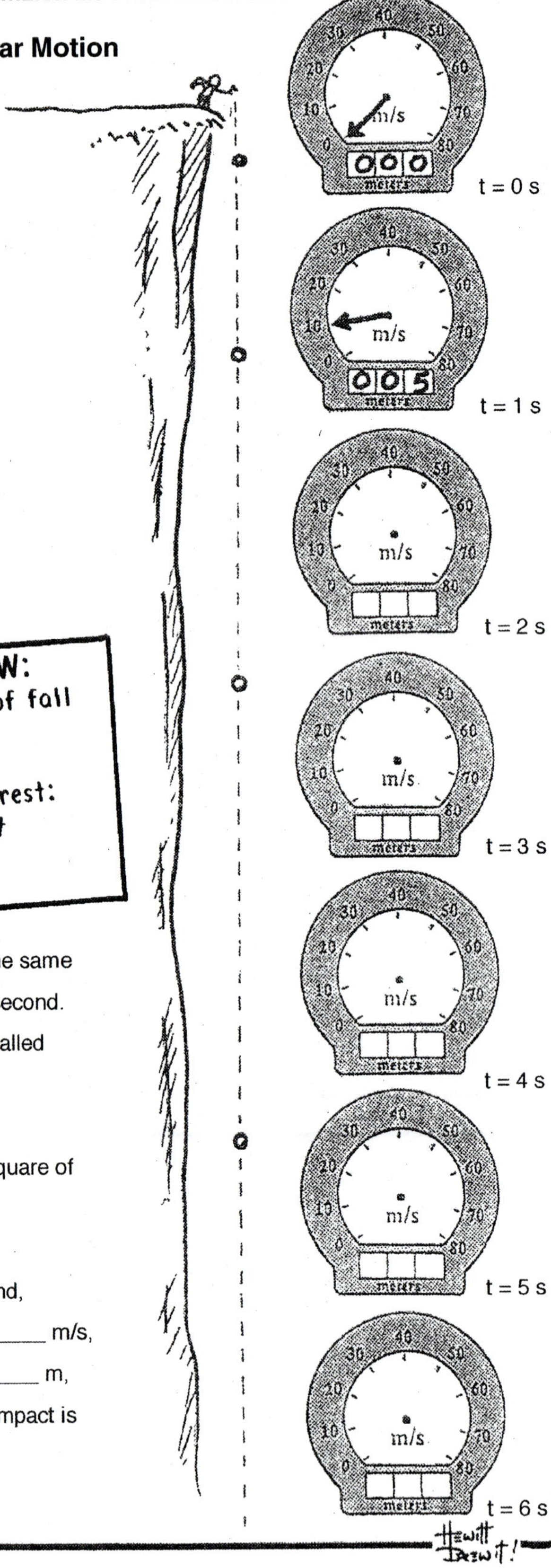

YOU NEED TO KNOW:

Instantaneous speed of fall from rest:

$$v = gt$$

Distance fallen from rest:

$$d = v_{average}t$$

or

$$d = \frac{1}{2}gt^2$$

1. The speedometer reading increases the same amount, _______________ m/s, each second. This increase in speed per second is called _______________.

2. The distance fallen increases as the square of the _______________.

3. If it takes 7 seconds to reach the ground, then its speed at impact is _______________ m/s, the total distance fallen is _______________ m, and its acceleration of fall just before impact is _______________ m/s².

m/s 000 meters t = 0 s

m/s 005 meters t = 1 s

m/s meters t = 2 s

m/s meters t = 3 s

m/s meters t = 4 s

m/s meters t = 5 s

m/s meters t = 6 s

Hewitt Drew it!

Name ______________________ Date ______________

CONCEPTUAL *Physics* **FUNDAMENTALS PRACTICE PAGE**

Chapter 3 Equilibrium and Linear Motion

Hang Time

Some athletes and dancers have great jumping ability. When leaping, they seem to momentarily "hang in the air" and defy gravity. The time that a jumper is airborne with feet off the ground is called hang time. Ask your friends to estimate the hang time of the great jumpers. They may say two or three seconds. But surprisingly, the hang time of the greatest jumpers is most always less than 1 second! A longer time is one of many illusions we have about nature.

To better understand this, find the answers to the following questions:

1. If you step off a table and it takes one-half second to reach the floor, what will be the speed when you meet the floor?

   ____

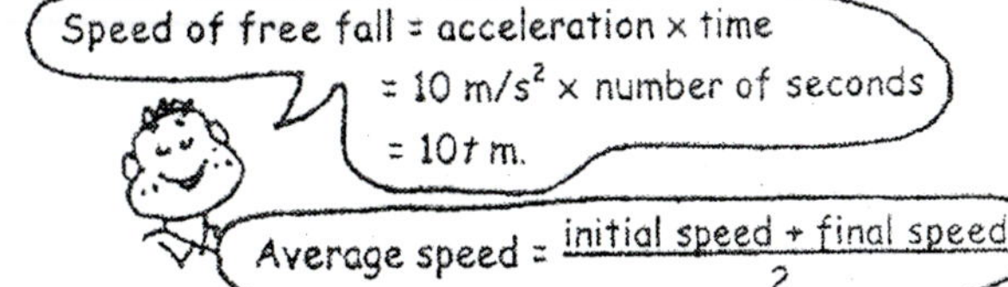

Speed of free fall = acceleration × time
= 10 m/s² × number of seconds
= 10t m.
Average speed = (initial speed + final speed) / 2

2. What will be your average speed of fall?

   ____

Distance = average speed × time.

3. What will be the distance of fall?

   ____

4. So how high is the surface of the table above the floor? ______________

Jumping ability is best measured by a standing vertical jump. Stand facing a wall with feet flat on the floor and arms extended upward. Make a mark on the wall at the top of your reach. Then make your jump and at the peak make another mark. The distance between these two marks measures your vertical leap. If it's more than 0.6 meters (2 feet), you're exceptional.

5. What is your vertical jumping distance? ______________

6. Calculate your personal hang time using the formula $d = 1/2\ gt^2$. (Remember that hang time is the time that you move upward + the time you return downward.)

Almost anybody can safely step off a 1.25-m (4-feet) high table. Can anybody in your school jump from the floor up onto the same table?

No way!

There's a big difference in how high you can reach and how high you raise your "center of gravity" when you jump. Even basketball star Michael Jordan in his prime couldn't quite raise his body 1.25 meters high, although he could easily reach higher than the more-than-3-meter high basket.

Here we're talking about vertical motion. How about running jumps? We'll see in Chapter 10 that the height of a jump depends only on the jumper's vertical speed at launch. While airborne, the jumper's horizontal speed remains constant while the vertical speed undergoes acceleration due to gravity. While airborne, no amount of leg or arm pumping or other bodily motions can change your hang time.

Hewitt Drew it!

Chapter 3 Equilibrium and Linear Motion

Non-Accelerated Motion

1. The sketch shows a ball rolling at constant velocity along a level floor. The ball rolls from the first position shown to the second in 1 second. The two positions are 1 meter apart. Sketch the ball at successive 1-second intervals all the way to the wall (neglect resistance).

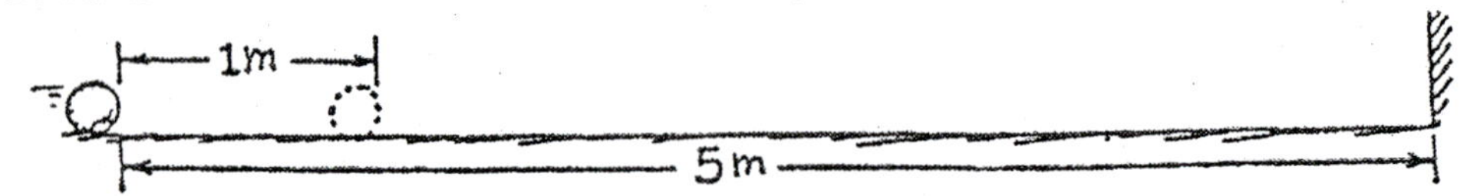

a. Did you draw successive ball positions evenly spaced, farther apart, or closer together? Why?

____

b. The ball reaches the wall with a speed of __________ m/s and takes a time of __________ seconds.

2. Table I shows data of sprinting speeds of some animals. Make whatever computations necessary to complete the table.

TABLE I

ANIMAL	DISTANCE	TIME	SPEED
CHEETAH	75 m	3 s	25 m/s
GREYHOUND	160 m	10 s	
GAZELLE	1 km		100 km/h
TURTLE		30 s	1 cm/s

Accelerated Motion

3. An object starting from rest gains a speed $v = at$ when it undergoes uniform acceleration. The distance it covers is $d = 1/2\ at^2$. Uniform acceleration occurs for a ball rolling down an inclined plane. The plane below is tilted so a ball picks up a speed of 2 m/s each second; then its acceleration $a = 2\ \text{m/s}^2$. The positions of the ball are shown at 1-second intervals. Complete the six blank spaces for distance covered and the four blank spaces for speeds.

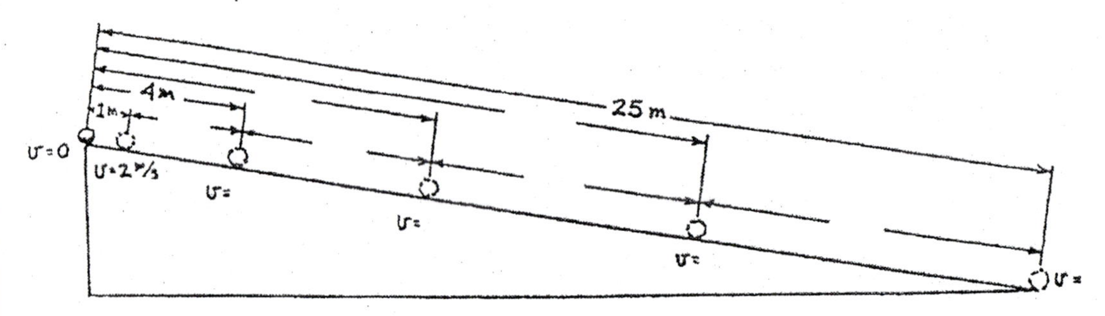

a. Do you see that the total distance from the starting point increases as the square of the time? This was discovered by Galileo. If the incline were to continue, predict the ball's distance from the starting point for the next 3 seconds.

____

b. Note the increase of distance between ball positions with time. Do you see an odd-integer pattern (also discovered by Galileo) for this increase? If the incline were to continue, predict the successive distances between ball positions for the next 3 seconds.

____

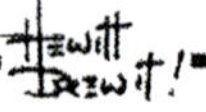

Name ______________________________ Date __________

CONCEPTUAL **Physics** FUNDAMENTALS PRACTICE PAGE

Chapter 4 Newton's Laws of Motion

Mass and Weight

Learning physics is learning the connections among concepts in nature, and also learning to distinguish between closely-related concepts. Velocity and acceleration, previously treated, are often confused. Similarly in this chapter, we find that mass and weight are often confused. They aren't the same! Please review the distinction between mass and weight in your textbook. To reinforce your understanding of this distinction, circle the correct answers below.

Comparing the concepts of mass and weight, one is basic—fundamental—depending only on the internal makeup of an object and the number and kind of atoms that compose it. The concept that is fundamental is [mass] [weight].

The concept that additionally depends on location in a gravitational field is [mass] [weight].

[Mass] [Weight] is a measure of the amount of matter in an object and only depends on the number and kind of atoms that compose it.

It can correctly be said that [mass] [weight] is a measure of "laziness" of an object.

[Mass] [Weight] is related to the gravitational force acting on the object.

[Mass] [Weight] depends on an object's location, whereas [mass] [weight] does not.

In other words, a stone would have the same [mass] [weight] whether it is on the surface of Earth or on the surface of the Moon. However, its [mass] [weight] depends on its location.

On the Moon's surface, where gravity is only about 1/6th Earth gravity [mass] [weight] [both the mass and the weight] of the stone would be the same as on Earth.

While mass and weight are not the same, they are [directly proportional] [inversely proportional] to each other. In the same location, twice the mass has [twice] [half] the weight.

The Standard International (SI) unit of mass is the [kilogram] [newton], and the SI unit of force is the [kilogram] [newton].

In the United States, it is common to measure the mass of something by measuring its gravitational pull to Earth, its weight. The common unit of weight in the U.S. is the [pound] [kilogram] [newton].

When I step on a weighing scale, two forces act on it: a downward pull of gravity, and an upward support force. These equal and opposite forces effectively compress a spring inside the scale that is calibrated to show weight. When in equilibrium, my weight = *mg*.

thanx to Daniela Taylor

Chapter 4 Newton's Laws of Motion
Converting Mass to Weight

Objects with mass also have weight (although they can be weightless under special conditions). If you know the mass of something in **kilograms** and want its weight in **newtons**, at Earth's surface, you can take advantage of the formula that relates weight and mass.

Weight = mass × acceleration due to gravity
$W = mg$

This is in accord with Newton's 2nd law, written as $F = ma$. When the force of gravity is the only force, the acceleration of any object of mass m will be g, the acceleration of free fall. Importantly, g acts as a proportionality constant, 9.8 N/kg, which is equivalent to 9.8 m/s^2.

Sample Question:
How much does a 1-kg bag of nails weigh on Earth?

$W = mg = (1\text{ kg})(9.8\text{ m/s}^2) = 9.8\text{ m/s}^2 = 9.8\text{ N}$.
or simply, $W = mg = (1\text{ kg})(9.8\text{ N/kg}) = 9.8\text{ N}$.

From $F = ma$, we see that the unit of force equals the units [kg x m/s^2]. Can you see the units [m/s^2] = [N/kg]?

Answer the following questions:
Felicia the ballet dancer has a mass of 45.0 kg.

1. What is Felicia's weight in newtons at Earth's surface? ______________

2. Given that 1 kilogram of mass corresponds to 2.2 pounds at Earth's surface, what is Felicia's weight in pounds on Earth? ______________

3. What would be Felicia's mass on the surface of Jupiter? ______________

4. What would be Felicia's weight on Jupiter's surface, where the acceleration due to gravity is 25.0 m/s^2? ______________

Different masses are hung on a spring scale calibrated in newtons.
The force exerted by gravity on 1 kg = 9.8 N.

9.8 N
1 kg

5. The force exerted by gravity on 5 kg = ______________ N.

6. The force exerted by gravity on ______________ kg = 98 N.

Make up your own mass and show the corresponding weight:
The force exerted by gravity on ____________ kg = ____________ N.

By whatever means (spring scales, measuring balances, etc.), find the mass of your physics book. Then complete the table.

OBJECT	MASS	WEIGHT
MELON	1 kg	
APPLE		1 N
BOOK		
A FRIEND	60 kg	

Hewitt Drew it!

Name ______________________________ Date ____________

Chapter 4 Newton's Laws of Motion
A Day at the Races with a = F/m

In each situation below, Cart A has a mass of **1 kg**. *Circle the correct answer* (A, B, or Same for both).

1. Cart A is pulled with a force of **1 N**.
   Cart B also has a mass of **1 kg** and is pulled with a force of **2 N**.
   Which undergoes the greater acceleration?

   [A] [B] [Same for both]

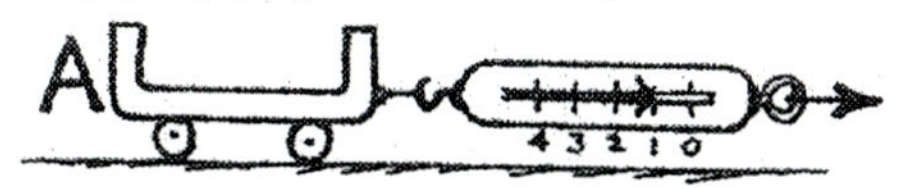

2. Cart A is pulled with a force of **1 N**.
   Cart B has a mass of **2 kg** and is also pulled with a force of **1 N**.
   Which undergoes the greater acceleration?

   [A] [B] [Same for both]

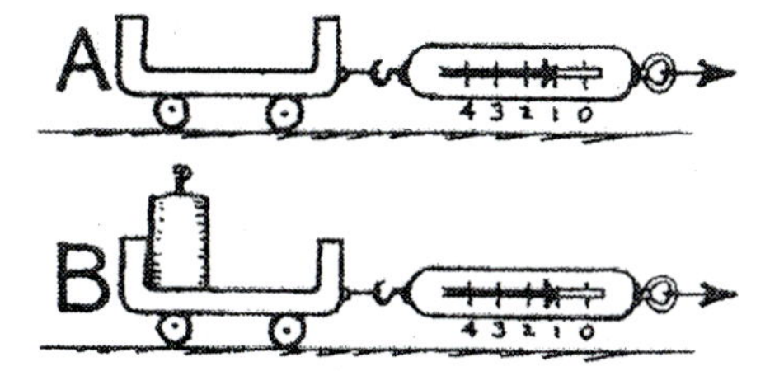

3. Cart A is pulled with a force of **1 N**.
   Cart B has a mass of **2 kg** and is pulled with a force of **2 N**.
   Which undergoes the greater acceleration?

   [A] [B] [Same for both]

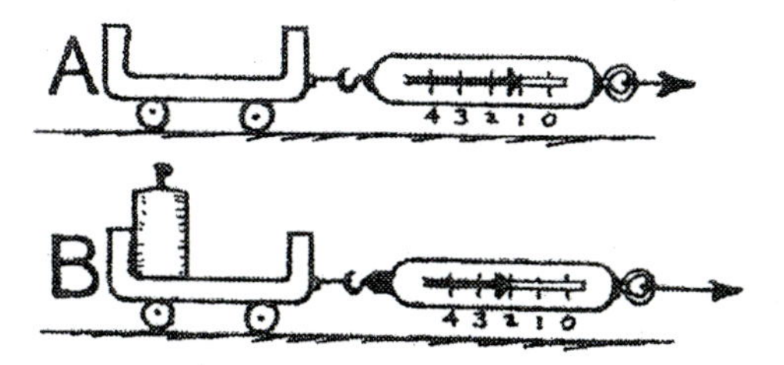

4. Cart A is pulled with a force of **1 N**.
   Cart B has a mass of **3 kg** and is pulled with a force of **3 N**.
   Which undergoes the greater acceleration?

   [A] [B] [Same for both]

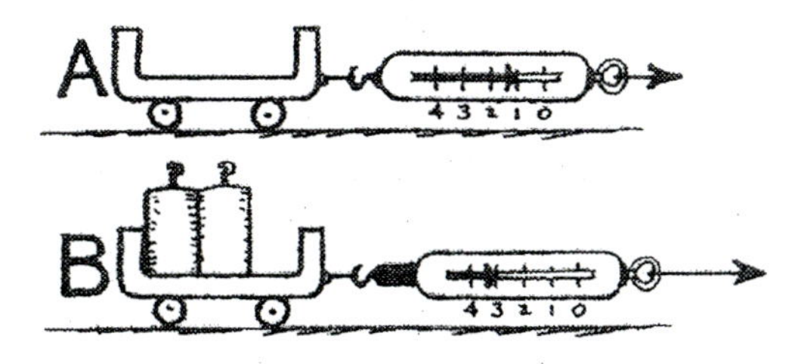

5. This time Cart A is pulled with a force of **4 N**.
   Cart B has a mass of **4 kg** and is pulled with a force of **4 N**.
   Which undergoes the greater acceleration?

   [A] [B] [Same for both]

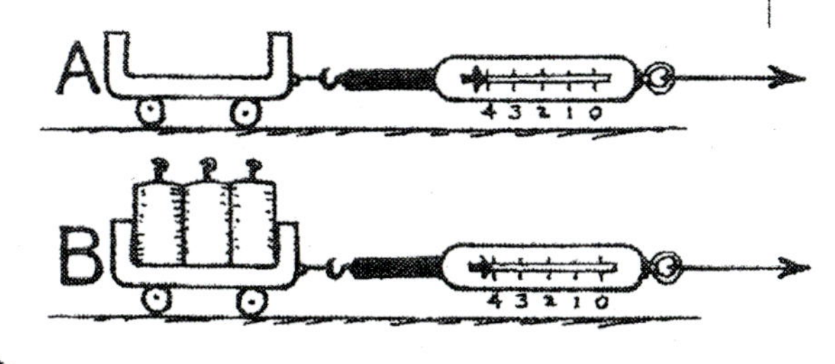

6. Cart A is pulled with a force of **2 N**.
   Cart B has a mass of **4 kg** and is pulled with a force of **3 N**.
   Which undergoes the greater acceleration?

   [A] [B] [Same for both]

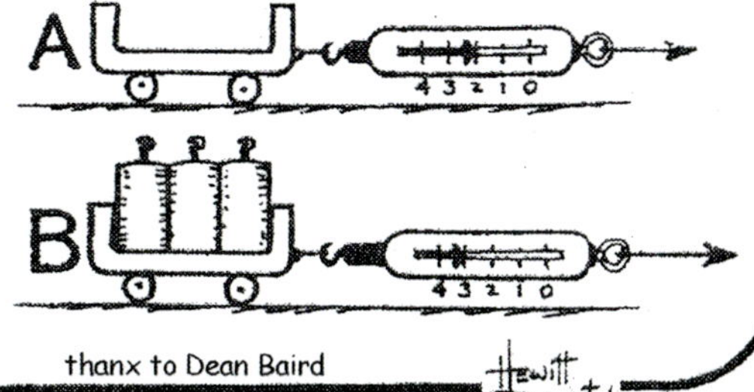

thanx to Dean Baird

Hewitt Drew it!

Chapter 4 Newton's Laws of Motion

Dropping Masses and Accelerating Cart

1. Consider a 1-kg cart being pulled by a 10-N applied force. According to Newton's 2nd law, acceleration of the cart is

$$a = \frac{F}{m} = \frac{10\ \text{N}}{1\ \text{kg}} = 10\ \text{m/s}^2.$$

This is the same as the acceleration of free fall, g—because a force equal to the cart's weight accelerates it.

2. Consider the acceleration of the cart when the applied force is due to a 10-N iron weight attached to a string draped over a pulley. Will the cart accelerate as before, at 10 m/s^2? The answer is no, because the mass being accelerated is the mass of the cart *plus* the mass of the piece of iron that pulls it. Both masses accelerate. The mass of the 10-N iron weight is 1 kg—so the total mass being accelerated (cart + iron) is 2 kg. Then,

The pulley changes only the direction of the force.

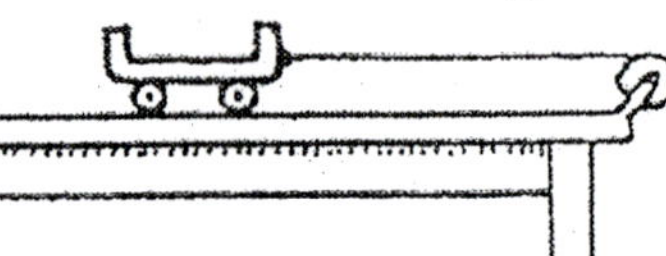

$$a = \frac{F}{m} = \frac{10\ \text{N}}{2\ \text{kg}} = 5\ \text{m/s}^2.$$

Don't forget: the total mass of a system includes the mass of the hanging iron.

Note this is half the acceleration due to gravity alone, g. So the acceleration of 2 kg produced by the weight of 1 kg is $g/2$.

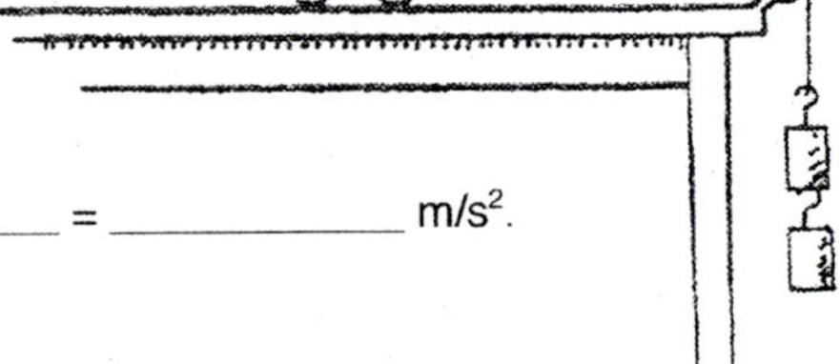

a. Find the acceleration of the 1-kg cart when two identical 10-N weights are attached to the string.

$$a = \frac{F}{m} = \frac{\text{applied force}}{\text{total mass}} = \_\_\_\_\_\_\_\_\_\_ = \_\_\_\_\_\_\_\_\_\_\ \text{m/s}^2.$$

Here we simplify and say $g = 10\ \text{m/s}^2$.

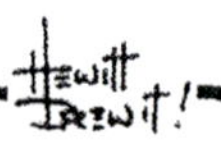

Chapter 4 Newton's Laws of Motion

Dropping Masses and Accelerating Cart*—continued*

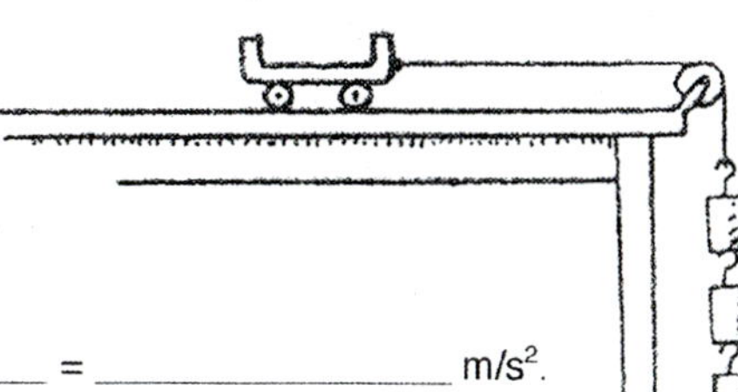

b. Find the acceleration of the 1-kg cart when the three identical 10-N weights are attach to the string.

$$a = \frac{F}{m} = \frac{\text{applied force}}{\text{total mass}} = \underline{\hspace{3cm}} = \underline{\hspace{3cm}} \text{ m/s}^2.$$

c. Find the acceleration of the 1-kg cart when four identical 10-N weights (not shown) are attached to the string.

$$a = \frac{F}{m} = \frac{\text{applied force}}{\text{total mass}} = \underline{\hspace{3cm}} = \underline{\hspace{3cm}} \text{ m/s}^2.$$

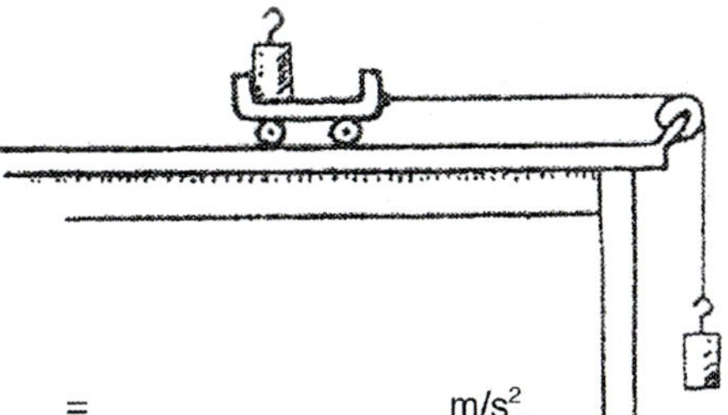

d. This time 1 kg of iron is added to the cart, and only one iron piece dangles from the pulley. Find the acceleration of the cart.

$$a = \frac{F}{m} = \frac{\text{applied force}}{\text{total mass}} = \underline{\hspace{3cm}} = \underline{\hspace{3cm}} \text{ m/s}^2.$$

The force due to gravity on a mass m is mg.
So gravitational force on 1 kg is (1 kg)(10 m/s^2) = 10 N.

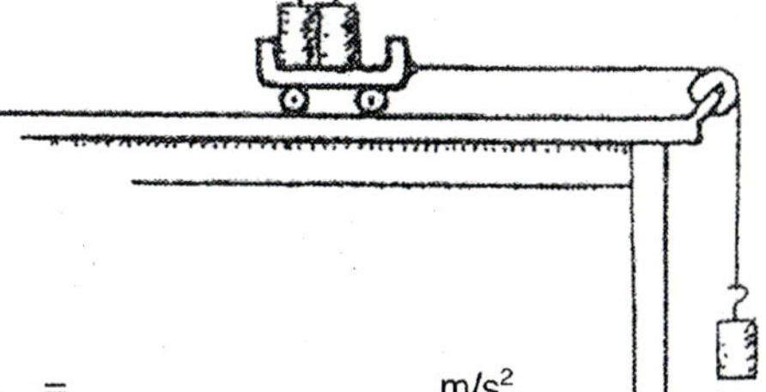

e. Find the acceleration of the cart when it carries two pieces of iron and only one iron piece dangles from the pulley.

$$a = \frac{F}{m} = \frac{\text{applied force}}{\text{total mass}} = \underline{\hspace{3cm}} = \underline{\hspace{3cm}} \text{ m/s}^2.$$

Hewitt Drew it!

Chapter 4 Newton's Laws of Motion

Dropping Masses and Accelerating Cart*—continued*

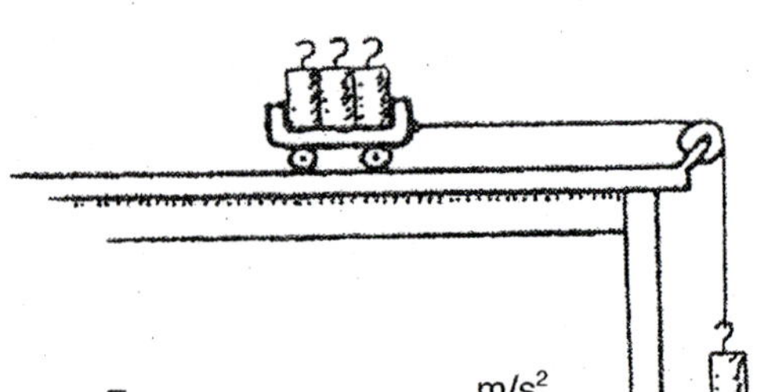

f. Find the acceleration of the cart when it carries 3 pieces of iron and only one iron piece dangles from the pulley.

$a = \frac{F}{m} = \frac{\text{applied force}}{\text{total mass}} =$ ________________ = ________________ m/s^2.

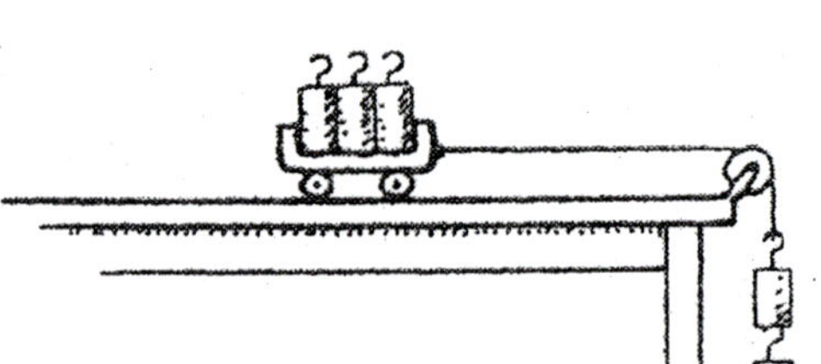

g. Find the acceleration of the cart when it carries 3 pieces of iron and 4 pieces of iron dangle from the pulley.

$a = \frac{F}{m} = \frac{\text{applied force}}{\text{total mass}} =$ ________________ = ________________ m/s^2.

Mass of cart is 1 kg. Mass of 10-N iron is also 1 kg.

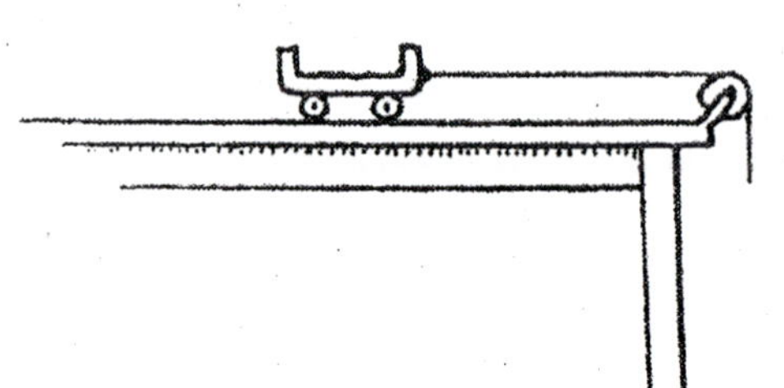

h. Draw your own combination of masses and find the acceleration.

$a = \frac{F}{m} = \frac{\text{applied force}}{\text{total mass}} =$ ________________ = ________________ m/s^2.

Hewitt Drew it!

Name ______________________________ Date __________

CONCEPTUAL Physics FUNDAMENTALS PRACTICE PAGE

Chapter 4 Newton's Laws of Motion

Force and Acceleration

1. Skelly the skater, total mass 25 kg, is propelled by rocket power.

a. Complete Table I (neglect resistance).

TABLE I

FORCE	ACCELERATION
100 N	
200 N	
	10 m/s²

b. Complete Table II for a constant 50-N resistance.

TABLE II

FORCE	ACCELERATION
50 N	0 m/s²
100 N	
200 N	

2. Block A on a horizontal friction-free table is accelerated by a force from a string attached to Block B of the same mass. Block B falls vertically and drags Block A horizontally. (Neglect the string's mass).

Circle the correct answers:

a. The *mass* of the system (A + B) is [*m*] [2 *m*].

b. The *force* that accelerates (A + B) is the weight of [A] [B] [A + B].

c. The weight of B is [*mg*/2] [*mg*] [2 *mg*].

d. Acceleration of (A + B) is [less than *g*] [*g*] [more than *g*].

e. Use $a = \frac{F}{m}$ to show the acceleration of (A + B) as a fraction of *g*. __________

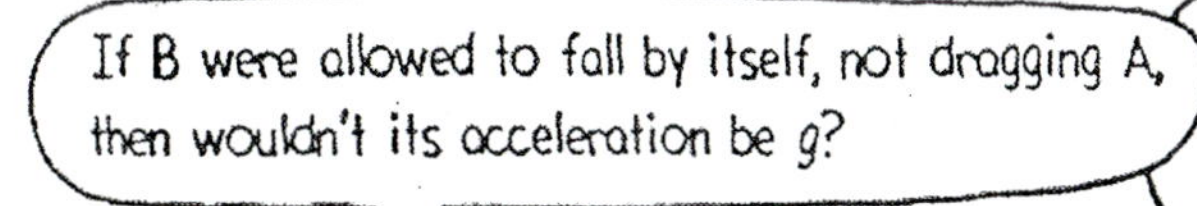

If B were allowed to fall by itself, not dragging A, then wouldn't its acceleration be *g*?

Yes, because the force that accelerates it would only be acting on its own mass – not twice the mass!

To better understand this, consider 3 and 4 on the other side!

Chapter 4 Newton's Laws of Motion

***Force and Acceleration**—continued*

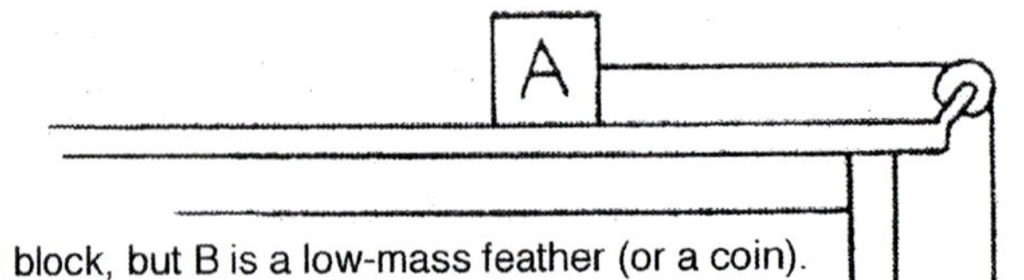

3. Suppose Block A is still a 1-kg block, but B is a low-mass feather (or a coin).

   a. Compared to the acceleration of the system of 2 equal-mass blocks

      the acceleration of (A + B) here is [less] [more]

      and is [close to zero] [close to *g*].

   b. In this case, the acceleration of B is

      [practically that of free fall] [nearly zero].

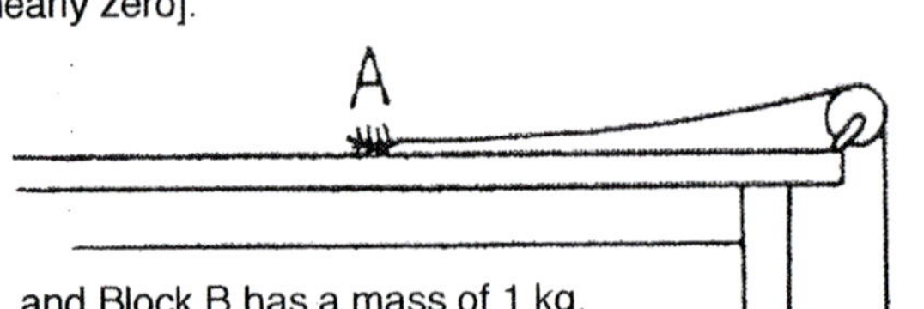

4. Suppose A is the feather or coin, and Block B has a mass of 1 kg.

   a. The acceleration of (A + B) here is [close to zero] [close to *g*].

   b. In this case, the acceleration of Block B is

      [practically that of free fall] [nearly zero].

5. Summarizing we see that when the weight of one object causes the acceleration of two objects, the range of possible accelerations is between [zero and *g*] [zero and infinity] [*g* and infinity].

6. For a change of pace, consider a ball that rolls down a uniform-slope ramp.

   a. Speed of the ball is [decreasing] [constant] [increasing].

   b. Acceleration is [decreasing] [constant] [increasing].

   c. If the ramp were steeper, acceleration would be [more] [the same] [less].

   d. When the ball reaches the bottom and rolls along the smooth level surface, it [continues to accelerate] [does not accelerate].

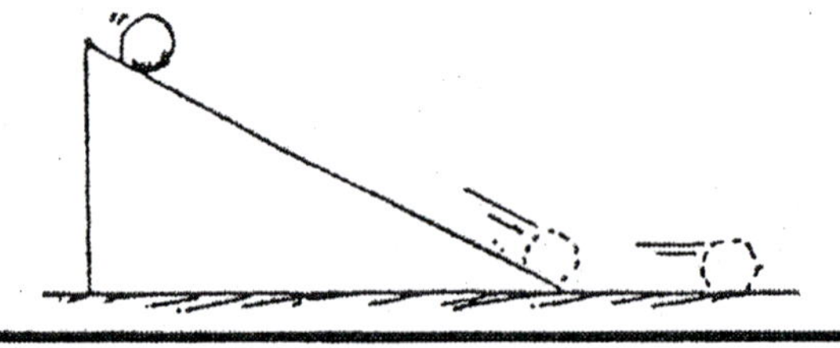

Hewitt Drew it!

Name ______________________________ Date __________

CONCEPTUAL **Physics** FUNDAMENTALS PRACTICE PAGE

Chapter 4 Newton's Laws of Motion

Friction

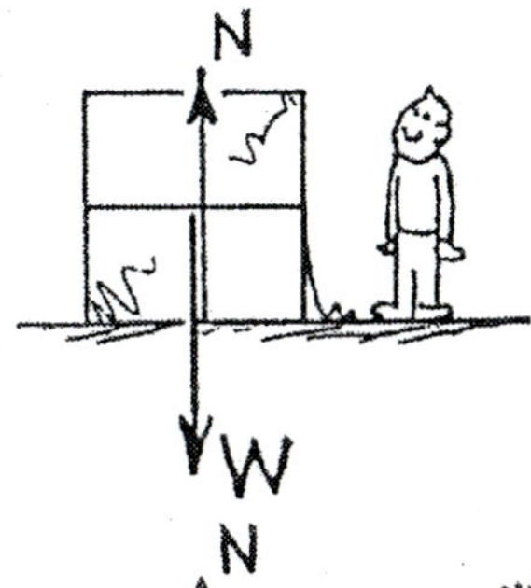

1. A crate filled with delicious junk food rests on a horizontal floor. Only gravity and the support force of the floor act on it, as shown by the vectors for weight **W** and normal force **N**.

   a. The net force on the crate is [zero] [greater than zero].

   b. Evidence for this is ______________________________.

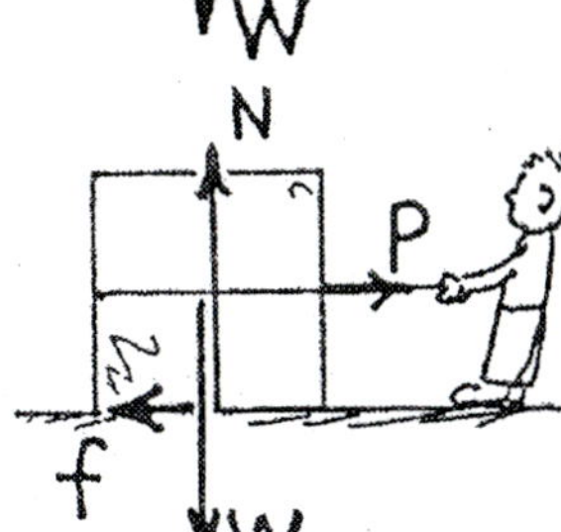

2. A slight pull **P** is exerted on the crate, not enough to move it. A force of friction ***f*** now acts,

   a. which is [less than] [equal to] [greater than] **P**.

   b. Net force on the crate is [zero] [greater than zero].

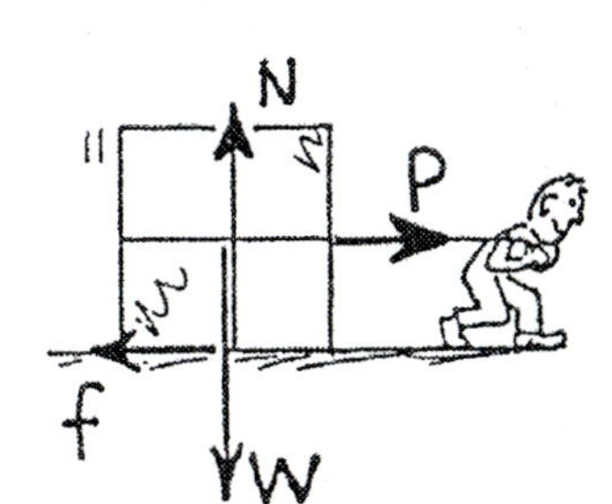

3. Pull **P** is increased until the crate begins to move. It is pulled so that it moves with constant velocity across the floor.

   a. Friction ***f*** is [;less than] [equal to] [greater than] **P**.

   b. Constant velocity means acceleration is [zero] [more than zero].

   c. Net force on the crate is [less than] [equal to] [more than] zero.

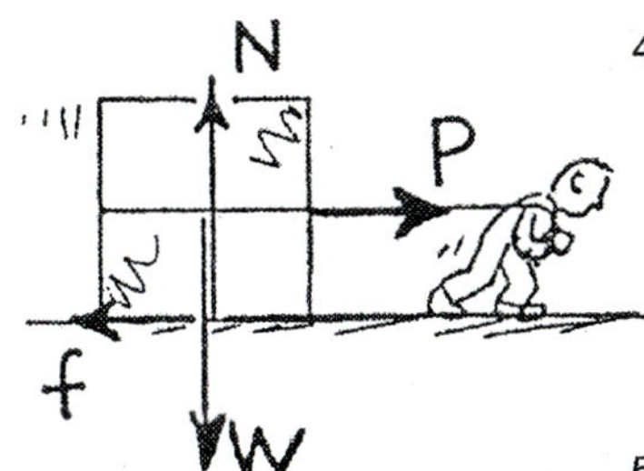

4. Pull **P** is further increased and is now greater than friction ***f***.

   a. Net force on the crate is [less than] [equal to] [greater than] zero.

   b. The net force acts toward the right, so acceleration acts toward the [left] [right].

5. If the pulling force **P** is 150 N and the crate doesn't move, what is the magnitude of ***f***? __________

6. If the pulling force **P** is 200 N and the crate doesn't move, what is the magnitude of ***f***? __________

7. If the force of sliding friction is 250 N, what force is necessary to keep the crate sliding at constant velocity? __________

8. If the mass of the crate is 50 kg and sliding friction is 250 N, what is the acceleration of the crate when the pulling force is 250 N? __________ 300 N? __________ 500 N? __________

Chapter 4 Newton's Laws of Motion
Falling and Air Resistance

Bronco skydives and parachutes from a stationary helicopter. Various stages of fall are shown in positions *a* through *f*. Using Newton's 2nd law,

$$a = \frac{F_{net}}{m} = \frac{W - R}{m}$$

find Bronco's acceleration at each position (answer in the blanks to the right). You need to know that Bronco's mass *m* is 100 kg so his weight is a constant 1000 N. Air resistance *R* varies with speed and cross-sectional area as shown.

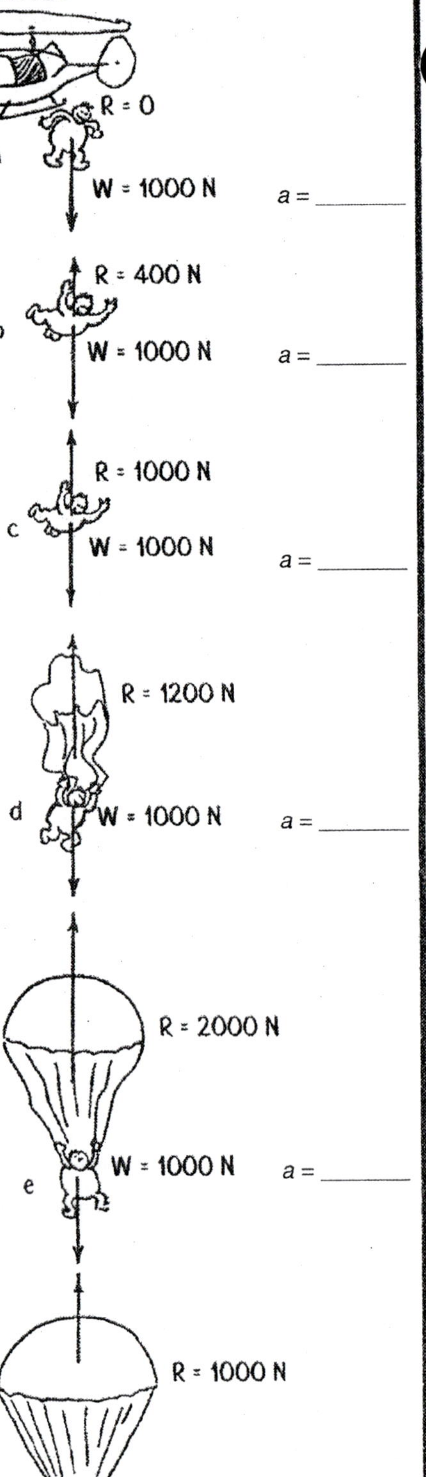

Circle the correct answers:

1. When Bronco's speed is least, his acceleration is

   [least] [most].

2. In which position(s) does Bronco experience a downward acceleration?

   [a] [b] [c] [d] [e] [f]

3. In which position(s) does Bronco experience an upward acceleration?

   [a] [b] [c] [d] [e] [f]

4. When Bronco experiences an upward acceleration,

   his velocity is [still downward] [upward also].

5. In which position(s) is Bronco's velocity constant?

   [a] [b] [c] [d] [e] [f]

6. In which position(s) does Bronco experience terminal velocity?

   [a] [b] [c] [d] [e]

7. In which position(s) is terminal velocity greatest?

   [a] [b] [c] [d] [e]

8. If Bronco were heavier, his terminal velocity would be

   [greater] [less] [the same].

Name ______________________ Date __________

CONCEPTUAL Physics FUNDAMENTALS PRACTICE PAGE

Chapter 4 Newton's Laws of Motion

Action and Reaction Pairs

1. In the example below, the action-reaction pair is shown by the arrows (vectors), and the action-reaction described in words. In *a* through *g*, draw the other arrow (vector) and state the reaction to the given action. Then make up your own example in *h*.

Example:

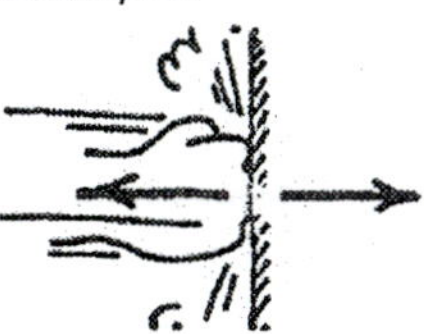

Fist hits wall.

Wall hits fist.

Head bumps ball.

a. ______________________

Windshield hits bug.

b. ______________________

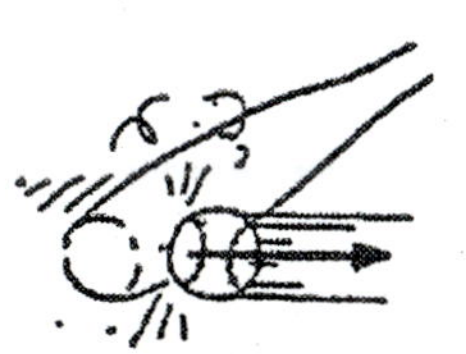

Bat hits ball.

c. ______________________

Hand touches nose.

d. ______________________

Hand pulls flower.

e. ______________________

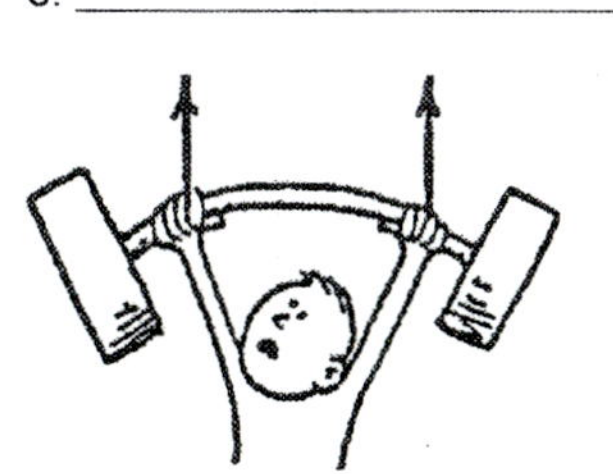

Athlete pushes bar upward.

f. ______________________

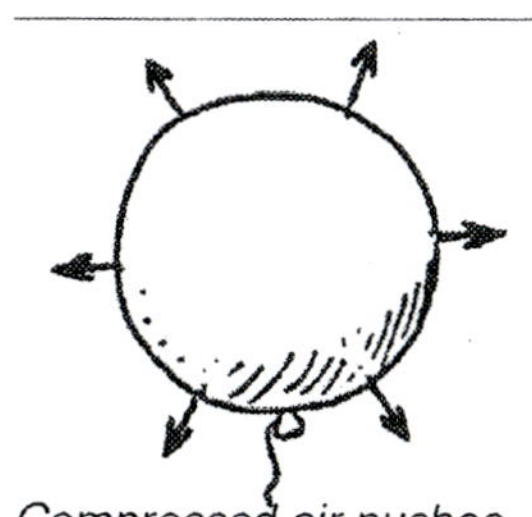

Compressed air pushes balloon surface outward.

g. ______________________

____

h. ______________________

____

2. Draw arrows to show the chain of at least six parts of action-reaction forces below.

YOU CAN'T TOUCH
WITHOUT BEING TOUCHED—
NEWTON'S THIRD LAW

Hewitt Drew it!

Chapter 4 Newton's Laws of Motion

Interactions

1. Nellie Newton holds an apple weighing 1 newton at rest on the palm of her hand. The force vectors shown are the forces that act on the apple.

   a. To say the weight of the apple is 1 N is to say that a downward gravitational force of 1 N is exerted on the apple by [Earth] [her hand].

   b. Nellie's hand supports the apple with normal force **N**, which acts in a direction opposite to **W**. We can say **N** [equals **W**] [has the same magnitude as **W**].

   c. Since the apple is at rest, the net force on the apple is [zero] [nonzero].

   d. Since **N** is equal and opposite to **W**, we [can] [cannot] say that **N** and **W** comprise an action-reaction pair. The reason is because action and reaction always [act on the same object] [act on different objects], and here we see **N** and **W** [both acting on the apple] [acting on different objects].

   e. In accord with the rule, "If ACTION is A acting on B, then REACTION is B acting on A," if we say *action* is Earth pulling down on the apple, then *reaction* is [the apple pulling up on Earth] [**N**, Nellie's hand pushing up on the apple].

   f. To repeat for emphasis, we see that **N** and **W** are equal and opposite to each other [and comprise an action-reaction pair] [but do not comprise an action-reaction pair].

To identify a pair of action-reaction forces in any situation, first identify the pair of interacting objects involved. Something is interacting with something else. In this case the whole Earth is interacting (gravitationally) with the apple. So Earth pulls downward on the apple (call it action), while the apple pulls upward on Earth (reaction).

Simply put, Earth pulls on apple (action); apple pulls on Earth (reaction).

Better put, apple and Earth *pull* on each other with equal and opposite forces that comprise a *single* interaction.

   g. Another pair of forces is **N** as shown, and the downward force of the apple against Nellie's hand, not shown. This force pair [is] [isn't] an action-reaction pair.

   h. Suppose Nellie now pushes upward on the apple with a force of 2 N. The apple [is still in equilibrium] [accelerates upward], and compared to **W**, the magnitude of **N** is [the same] [twice] [not the same, and not twice].

   i. Once the apple leaves Nellie's hand, **N** is [zero] [still twice the magnitude of **W**], and the net force on the apple is [zero] [only **W**] [still **W** - **N**, a negative force].

Hewitt Drew it!

Name ______________________________ Date __________

CONCEPTUAL **Physics** FUNDAMENTALS PRACTICE PAGE

Chapter 4 Newton's Laws of Motion

Vectors and the Parallelogram Rule

1. When two vectors **A** and **B** are at an angle to each other, they add to produce the resultant **C** by the *parallelogram rule.* Note that **C** is the diagonal of a parallelogram where **A** and **B** are adjacent sides. Resultant **C** is shown in the first two diagrams, *a* and *b.* Construct resultant **C** in diagrams *c* and *d.* Note that in diagram *d* you form a rectangle (a special case of a parallelogram).

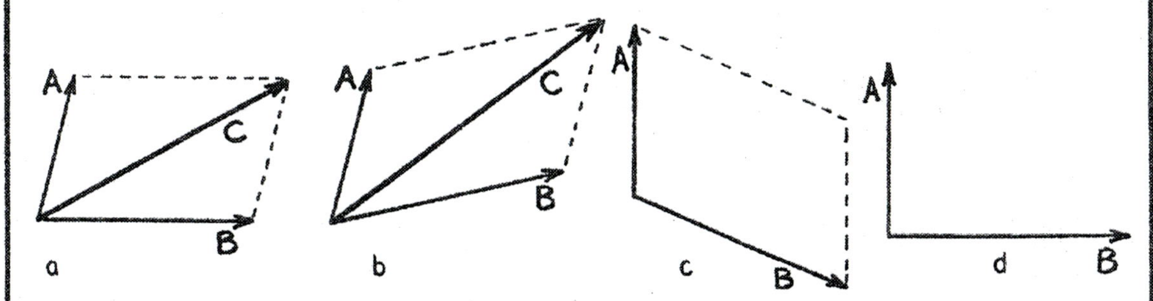

2. Below we see a top view of an airplane being blown off course by wind in various directions. Use the parallelogram rule to show the resulting speed and direction of travel for each case. In which case does the airplane travel fastest across the ground? __________ Slowest? __________

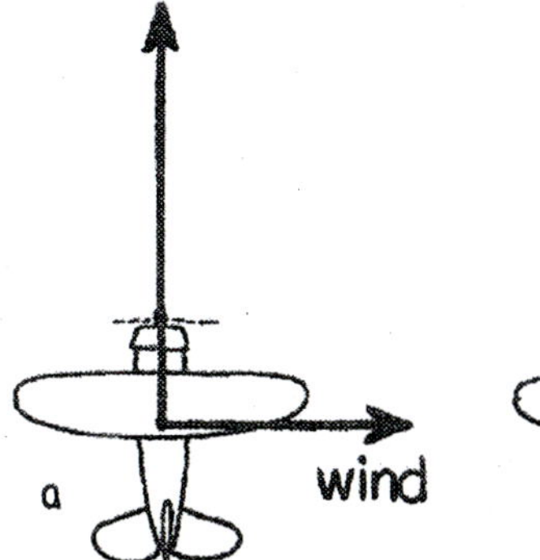

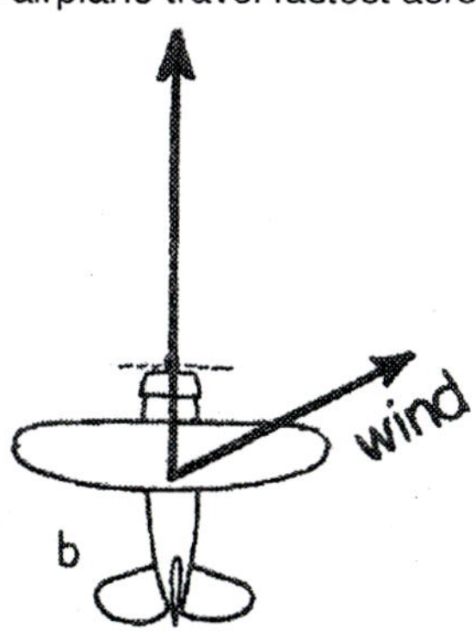

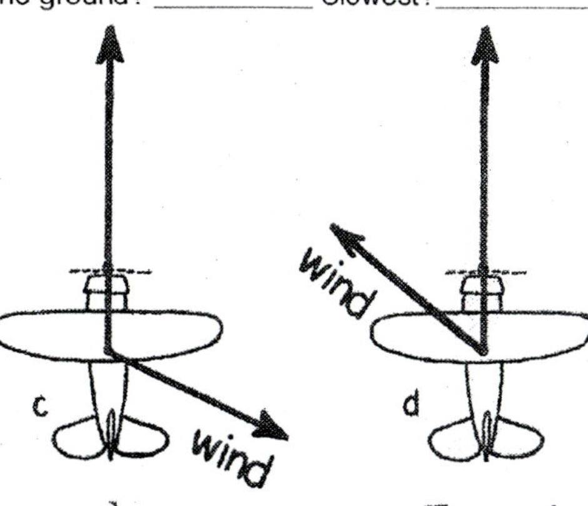

3. To the right we see the top views of 3 motorboats crossing a river. All have the same speed relative to the water, and all experience the same water flow.

Construct resultant vectors showing the speed and direction of the boats.

a. Which boat takes the shortest path to the opposite shore?

____

b. Which boat reaches the opposite shore first?

____

c. Which boat provides the fastest ride?

____

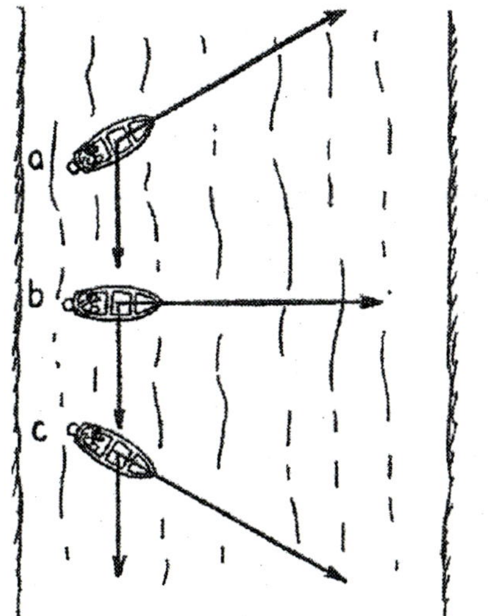

Hewitt Drew it!

Chapter 4 Newton's Laws of Motion

Velocity Vectors and Components

1. Draw the resultants of the four sets of vectors below.

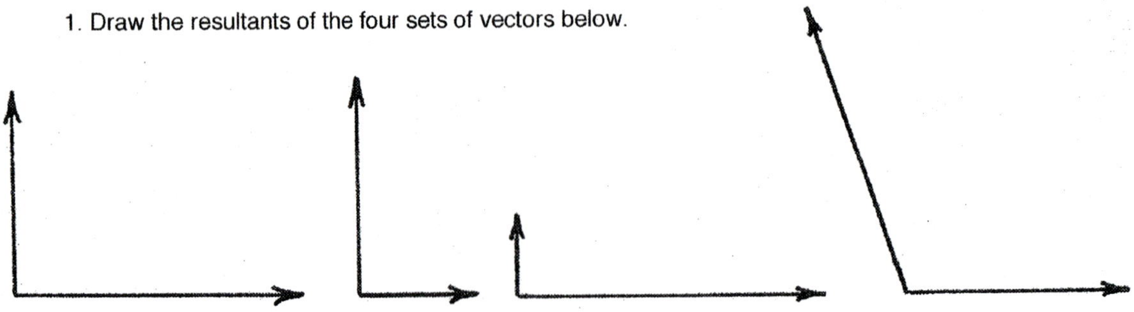

2. Draw the horizontal and vertical components of the four vectors below.

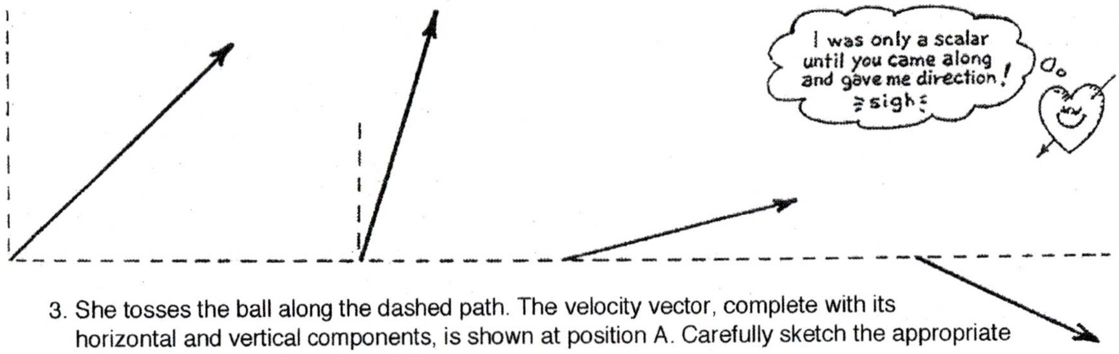

I was only a scalar until you came along and gave me direction! ≑sigh≑

3. She tosses the ball along the dashed path. The velocity vector, complete with its horizontal and vertical components, is shown at position A. Carefully sketch the appropriate components for positions B and C.

   a. Since there is no acceleration in the horizontal direction, how does the horizontal component of velocity compare for positions A, B, and C? ______________________

   b. What is the value of the vertical component of velocity at position B? ______________

   c. How does the vertical component of velocity at position C compare with that of position A?

____

____

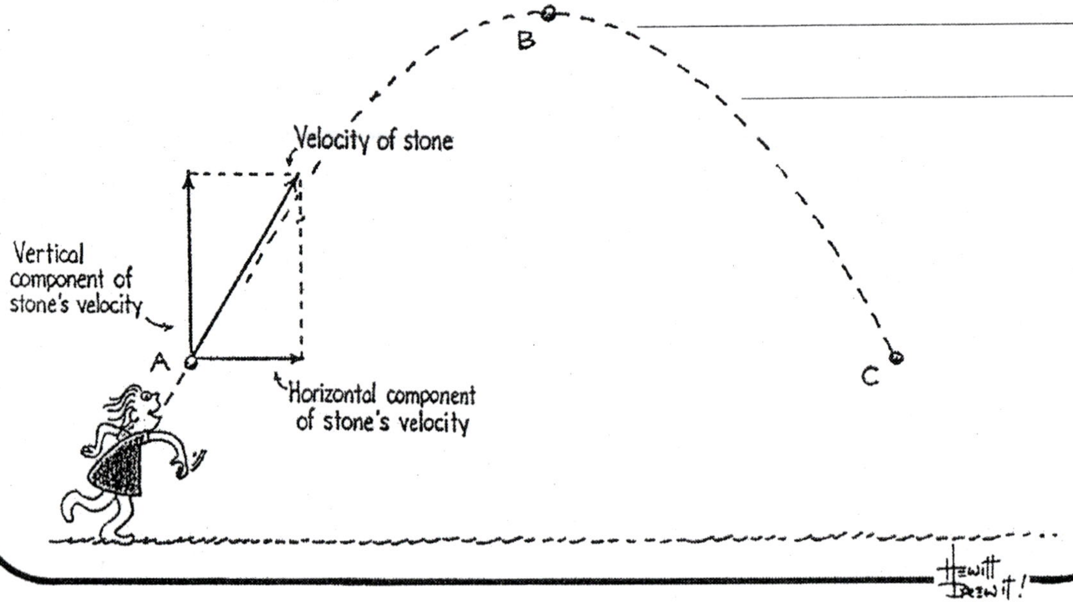

B

Velocity of stone

Vertical component of stone's velocity

A

Horizontal component of stone's velocity

C

Hewitt Drew it!

Name ______________________ Date __________

CONCEPTUAL Physics FUNDAMENTALS PRACTICE PAGE

Chapter 4 Newton's Laws of Motion

Force and Velocity Vectors

1. Draw sample vectors to represent the force of gravity on the ball in the positions shown below after it leaves the thrower's hand. (Neglect air resistance.)

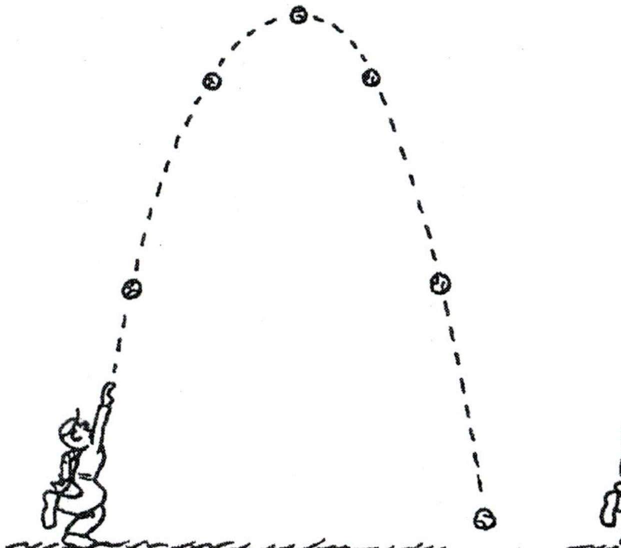

2. Draw sample bold vectors to represent the velocity of the ball in the positions shown below. With lighter vectors, show the horizontal and vertical components of velocity for each position.

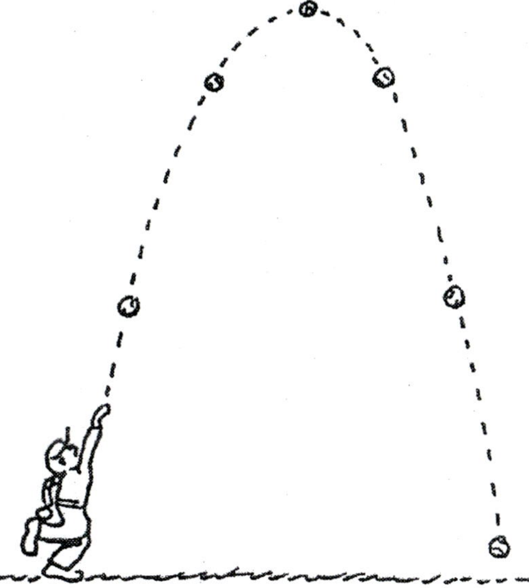

3.a. Which velocity component in the previous question remains constant? Why?

____

b. Which velocity component changes along the path? Why?

____

4. It is important to distinguish between force and velocity vectors. Force vectors combine with other force vectors, and velocity vectors combine with other velocity vectors. Do velocity vectors combine with force vectors?

____

5. All forces on the bowling ball, weight (down) and support of alley (up), are shown by vectors at its center before it strikes the pin *a*.
Draw vectors of all the forces that act on the ball *b* when it strikes the pin, and *c* after it strikes the pin.

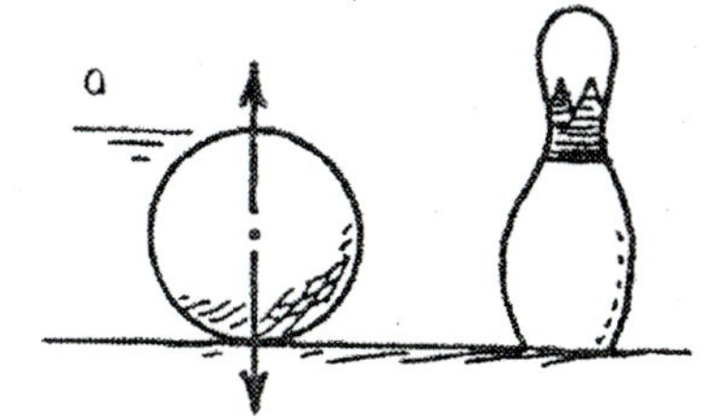

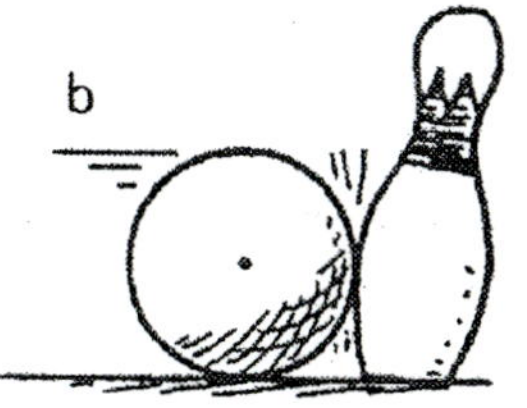

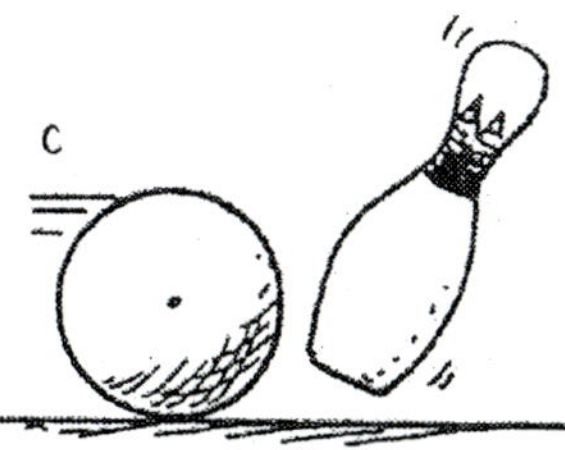

thanx to Howie Brand

Hewitt Drew it!

Chapter 4 Newton's Laws of Motion

Force Vectors and the Parallelogram Rule

1. The heavy ball is supported in each case by two strands of rope. The tension in each strand is shown by the vectors. Use the parallelogram rule to find the resultant of each vector pair.

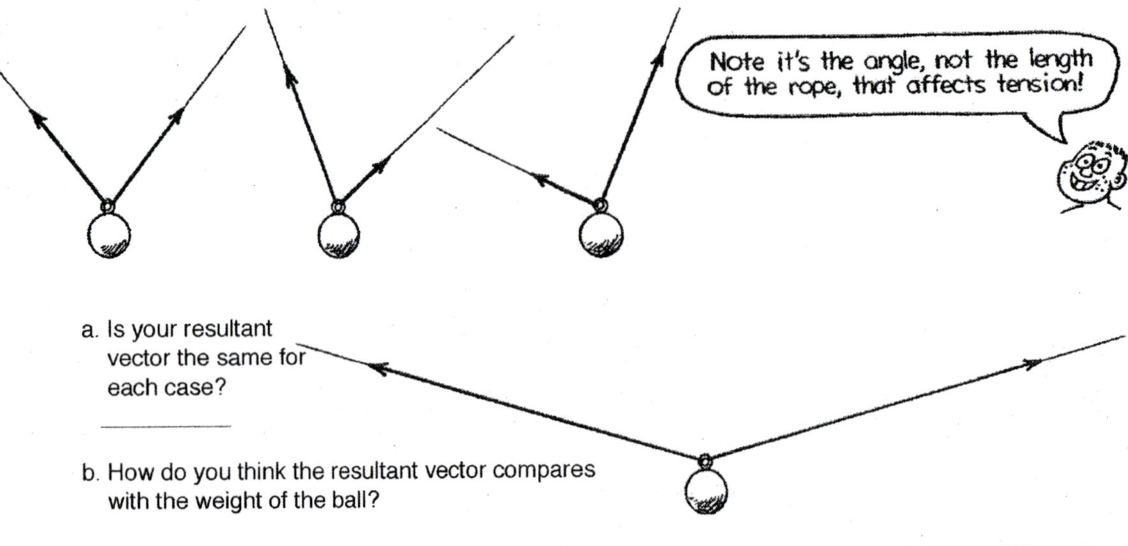

Note it's the angle, not the length of the rope, that affects tension!

a. Is your resultant vector the same for each case?

____

b. How do you think the resultant vector compares with the weight of the ball?

____

2. Now let's do the opposite of what we've done above. More often, we know the weight of the suspended object, but we don't know the rope tensions. In each case below, the weight of the ball is shown by the vector **W**. Each dashed vector represents the resultant of the pair of rope tensions. Note that each is equal and opposite to vectors **W** (they must be; otherwise the ball wouldn't be at rest).

a. Construct parallelograms where the ropes define adjacent sides and the dashed vectors are the diagonals.

b. How do the relative lengths of the sides of each parallelogram compare to rope tension?

c. Draw rope-tension vectors, clearly showing their relative magnitudes.

No wonder that hanging from a horizontal tightly-stretched clothesline breaks it!

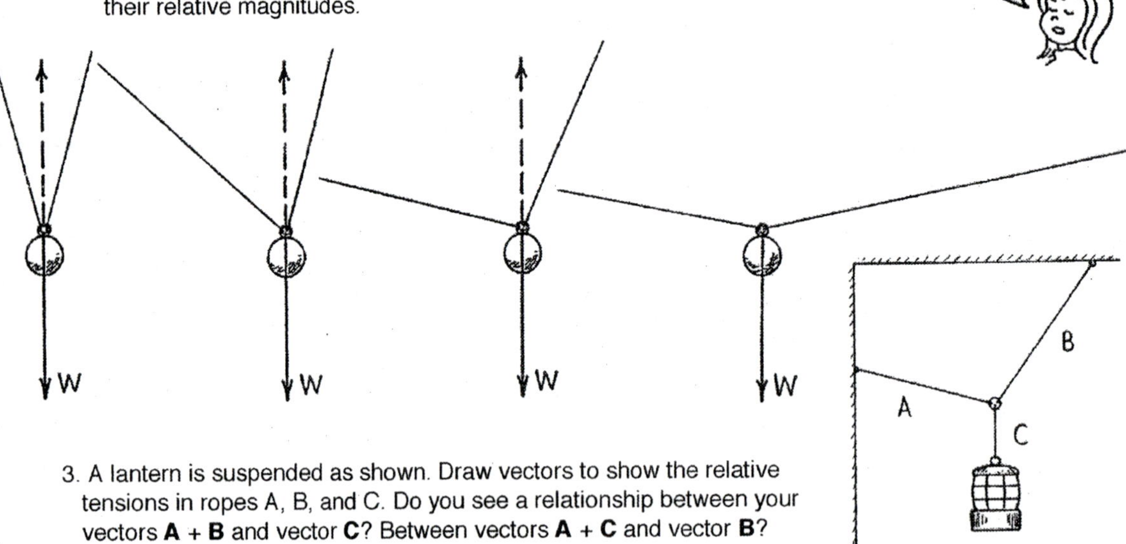

3. A lantern is suspended as shown. Draw vectors to show the relative tensions in ropes A, B, and C. Do you see a relationship between your vectors **A** + **B** and vector **C**? Between vectors **A** + **C** and vector **B**?

____

Hewitt Drew it!

Name ______________________ Date __________

CONCEPTUAL **Physics** FUNDAMENTALS PRACTICE PAGE

Chapter 4 Newton's Laws of Motion

Force-Vector Diagrams

In each case, a rock is acted on by one or more forces. Using a pencil and a ruler, draw an accurate vector diagram showing all forces acting on the rock, and no other forces. The first two cases are done as examples. The parallelogram rule in case 2 shows that the vector sum of **A** + **B** is equal and opposite to **W** (that is, **A** + **B** = -**W**). Do the same for cases 3 and 4. Draw and label vectors for the weight and normal support forces in cases 5 to 10, and for the appropriate forces in cases 11 and 12.

1. Static

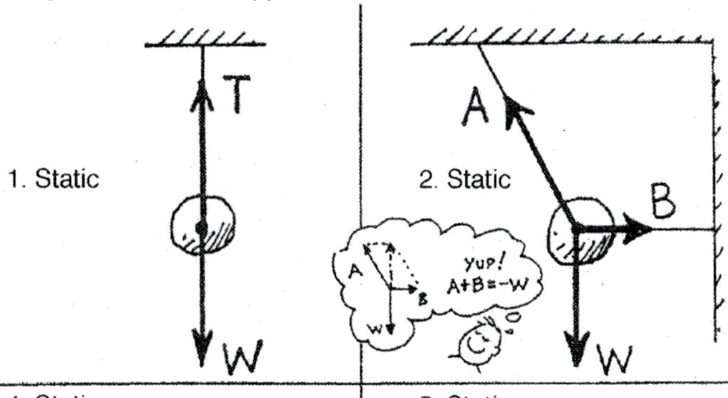

2. Static

3. Static

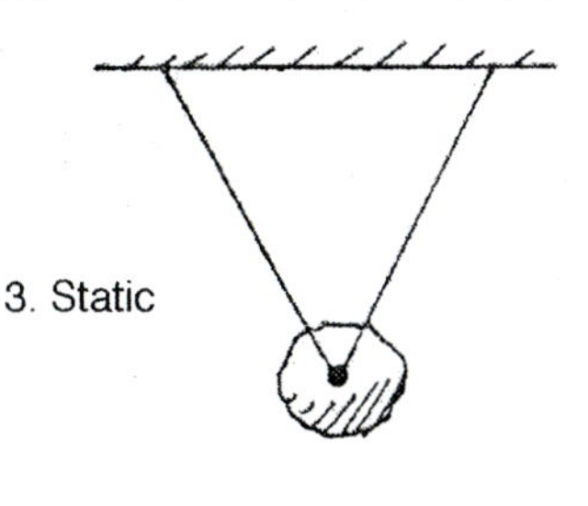

4. Static

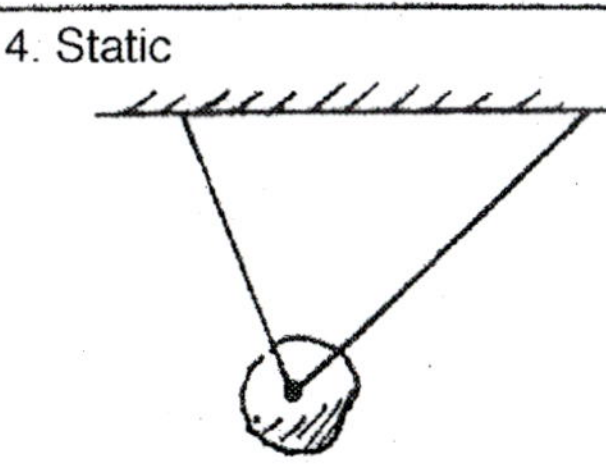

5. Static

6. Sliding at constant speed without friction

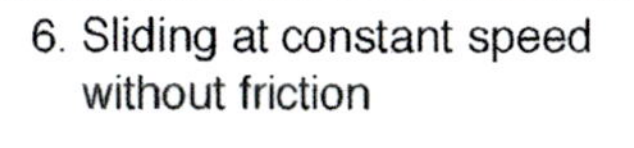

7. Decelerating due to friction

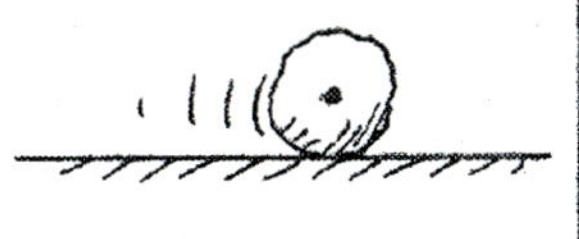

8. Static (Friction prevents sliding)

9. Rock slides (No friction)

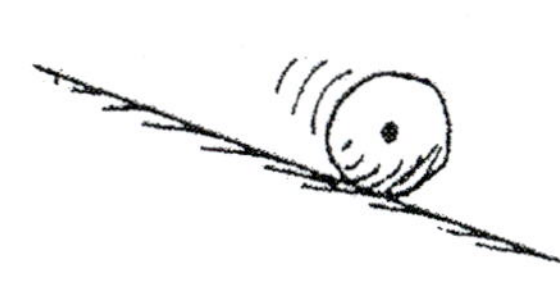

10. Static

11. Rock in free fall

12. Falling at terminal velocity

thanx to Jim Court

Hewitt Drew it!

Chapter 4 Newton's Laws of Motion

More on Vectors

A

B

T

C

W

1. Each of the vertically-suspended blocks has the same weight **W**. The two forces acting on Block C (**W** and rope tension **T**) are shown. Draw vectors to a reasonable scale for rope tensions acting on Blocks A and B.

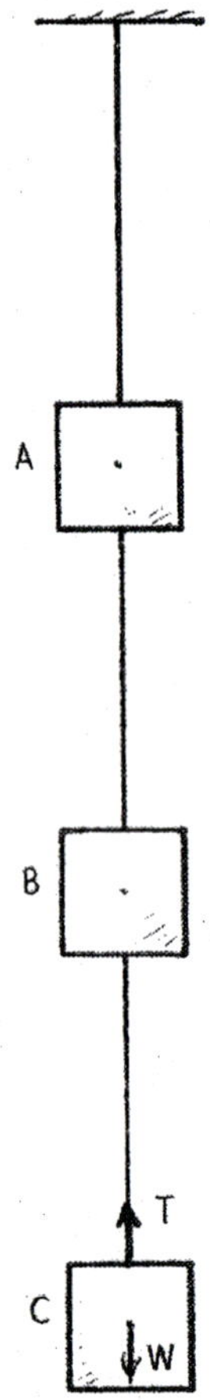

F_y F θ F_x

2. The cart is pulled with force **F** at angle θ as shown. F_x and F_y are components of **F**.

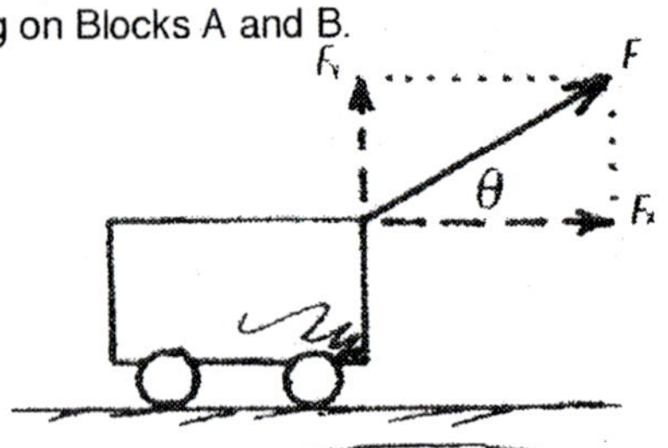

   a. How will the magnitude of F_x change if the angle θ is increased by a few degrees?

   [more] [less] [no change]

   b. How will the magnitude of F_y change if the angle θ is increased by a few degrees?

   [more] [less] [no change]

   c. What will be the value of F_x if angle θ is 90°?

   [more than **F**] [zero] [no change]

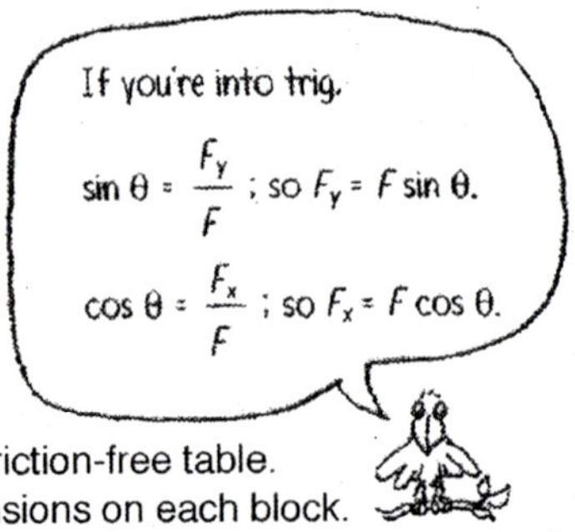

If you're into trig.

$\sin\theta = \frac{F_y}{F}$; so $F_y = F\sin\theta$.

$\cos\theta = \frac{F_x}{F}$; so $F_x = F\cos\theta$.

3. Force **F** pulls three blocks of equal mass across a friction-free table. Draw vectors of appropriate lengths for the rope tensions on each block.

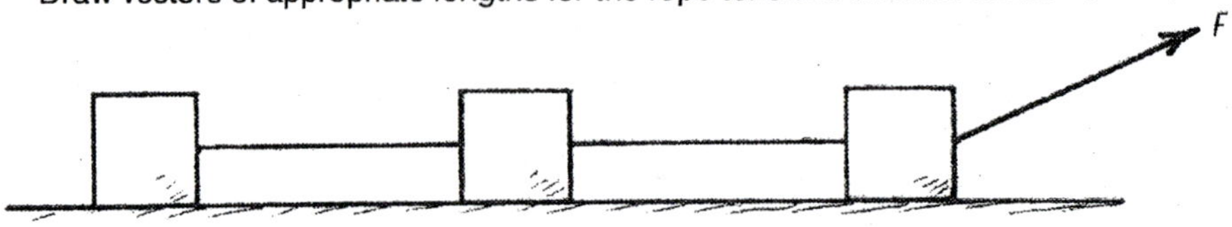

F

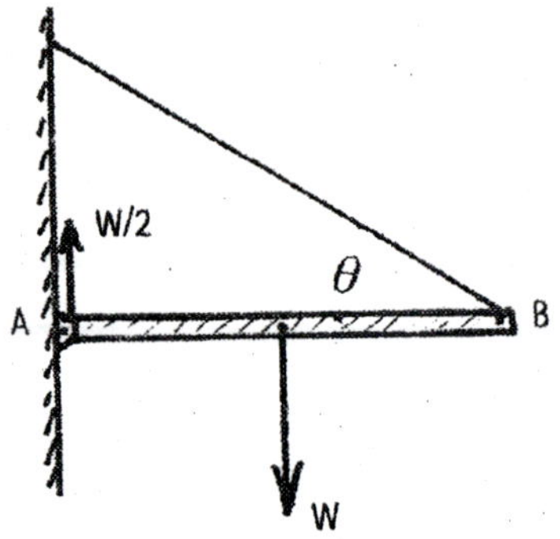

4. Consider the boom supported by hinge A and by a cable B. Vectors are shown for the weight **W** of the boom at its center, and **W**/2 for vertical component of upward force supplied by the hinge.

W/2 θ A B W

   a. Draw a vector representing the cable tension **T** at B. Why is it correct to draw its length so that the vertical component of **T** = **W**/2?

   ____

   b. Draw component T_x at B. Then draw the horizontal component of the force at A. How do these horizontal components compare, and why?

   ____

5. The block rests on the inclined plane. The vector for its weight **W** is shown. How many other forces act on the block, including static friction? __________ . Draw them to a reasonable scale.

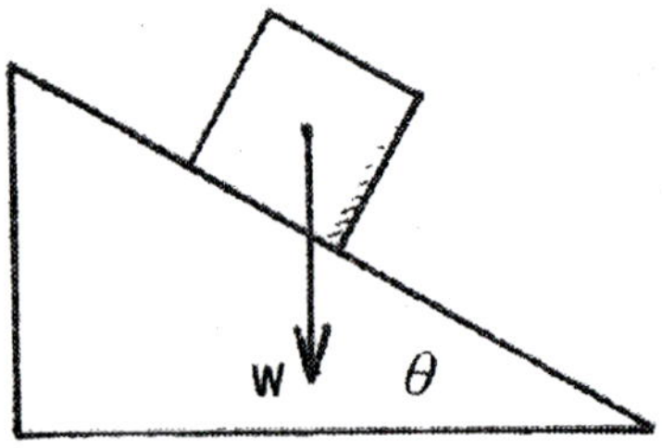

W θ

   a. How does the component of **W** parallel to the plane compare with the force of friction? ______________________________

   b. How does the component of **W** perpendicular to the plane compare with the normal force?

   ____

Hewitt Drew it!

Name ______________________________________ Date __________

CONCEPTUAL **Physics** FUNDAMENTALS PRACTICE PAGE

Chapter 5 Momentum and Energy

Changing Momentum

1. A moving car has momentum. If it moves twice as fast, its momentum is

   ________________ as much.

2. Two cars, one twice as heavy as the other, move down a hill at the same speed. Compared

   with the lighter car, the momentum of the heavier car is ______________ as much.

3. The recoil momentum of a cannon that kicks is

   [more than] [less than] [the same as]

   the momentum of the cannonball it fires. (Here we neglect friction and the momentum of the gases.)

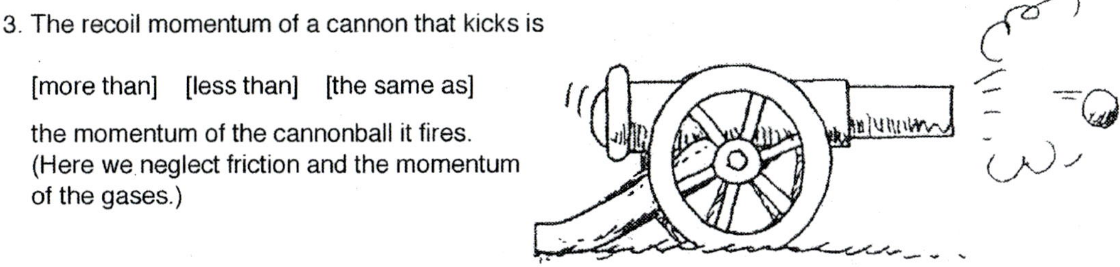

4. Suppose you are traveling in a bus at highway speed on a nice summer day and the momentum of an unlucky bug is suddenly changed as it splatters onto the front window.

   a. Compared to the force that acts on the bug, how much force acts on the bus?

      [more] [less] [the same]

   b. The time of impact is the same for both the bug and the bus. Compared to the impulse on the bug, this means the impulse on the bus is

      [more] [less] [the same].

   c. Although the momentum of the bus is very large compared to the momentum of the bug, the *change* in momentum of the bus, compared to the *change* of momentum of the bug is

      [more] [less] [the same].

   d. Which undergoes the greater acceleration?

      [bus] [both the same] [bug]

   e. Which therefore, suffers the greater damage?

      [bus] [both the same] [bug of course!]

Isn't it amazing, that in a collision between two very different entities — a bug and a bus, that three opposite quantities remain equal: impact forces, impulses, and changes in momentum!

Chapter 5 Momentum and Energy
***Changing Momentum**—continued*

5. Granny whizzes around the rink and is suddenly confronted with Ambrose at rest directly in her path. Rather than knock him over, she picks him up and continues in motion without "braking."

   Consider both Granny and Ambrose as two parts of one system. Since no outside forces act on the system, the momentum of the system before collision equals the momentum of the system after collision.

   a. Complete the before-collision data in the table below.

BEFORE COLLISION	
Granny's mass	80 kg
Granny's speed	3 m/s
Granny's momentum	________
Ambrose's mass	40 kg
Ambrose's speed	0 m/s
Ambrose's momentum	________
Total momentum	________

   b. After collision, Granny's speed [increases] [decreases].

   c. After collision, Ambrose's speed [increases] [decreases].

   d. After collision, the total mass of Granny + Ambrose is ________.

   e. After collision, the total momentum of Granny + Ambrose is

   ____

   f. Use the conservation of momentum law to find the speed of Granny and Ambrose together after collision. (Show your work in the space below.)

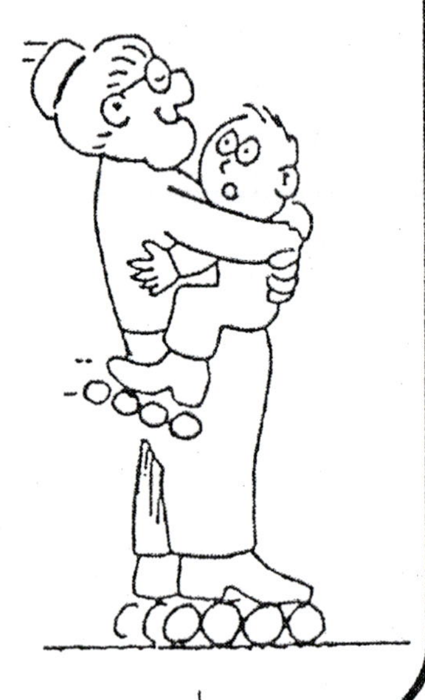

New speed ____________

Chapter 5 Momentum and Energy
Systems

1. When the compressed spring is released, Blocks A and B will slide apart. There are 3 systems to consider, indicated by the closed dashed lines below—A, B, and A + B. Ignore the vertical forces of gravity and the support force of the table.

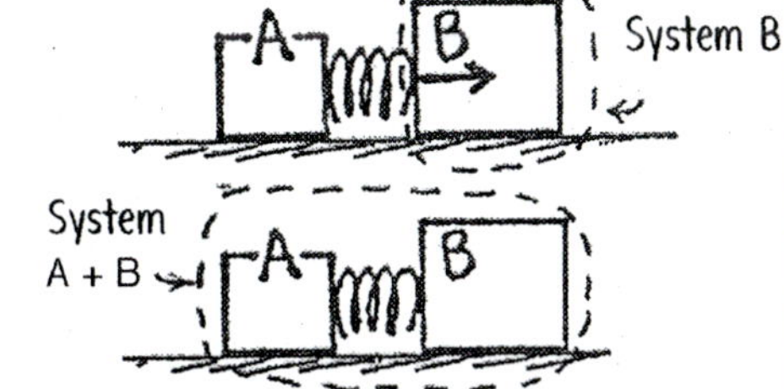

a. Does an external force act on System A? [Y] [N]

Will the momentum of System A change? [Y] [N]

b. Does an external force act on System B? [Y] [N]

Will the momentum of System B change? [Y] [N]

c. Does an external force act on System A + B? [Y] [N]

Will the momentum of System A + B change? [Y] [N]

2. Billiard ball A collides with billiard ball B at rest. Isolate each system with a closed dashed line. Draw only the external force vectors that act on each system.

a. Upon collision, the momentum of System A [increases] [decreases] [remains unchanged].

b. Upon collision, the momentum of System B [increases] [decreases] [remains unchanged].

c. Upon collision, the momentum of System A + B [increases] [decreases] [remains unchanged].

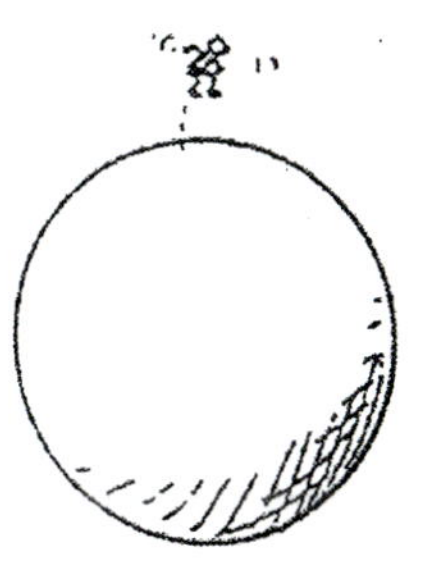

3.a. A girl jumps upward. In the left sketch, draw a closed dashed line to indicate the system of the girl.
Is there an external force acting on her? [Y] [N]

Does her momentum change? [Y] [N]

Is the girl's momentum conserved? [Y] [N]

b. In the right sketch, draw a closed dashed line to indicate the system (girl + Earth). Is there an external force acting on the system due to the interaction between the girl and Earth? [Y] [N]

4. A block strikes a blob of jelly. Isolate 3 systems with a closed dashed line and show the external force on each. In which system is momentum conserved?

5. A truck crashes into a wall. Isolate 3 systems with a closed dashed line and show the external force on each. In which system is momentum conserved?

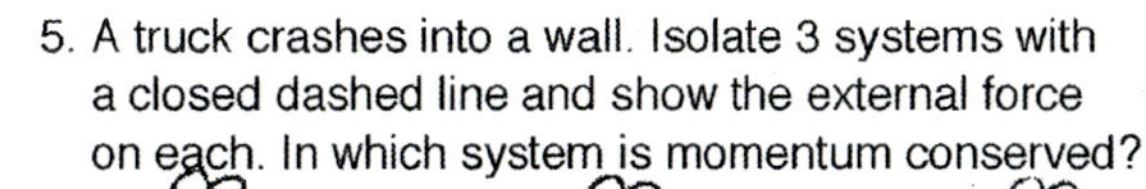

thanx to Cedric Linder

Hewitt Drew it!

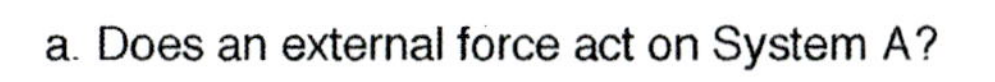

Name ______________________ Date __________

CONCEPTUAL **Physics** FUNDAMENTALS PRACTICE PAGE

Chapter 5 Momentum and Energy

Work and Energy

1. How much work (energy) is needed to lift an object that weighs 200 N to a height of 4 meters?

   ____

2. How much power is needed to lift the 200-N object to a height of 4 m in 4 seconds? ________

3. What is the power output of an engine that does 60,000 J of work in 10 seconds? ________

4. The block of ice weighs 500 newtons. (Neglect friction.)

   a. How much force parallel to the incline is needed to push it to the top? ________

   b. How much work is required to push it to the top of the incline? ________

   c. What is the potential energy of the block relative to ground level? ________

   d. What would be the potential energy if the block were simply lifted vertically 3 m? ________

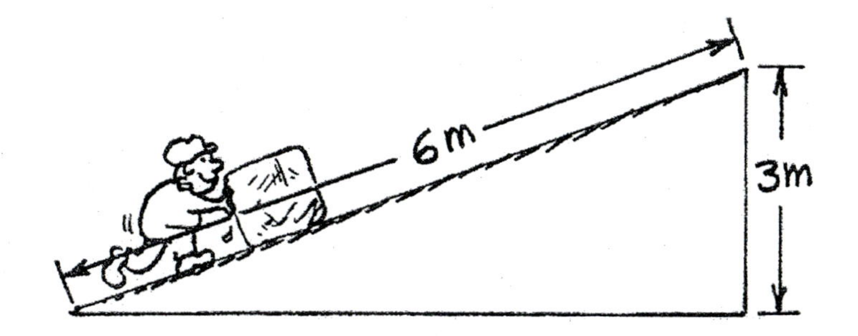

5. All the ramps below are 5 meters high. We know that the KE of the block at the bottom of each ramp will be equal to the loss of PE (conservation of energy).
   Find the speed of the block at ground level in each case. (Hint: Do you recall from earlier chapters how much time it takes something to fall a vertical distance of 5 m from a position of rest assuming $g = 10\ m/s^2$ and how much speed a falling object acquires in this time?) This gives you the answer to Case 1.
   Discuss with your classmates how energy conservation provides the answers to Cases 2 and 3.

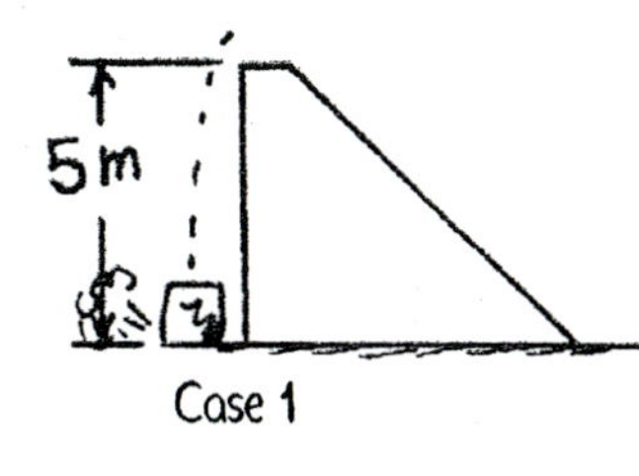

Case 1

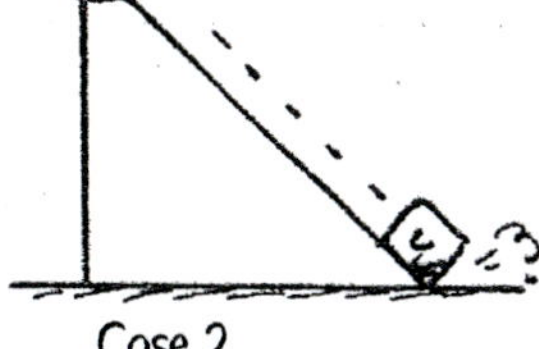

Case 2

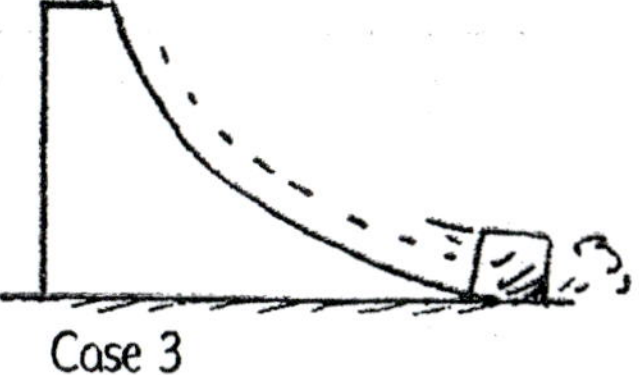

Case 3

Speed ________ m/s Speed ________ m/s Speed ________ m/s

Hewitt Drew it!

Chapter 5 Momentum and Energy

Work and Energy—continued

6. Which block reaches the bottom of the incline first? Assume no friction. (Be careful!) Explain your answer.

____

____

____

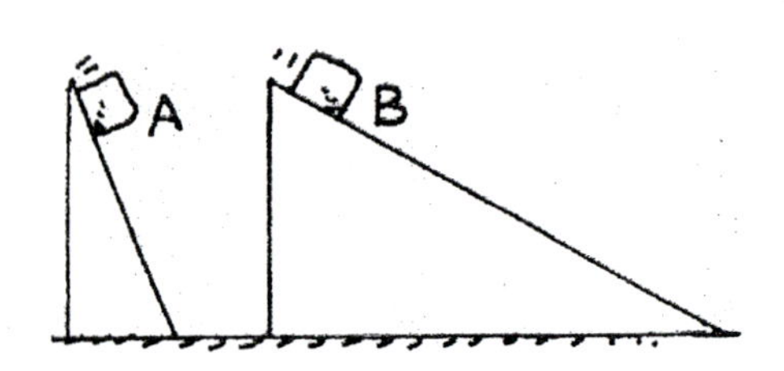

7. Both the KE and PE of a block freely sliding down a ramp are shown below only at the bottom position in the sketch. Fill in the missing values for the other positions.

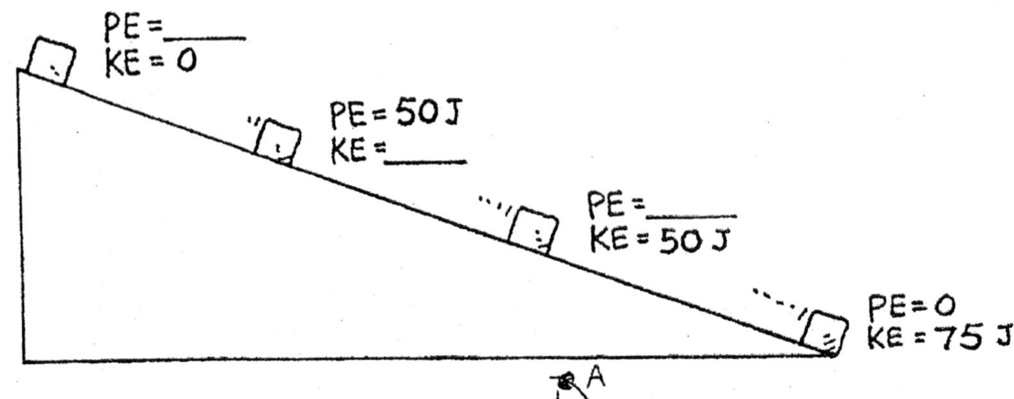

8. A big metal bead slides due to gravity along an upright friction-free wire. It starts from rest at the top of the wire as shown in the sketch.

How fast is it traveling as it passes

Point B? __________

Point D? __________

Point E? __________

Maximum speed at Point __________

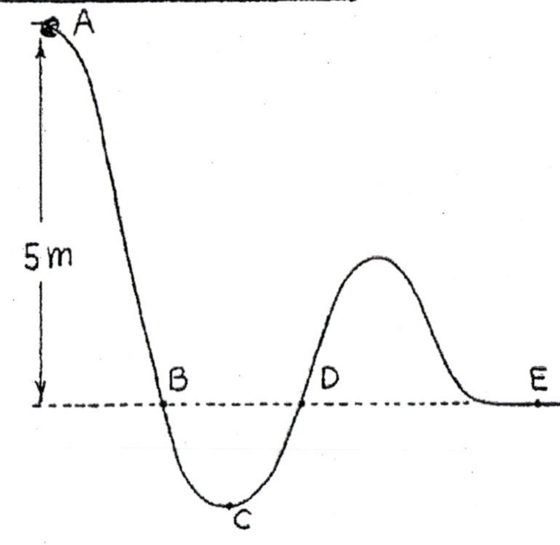

9. Rows of wind-powered generators are used in various windy locations to generate electric power. Does the power generated affect the speed of the wind? Would locations behind the "windmills" be windier if they weren't there. Discuss this in terms of energy conservation with your classmates.

____

____

____

Hewitt Drew it!

Name ______________________ Date __________

CONCEPTUAL Physics FUNDAMENTALS PRACTICE PAGE

Chapter 5 Momentum and Energy

Conservation of Energy

1. Fill in the blanks for the six systems.

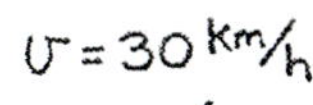

v = 60 km/h
KE = ______

v = 90 km/h
KE = ______

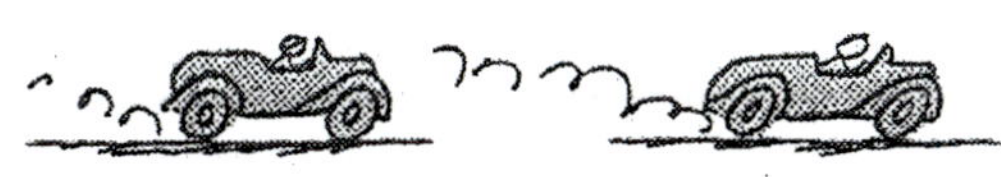

PE = 15000 J
KE = 0

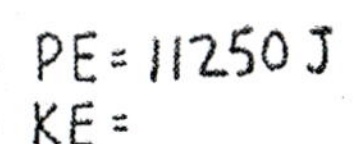

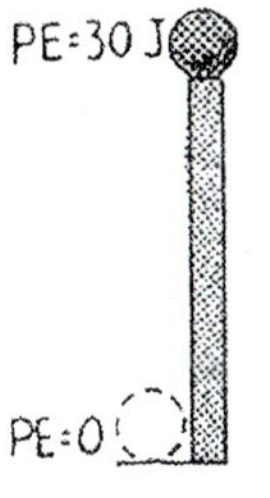

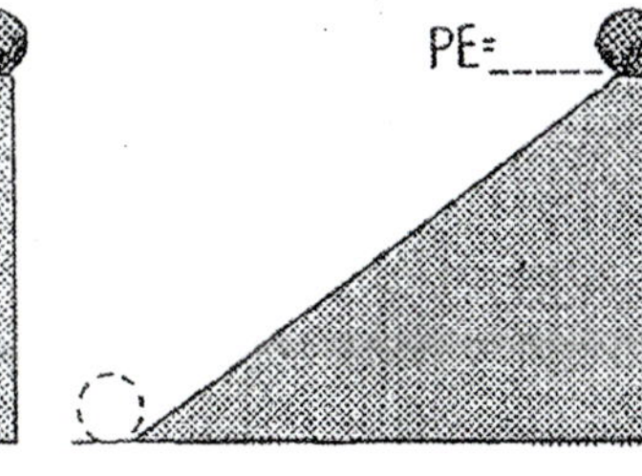

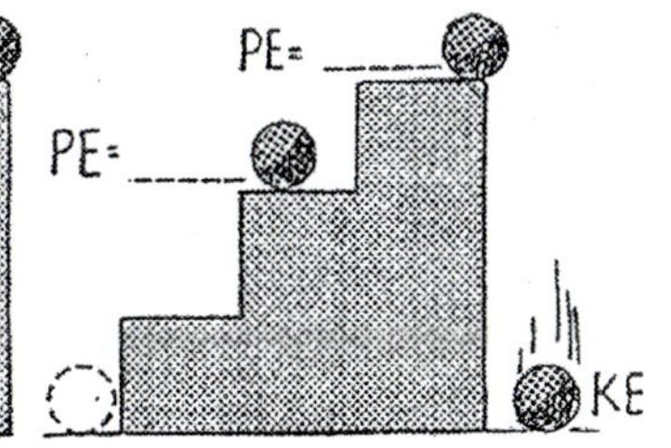

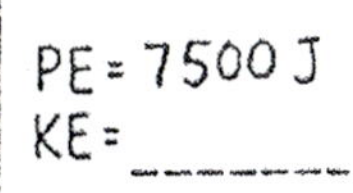

PE = 3750 J
KE = ______

WORK DONE = ______

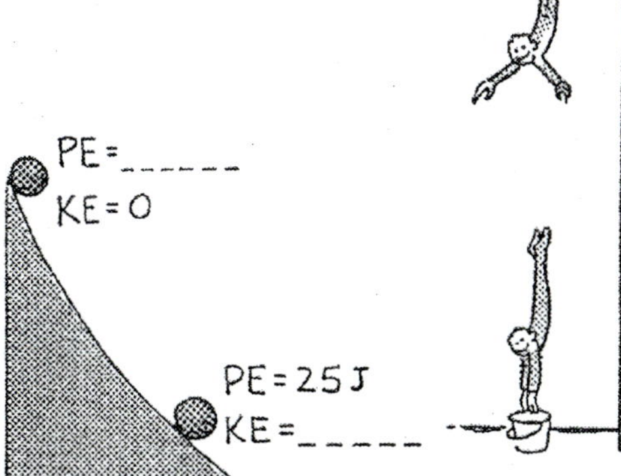

PE = 0
KE = 50 J

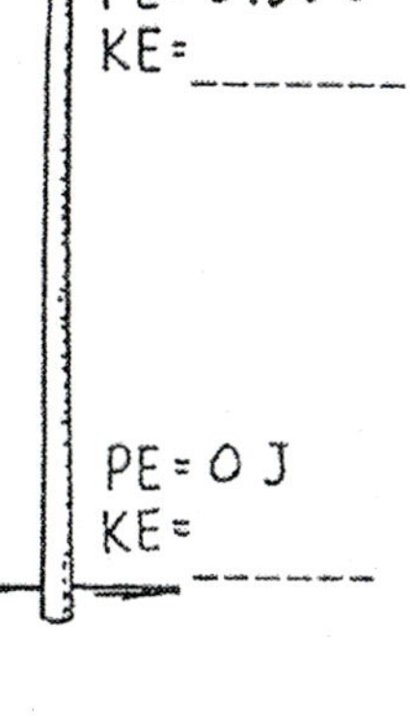

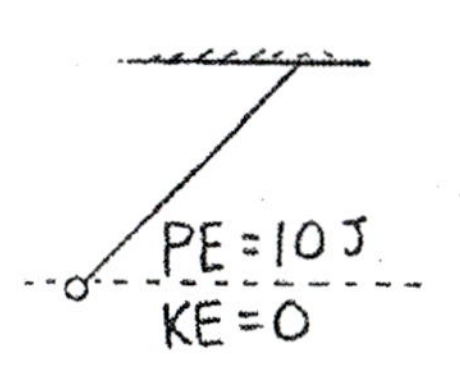

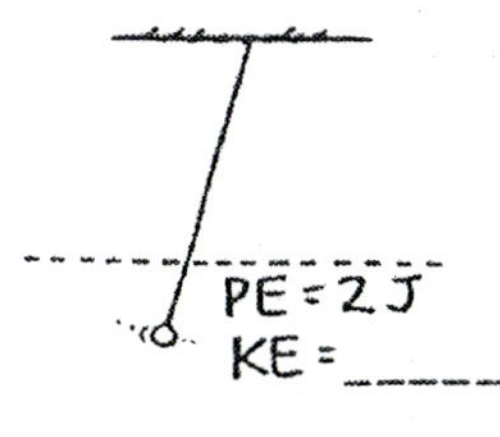

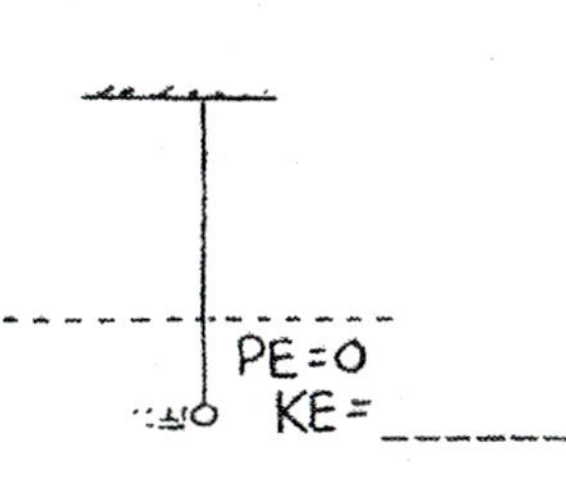

PE = ______
KE = ______

Hewitt Drew it!

Chapter 5 Momentum and Energy

Conservation of Energy—continued

2. The woman supports a 100-N load with the friction-free pulley systems shown below. Fill in the spring-scale readings that show how much force she must exert.

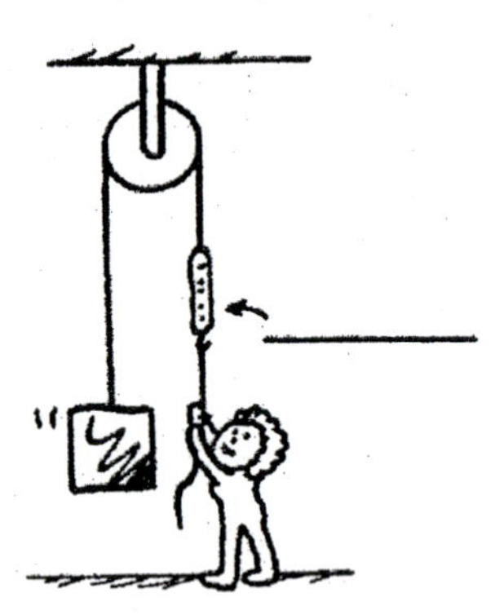

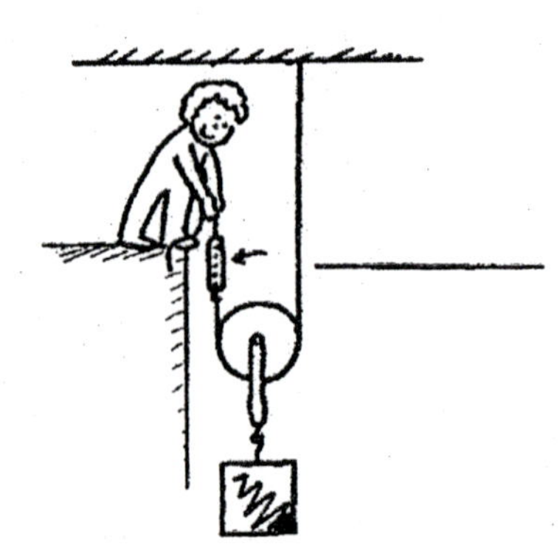

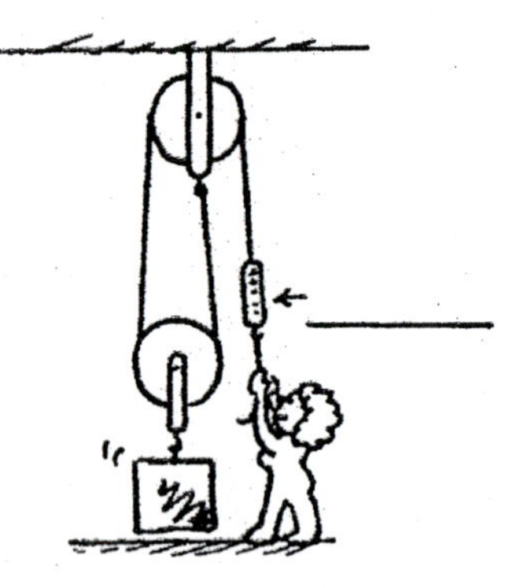

3. A 600-N block is lifted by the friction-free pulley system shown.

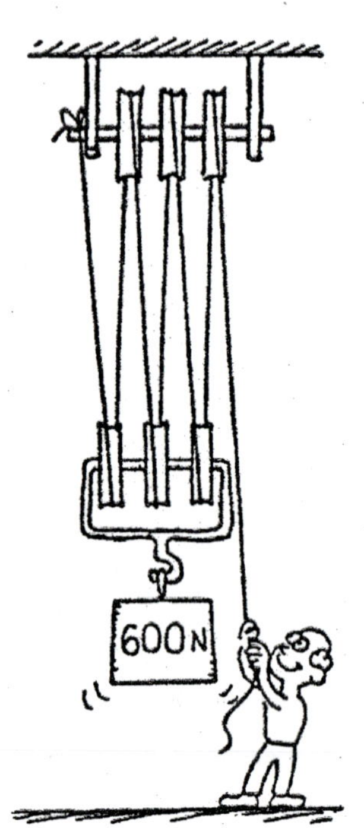

a. How many strands of rope support the 600-N weight?

____

b. What is the tension in each strand?

____

c. What is the tension in the end held by the man?

____

d. If the man pulls his end down 60 cm, how many cm will the weight rise?

____

e. If the man does 60 joules of work, what will be the increase of PE of the 600-N weight?

____

4. Why don't balls bounce as high during the second bounce as they do in the first bounce?

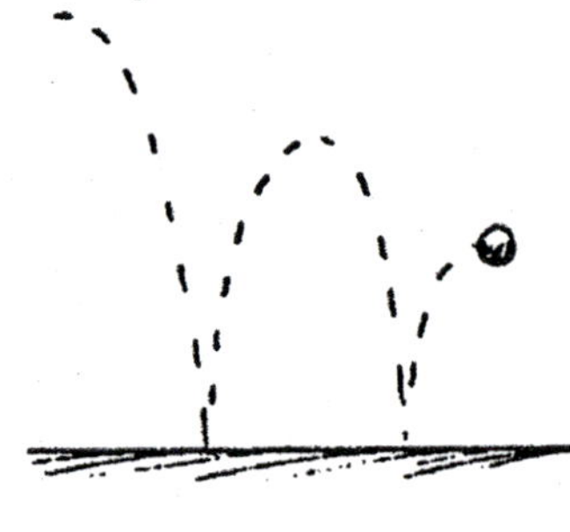

____

____

____

Can you see how the conservation of energy applies to all changes in nature?

Hewitt Drew it!

Name ______________________________ Date __________

CONCEPTUAL *Physics* FUNDAMENTALS PRACTICE PAGE

Chapter 5 Momentum and Energy

Momentum and Energy

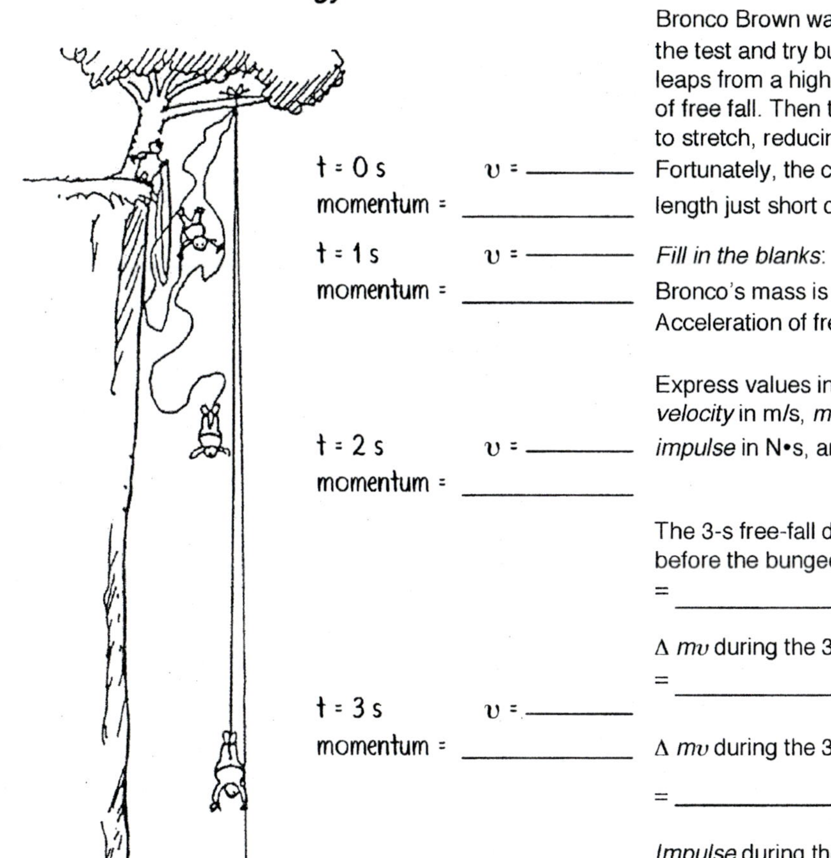

t = 0 s v = ________
momentum = ________

t = 1 s v = ________
momentum = ________

t = 2 s v = ________
momentum = ________

t = 3 s v = ________
momentum = ________

t = 5 s v = ________
momentum = ________

Bronco Brown wants to put $Ft = \Delta\, mv$ to the test and try bungee jumping. Bronco leaps from a high cliff and experiences 3 s of free fall. Then the bungee cord begins to stretch, reducing his speed to zero in 2 s. Fortunately, the cord stretches to its maximum length just short of the ground below.

Fill in the blanks:

Bronco's mass is 100 kg.
Acceleration of free fall is 10 m/s^2.

Express values in SI units (*distance* in m, *velocity* in m/s, *momentum* in kg•m/s, *impulse* in N•s, and *deceleration* in m/s^2).

The 3-s free-fall distance of Bronco just before the bungee cord begins to stretch
= ________

$\Delta\, mv$ during the 3 to 5-s interval of free fall
= ________

$\Delta\, mv$ during the 3 to 5-s of slowing down
= ________

Impulse during the 3 to 5-s of slowing down
= ________

Average force exerted by the cord during the 3 to 5-s interval of slowing down
= ________

How about *work* and *energy*? How much KE does Bronco have 3 s after he first jumps?

____

How much does gravitational PE decrease during this 3 s?

____

What two kinds of PE are changing during the 3 to 5-s slowing-down interval?

____

____

Hewitt Drew it!

Chapter 5 Momentum and Energy

Energy and Momentum

A MiniCooper and a Lincoln Town Car are initially at rest on a horizontal parking lot at the edge of a steep cliff. For simplicity, we assume that the Town Car has twice as much mass as the MiniCooper. Equal constant forces are applied to each car and they accelerate across equal distances (we ignore the effects of friction). When they reach the far end of the lot, the force is suddenly removed, whereupon they sail through the air and crash to the ground below. (The cars are wrecks to begin with, and this is a scientific experiment!)

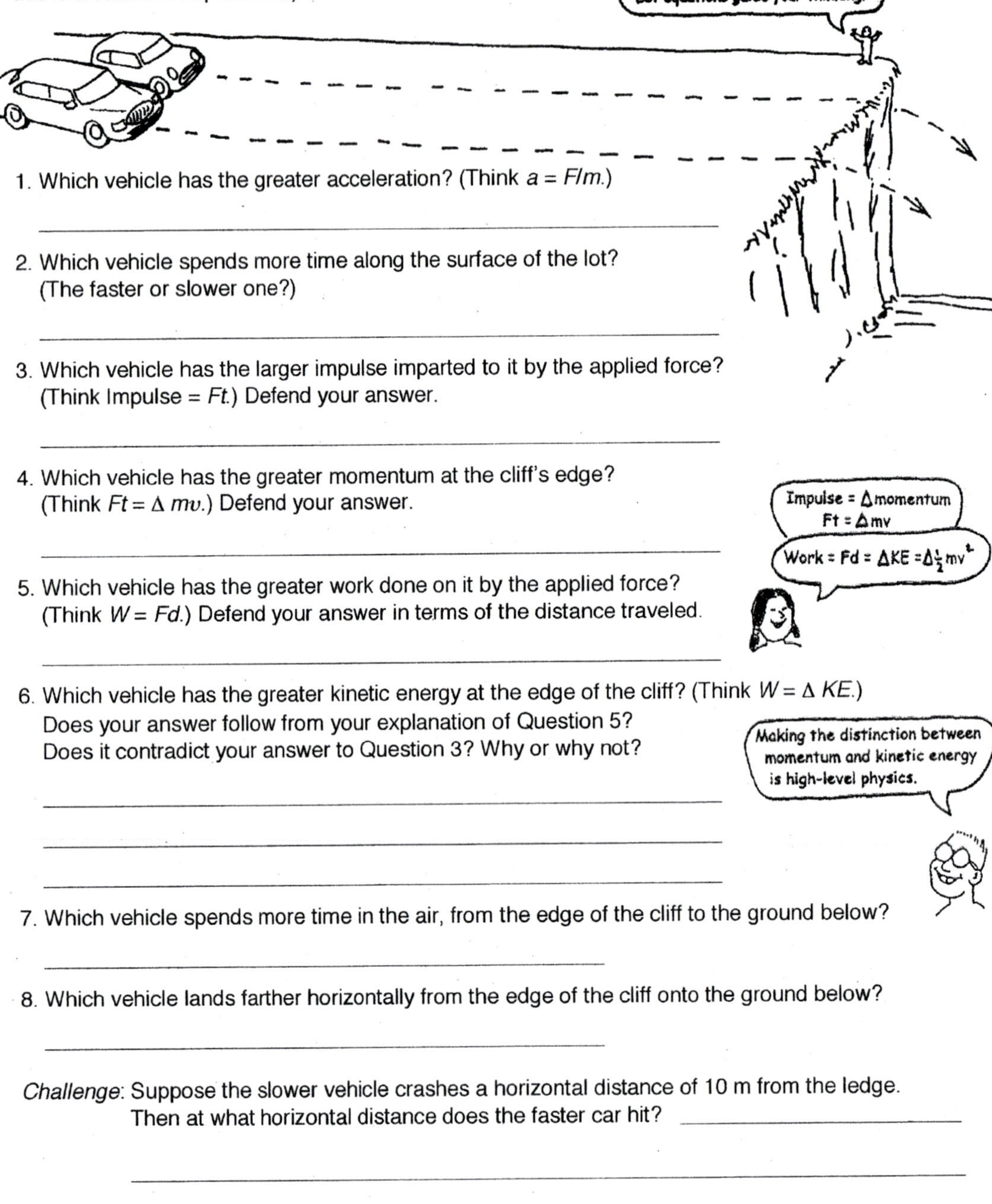

Let equations guide your thinking!
Impulse = Δmomentum
Ft = Δmv
Work = Fd = ΔKE = Δ½mv²
Making the distinction between momentum and kinetic energy is high-level physics.
Hewitt Drew it!

1. Which vehicle has the greater acceleration? (Think $a = F/m$.)

____

2. Which vehicle spends more time along the surface of the lot? (The faster or slower one?)

____

3. Which vehicle has the larger impulse imparted to it by the applied force? (Think Impulse = Ft.) Defend your answer.

____

4. Which vehicle has the greater momentum at the cliff's edge? (Think $Ft = \Delta\, mv$.) Defend your answer.

____

5. Which vehicle has the greater work done on it by the applied force? (Think $W = Fd$.) Defend your answer in terms of the distance traveled.

____

6. Which vehicle has the greater kinetic energy at the edge of the cliff? (Think $W = \Delta\, KE$.) Does your answer follow from your explanation of Question 5? Does it contradict your answer to Question 3? Why or why not?

____

____

____

7. Which vehicle spends more time in the air, from the edge of the cliff to the ground below?

____

8. Which vehicle lands farther horizontally from the edge of the cliff onto the ground below?

____

Challenge: Suppose the slower vehicle crashes a horizontal distance of 10 m from the ledge. Then at what horizontal distance does the faster car hit? ____________________

____

____

Name ______________________ Date __________

CONCEPTUAL Physics FUNDAMENTALS PRACTICE PAGE

Chapter 6 Gravity, Projectiles, and Satellites

Inverse-Square Law

1. Paint spray travels radially away from the nozzle of the can in straight lines. Like gravity, the strength (intensity) of the spray obeys an inverse-square law. Complete the diagram by filling in the blank spaces.

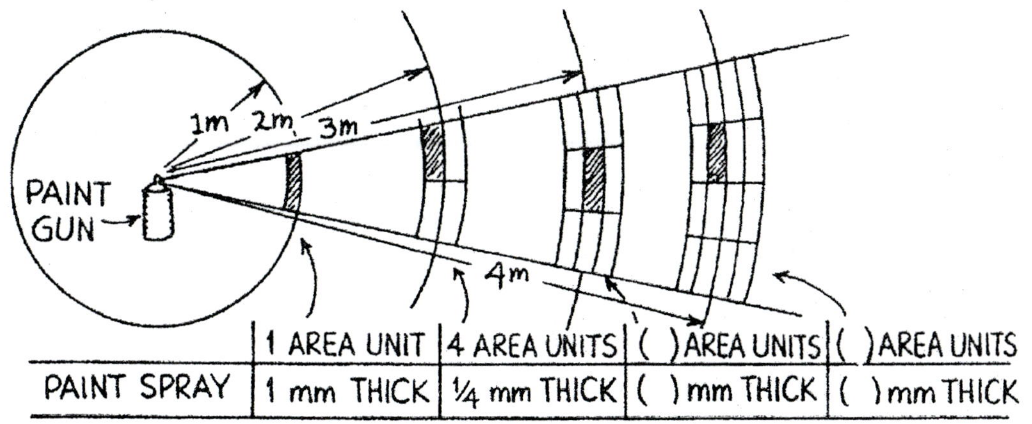

	1 AREA UNIT	4 AREA UNITS	( ) AREA UNITS	( ) AREA UNITS
PAINT SPRAY	1 mm THICK	¼ mm THICK	( ) mm THICK	( ) mm THICK

2. A small light source located 1 m in front of an opening of area 1 m^2 illuminates a wall behind. If the wall is 1 m behind the opening (2 m from the light source), the illuminated area covers 4 m^2. How many square meters will be illuminated if the wall is

5 m from the source? __________

10 m from the source? __________

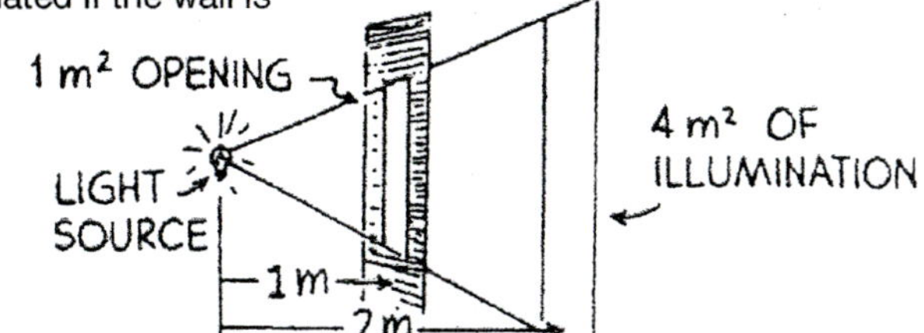

3. If you stand at rest on a weighing scale and find that you are pulled toward Earth with a force of 500 N, then the normal force on the scale is also __________ N and you weigh __________ N. How much does Earth weigh? If you tip the scale upside down and repeat the weighing process, you and Earth are still pulled together with a force of __________ N, and therefore, relative to you, the whole 6,000,000,000,000,000,000,000,000-kg Earth weighs __________ N! Weight, unlike mass, is a relative quantity.

You are pulled to Earth with a force of 500 N, so you weigh 500 N.

DO YOU SEE WHY IT MAKES SENSE TO DISCUSS THE EARTH'S MASS, BUT NOT ITS WEIGHT?

Earth is pulled toward you with a force of 500 N, so it weighs 500 N.

Hewitt Drew it!

Chapter 6 Gravity, Projectiles, and Satellites

***Inverse-Square Law**—continued*

4. The spaceship is attracted to both the planet and the planet's moon. The planet has four times the mass of its moon. The force of attraction of the spaceship to the planet is shown by the vector.

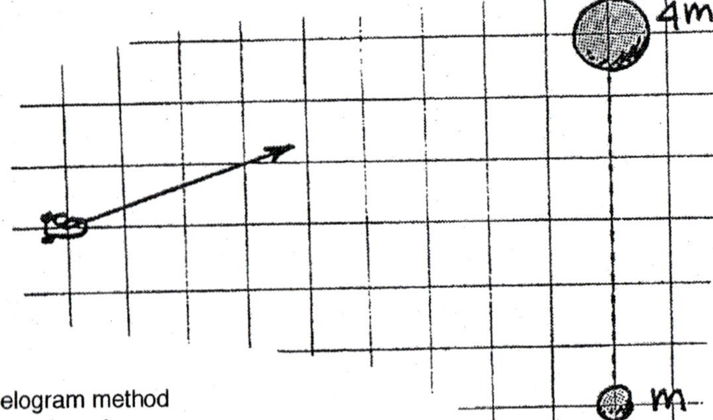

   a. Carefully sketch another vector to show the spaceship's attraction to the moon. Then apply the parallelogram method of Chapter 3 and sketch the resultant force.

   b. Determine the location between the planet and its moon (along the dotted line) where gravitational forces cancel. Make a sketch of the spaceship there.

5. Consider a planet of uniform density that has a straight tunnel from the North Pole through the center to the South Pole. At the surface of the planet, an object weighs 1 ton.

   a. Fill in the gravitational force on the object when it is halfway to the center, then at the center.

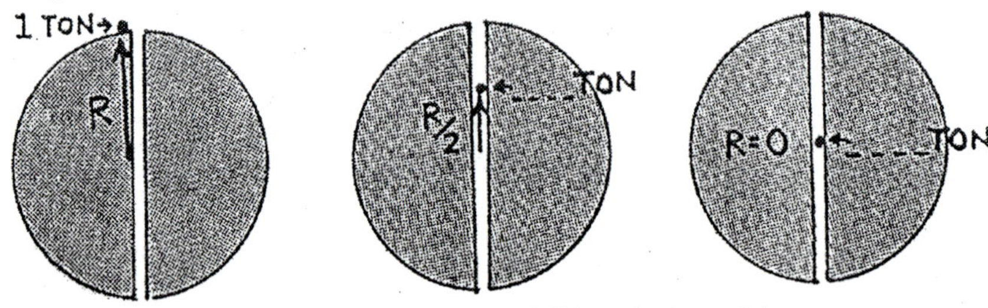

   b. Describe the motion you would experience if you fell into the tunnel.

____

6. Consider an object that weighs 1 ton at the surface of a planet, just before the planet gravitationally collapses.

   a. Fill in the weights of the object on the planet's shrinking surface at the radial values shown.

   b. When the planet has collapsed to 1/10 of its initial radius, a ladder is erected that puts the object as far from its center as the object was originally. Fill in its weight at this position.

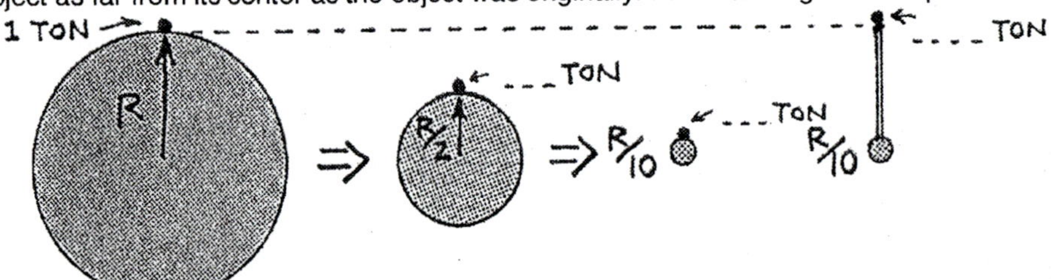

Hewitt Drew it!

Name ______________________________ Date __________

CONCEPTUAL **Physics** FUNDAMENTALS PRACTICE PAGE

Chapter 6 Gravity, Projectiles, and Satellites

Our Ocean Tides

1. Consider two equal-mass blobs of water, A and B, initially at rest in the Moon's gravitational field. The vector shows the gravitational force of the Moon on A.

a. Draw a force vector on B due to the Moon's gravity.

b. Is the force on B more or less than the force on A? __________

c. Why? ______________________________

d. The blobs accelerate toward the Moon. Which has the greater acceleration? [A] [B]

e. Because of the different accelerations, with time

[A gets farther ahead of B] [A and B gain identical speeds] and the distance between A and B

[increases] [stays the same] [decreases].

f. If A and B were connected by a rubber band, with time the rubber band would

[stretch] [not stretch].

g. This [stretching] [nonstretching] is due to the [difference] [nondifference] in the Moon's gravitational pulls.

h. The two blobs will eventually crash into the Moon. To orbit around the Moon instead of crashing into it, the blobs should move

[away from the Moon] [tangentially]. Then their accelerations will consist of changes in

[speed] [direction].

2. Now consider the same two blobs located on opposite sides of Earth.

a. Because of difference in the Moon's pull on the blobs,

they tend to [spread away from each other] [approach each other].

b. Does this spreading produce ocean tides? [Yes] [No]

c. If Earth and Moon were closer, gravitational force between them would be

[more] [the same] [less], and the difference in gravitational forces on the near and far parts

of the ocean would be [more] [the same] [less].

d. Because Earth's orbit about the Sun is slightly elliptical, Earth and Sun are closer in December than in June. Taking the Sun's tidal force into account, on a world average, ocean tides are

greater in [December] [June] [no difference].

Hewitt Drew it!

Name ______________________ Date __________

CONCEPTUAL **Physics** FUNDAMENTALS PRACTICE PAGE

Chapter 6 Gravity, Projectiles, and Satellites

Independence of Horizontal and Vertical Components of Motion

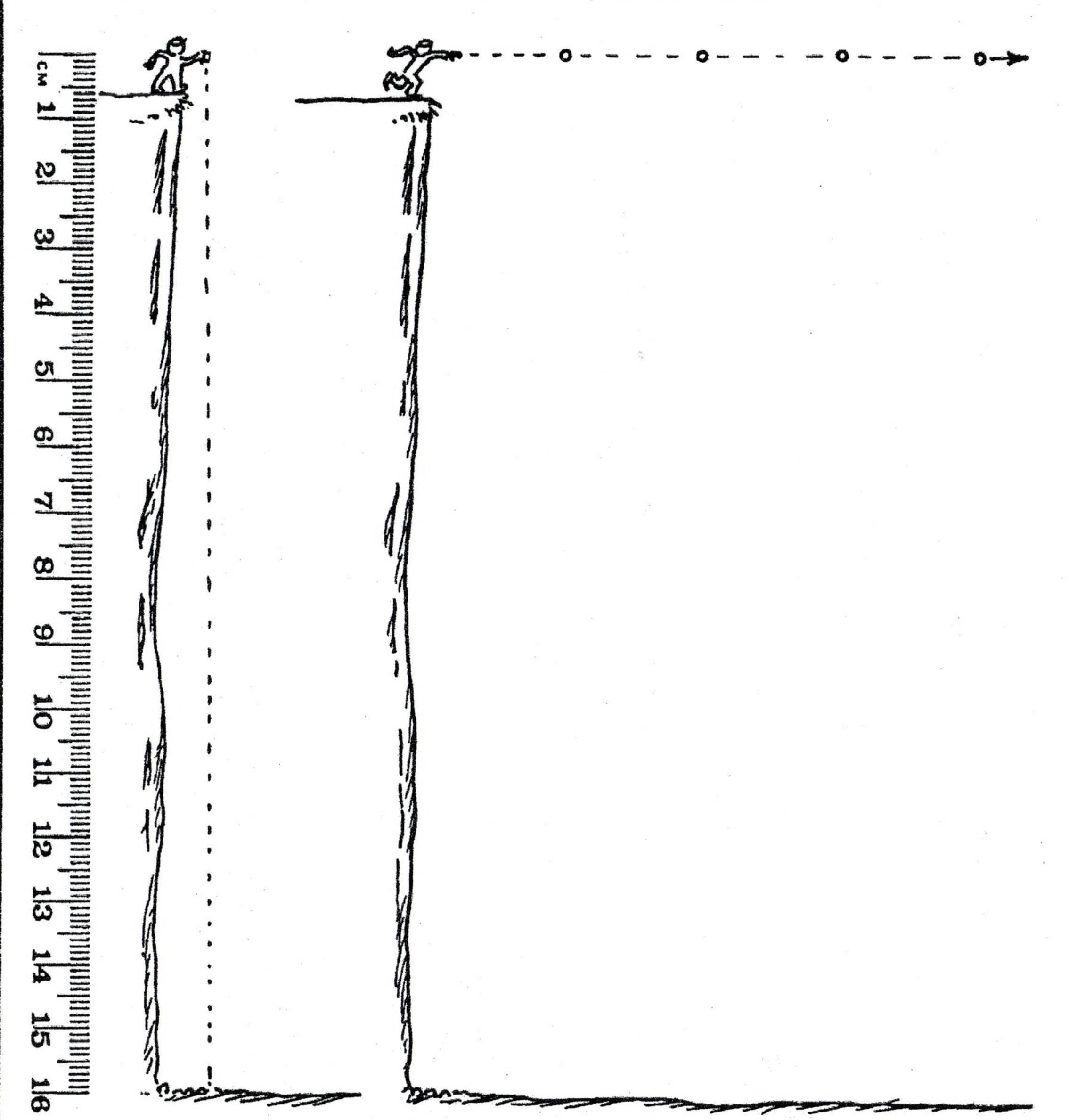

1. Above left: Use the scale 1 cm: 5 m and draw the positions of the dropped ball at 1-second intervals. Neglect air resistance and assume $g = 10\ m/s^2$.
   Estimate the number of seconds the ball is in the air. __________ seconds

2. Above right: The four positions of the thrown ball with no gravity are at 1-second intervals. At 1 cm: 5 m, carefully draw the positions of the ball with gravity. Connect your positions with a smooth curve to show the path of the ball.
   How is the motion in the vertical direction affected by motion in the horizontal direction?

Chapter 6 Gravity, Projectiles, and Satellites

***Independence of Horizontal and Vertical Components of Motion**—continued*

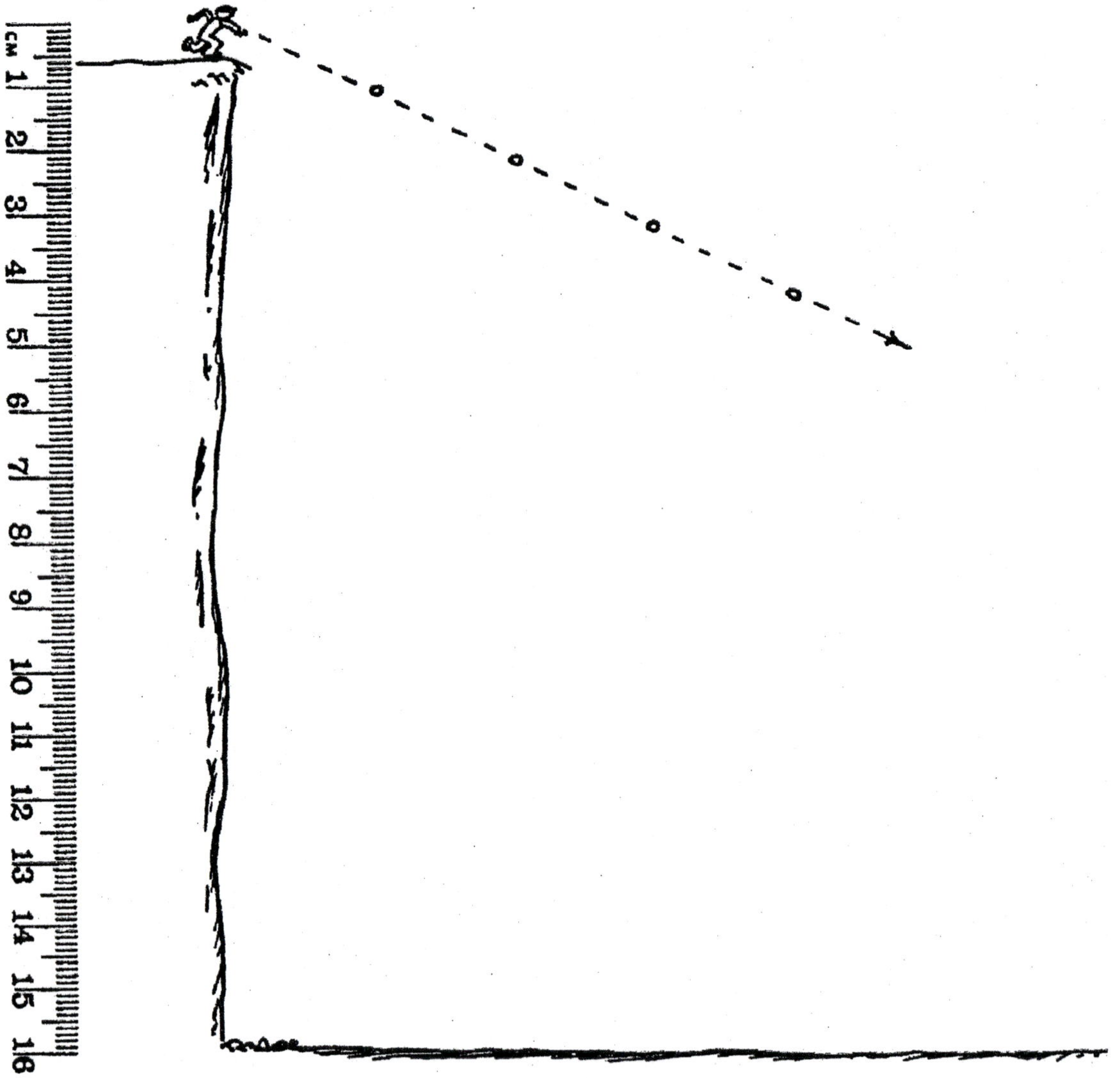

CM 1 2 3 4 5 6 7 8 9 10 11 12 13 14 15 16

3. This time the ball is thrown below the horizontal. Use the same scale 1 cm: 5m and carefully draw the positions of the ball as it falls beneath the dashed line. Connect your positions with a smooth curve. Estimate the number of seconds the ball remains in the air. __________

4. Suppose that you are an accident investigator on site to determine whether or not a car was speeding before it crashed through the rail of the bridge and into the mudbank. The speed limit on the bridge is 55 mph = 24 m/s. What is your conclusion?

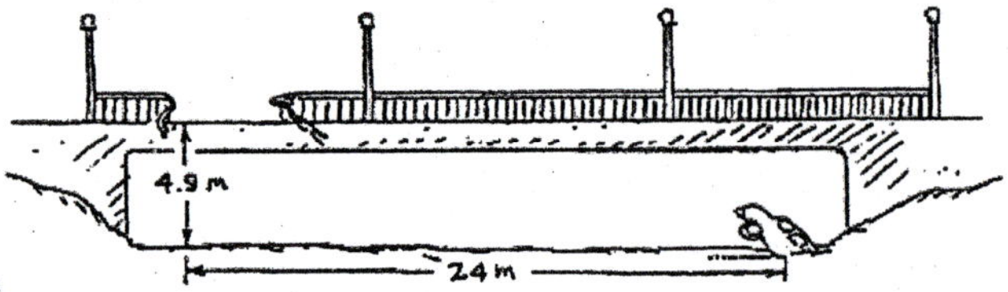

4.9 m

24 m

____

____

____

____

Hewitt Drew it!

Name ______________________________ Date ____________

CONCEPTUAL *Physics* FUNDAMENTALS PRACTICE PAGE

Chapter 6 Gravity, Projectiles, and Satellites
Tossed Ball

A ball tossed upward has initial velocity components 30 m/s vertical, and 5 m/s horizontal. The location of the ball is shown at 1-second intervals. Air resistance is negligible, and $g = 10\ \text{m/s}^2$. Write the values in the boxes for ascending velocity components, and your calculated resultant descending velocities.

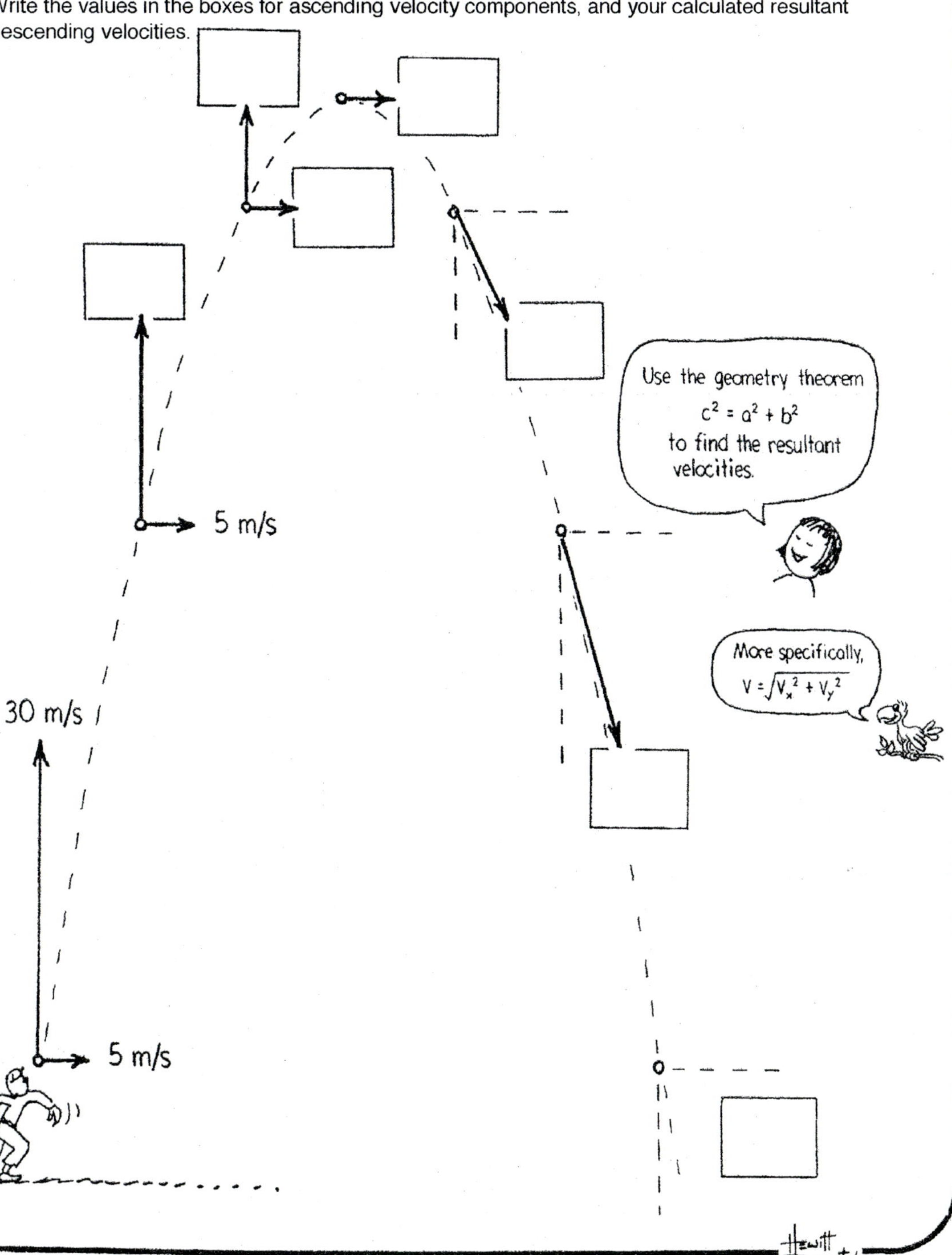

5 m/s
30 m/s
5 m/s
Use the geometry theorem
$c^2 = a^2 + b^2$
to find the resultant velocities.
More specifically,
$V = \sqrt{V_x^2 + V_y^2}$

Name ______________________________ Date __________

CONCEPTUAL **Physics** FUNDAMENTALS PRACTICE PAGE

Chapter 6 Gravity, Projectiles, and Satellites

Satellite in Circular Orbit

1. Figure A shows "Newton's Mountain," so high that its top is above the drag of the atmosphere. The cannonball is fired and hits the ground.

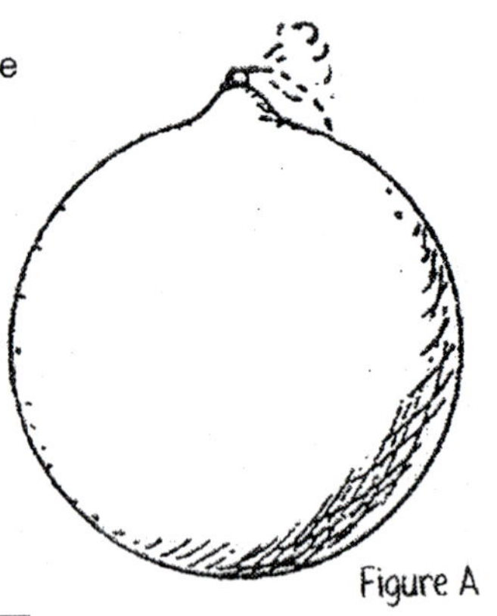

Figure A

a. Draw a likely path that the cannonball might take if it were fired a little bit faster.

b. Repeat for a still greater speed, but less than 8 km/s.

c. Then draw its orbital path for a speed of 8 km/s.

d. What is the shape of the 8-km/s curve?

____

e. What would be the shape of the orbital path if the cannonball were fired at a speed of 9 km/s?

____

2. Figure B shows a satellite in circular orbit.

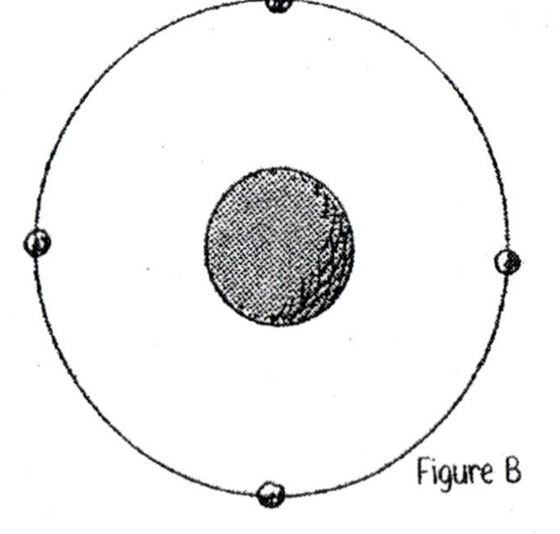

Figure B

a. At each of the four positions, draw a vector that represents the gravitational *force* exerted on the satellite.

b. Label the force vectors ***F***.

c. Then draw a vector at each location to represent the v*elocity* of the satellite, and label it ***V***.

d. Are all four ***F*** vectors the same length? Why or why not?

____

e. Are all four ***V*** vectors the same length? Why or why not?

____

f. What is the angle between your ***F*** and ***V*** vectors? __________

g. Is there any component of ***F*** parallel to ***V***? __________

h. What does this indicate about the work the force of gravity can do on the satellite?

____

i. Does the KE of the satellite in Figure B remain constant or does it vary? __________

j. Does the PE of the satellite remain constant or does it vary? __________

Hewitt Drew it!

Chapter 6 Gravity, Projectiles, and Satellites

Satellite in Elliptical Orbit

3. Repeat the procedure you used for the circular orbit, drawing vectors ***F*** and ***V*** for each position in Figure C, including proper labeling. Show greater magnitudes with greater lengths. Don't bother making the scale accurate.

a. Are your vectors ***F*** all the same magnitude? Why or why not?

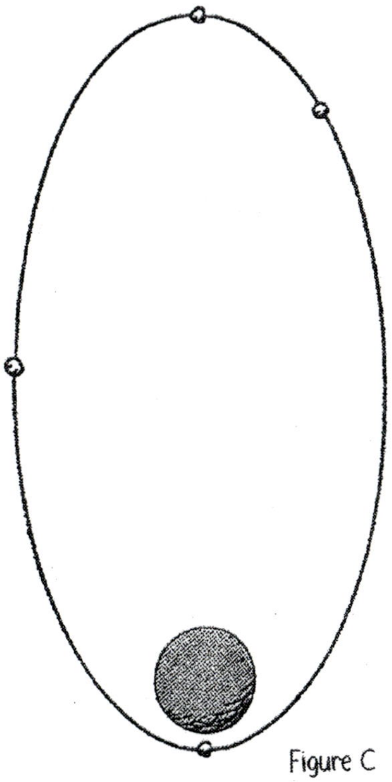

____

____

b. Are your vectors ***V*** all the same magnitude? Why or why not?

____

____

c. Is the angle between vectors ***F*** and ***V*** everywhere the same, or does it vary?

____

____

d. Are there places where there is a component of ***F*** parallel to ***V***?

____

____

Figure C

e. Is work done on the satellite where there is a component of ***F*** parallel to ***V***? If so, does this change the KE of the satellite?

____

f. Where there is a component of ***F*** parallel to or in the direction of ***V***, does this increase or decrease the KE of the satellite?

____

____

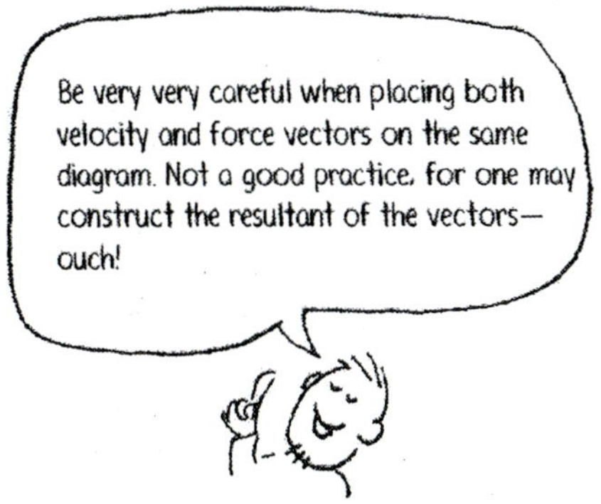

g. What can you say about the sum of KE + PE along the orbit?

____

____

Be very very careful when placing both velocity and force vectors on the same diagram. Not a good practice, for one may construct the resultant of the vectors—ouch!

Hewitt Drew it!

Name ______________________ Date ____________

CONCEPTUAL **Physics** FUNDAMENTALS PRACTICE PAGE

Chapter 7 Fluid Mechanics

Archimedes' Principle I

1. Consider a balloon filled with 1 liter of water (1000 cm^3) in equilibrium in a container of water, as shown in Figure 1.

   a. What is the mass of the 1 liter of water?

   ____

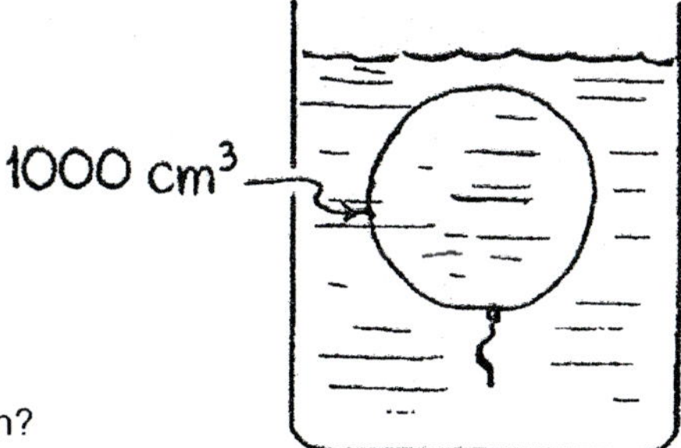

Figure 1

   b. What is the weight of the 1 liter of water?

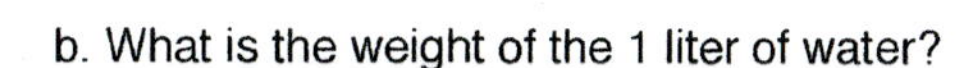

   ____

   c. What is the weight of water displaced by the balloon?

   ____

   d. What is the buoyant force on the balloon?

   ____

WATER DOES NOT SINK IN WATER!

   e. Sketch a pair of vectors in Figure 1: one for the weight of the balloon and the other for the buoyant force that acts on it. How do the size and directions of your vectors compare?

   ____

2. As a thought experiment, pretend we could remove the water from the balloon but still retain the same size of 1 liter. Then inside the balloon is a vacuum.

   a. What is the mass of the liter of nothing?

   ____

   b. What is the weight of the liter of nothing?

   ____

ANYTHING THAT DISPLACES 9.8 N OF WATER EXPERIENCES 9.8 N OF BUOYANT FORCE.

   c. What is the weight of water displaced by the nearly massless 1-liter balloon?

   ____

   d. What is the buoyant force on the nearly massless balloon?

   ____

CUZ IF YOU PUSH 9.8 N OF WATER ASIDE THE WATER PUSHES BACK ON YOU WITH 9.8 N!

   e. In which direction would the nearly massless balloon accelerate?

   ____

Hewitt Drew It!

Chapter 7 Fluid Mechanics

***Archimedes' Principle I**—continued*

3. Assume the balloon is replaced by a 0.5-kilogram piece of wood that has exactly the same volume (1000 cm^3), as shown in Figure 2. The wood is held in the same submerged position beneath the surface of the water.

1000 cm³

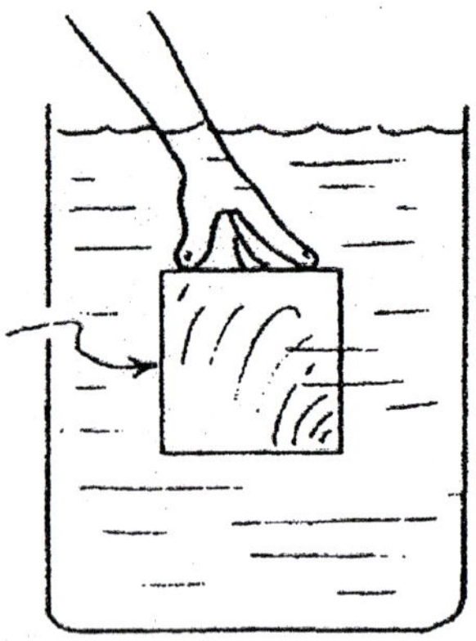

Figure 2

   a. What volume of water is displaced by the wood?

   ____

   b. What is the mass of the water displaced by the wood?

   ____

   c. What is the weight of the water displaced by the wood? ______________

   d. How much buoyant force does the surrounding water exert on the wood? ______________

   e. When the hand is removed, what is the net force on the wood?

   ____

   f. In which direction does the wood accelerate when released? ______________

THE BUOYANT FORCE ON A SUBMERGED OBJECT EQUALS THE WEIGHT OF WATER DISPLACED
... NOT THE WEIGHT OF THE OBJECT ITSELF!
... UNLESS IT IS FLOATING!

4. Repeat parts *a* through *f* in the previous question for a 5-kg rock that has the same volume (1000 cm^3), as shown in Figure 3. Assume the rock is suspended in the container of water by a string.

1000 cm³
WHEN THE WEIGHT OF AN OBJECT IS GREATER THAN THE BUOYANT FORCE EXERTED ON IT, IT SINKS!

Figure 3

a. ______________

b. ______________

c. ______________

d. ______________

e. ______________

f. ______________

Hewitt Drew it!

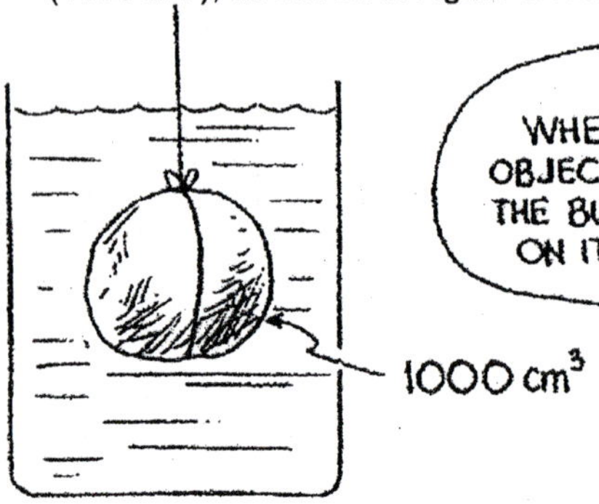

Name ______________________________ Date __________

CONCEPTUAL **Physics** FUNDAMENTALS PRACTICE PAGE

Chapter 7 Fluid Mechanics

Archimedes' Principle II

1. The water lines for the first three cases are shown. Sketch in the appropriate water lines for cases *d* and *e*, and make up your own for case *f*.

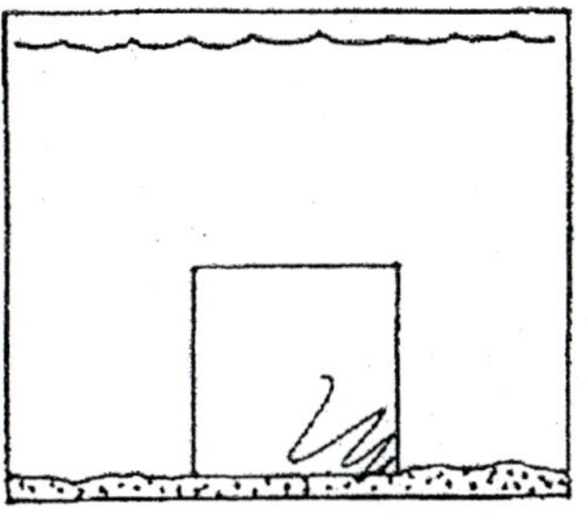

a. DENSER THAN WATER

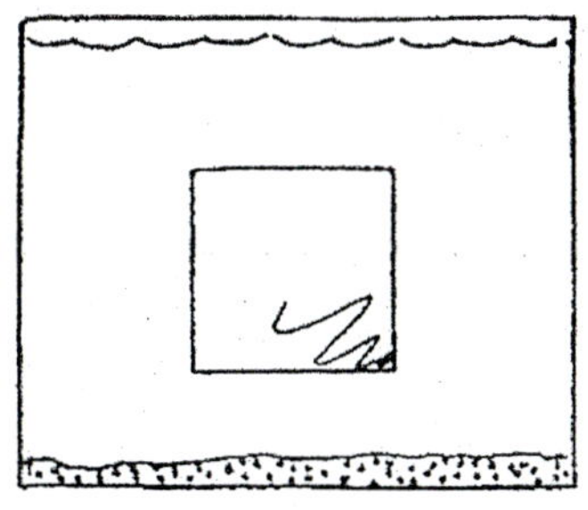

b. SAME DENSITY AS WATER

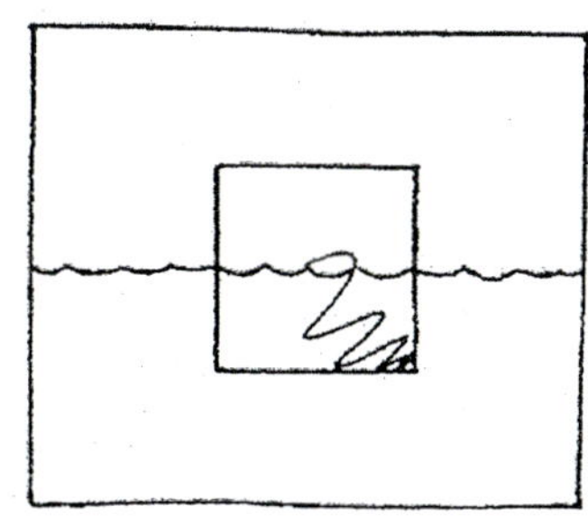

c. 1/2 AS DENSE AS WATER

d. 1/4 AS DENSE AS WATER

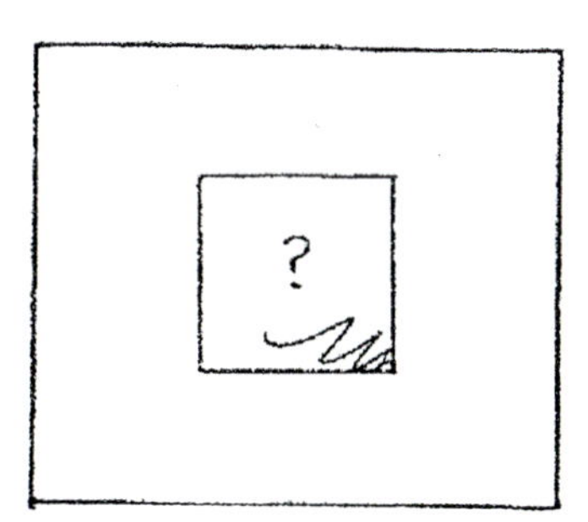

e. 3/4 AS DENSE AS WATER

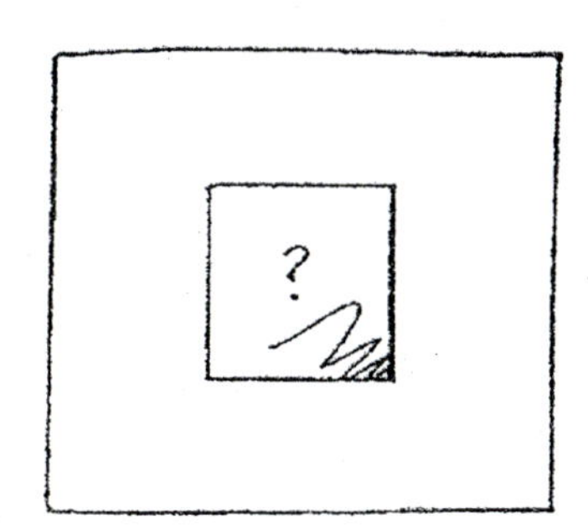

f. ____ AS DENSE AS WATER

2. If the weight of a ship is 100 million N, then the water it displaces weighs ______________.

If a cargo weighing 1000 N is put on board, then the ship will sink down until an extra

______________ of water is displaced.

3. The first two sketches below show the water line for an empty and a loaded ship. Draw the appropriate water line for the third sketch.

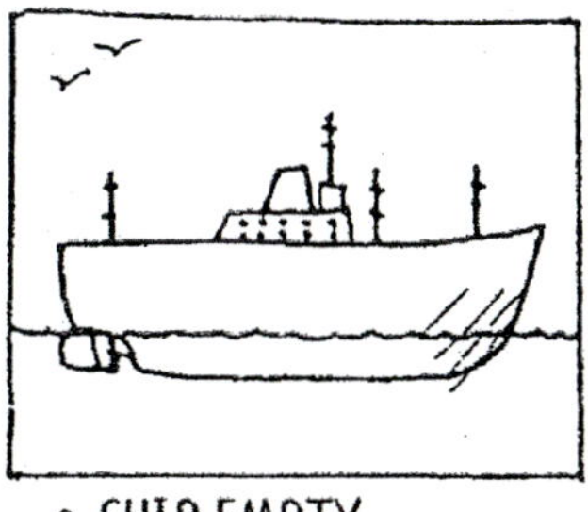

a. SHIP EMPTY

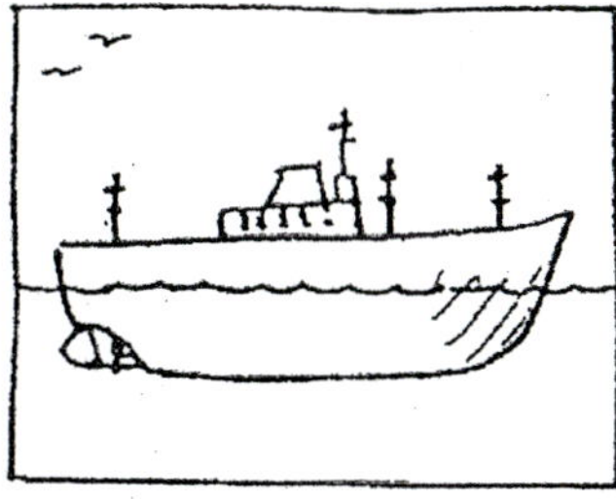

b. SHIP LOADED WITH 50 TONS OF IRON

c. SHIP LOADED WITH 50 TONS OF STYROFOAM

Hewitt Drew it!

Chapter 7 Fluid Mechanics

Archimedes' Principles II—continued

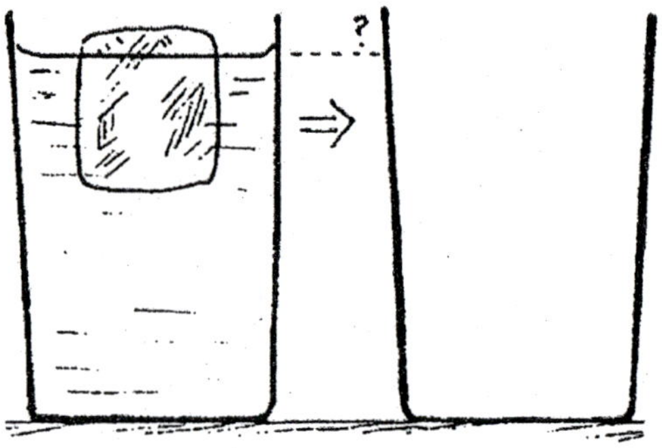

4. Here is an ice cube floating in a glass of ice water. Draw the water line after the ice cube melts. (Will the water line rise, fall, or remain the same?)

____

____

____

5. The air-filled balloon is weighted so it sinks in water. Near the surface, the balloon has a certain volume. Draw the balloon at the bottom (inside the dashed square) and show whether it is bigger, smaller, or the same size.

   a. Since the weighted balloon sinks, how does its overall density compare to the density of water?

   ____

   b. As the weighted balloon sinks, does its density increase, decease, or remain the same?

   ____

   c. Since the weighted balloon sinks, how does the buoyant force on it compare to its weight?

   ____

   d. As the weighted balloon sinks deeper, does the buoyant force on it increase, decrease, or remain the same?

   ____

6. What would your answers be to the above questions (5.a to d) for a rock instead of an air-filled balloon?

   a. _______________

   b. _______________

   c. _______________

   d. _______________

Hewitt Drew it!

Name ______________________ Date __________

CONCEPTUAL Physics FUNDAMENTALS PRACTICE PAGE

Chapter 7 Fluid Mechanics

Gas Pressure

1. A principle difference between a liquid and a gas is that when a liquid is under pressure, its volume

   [increases] [decreases] [doesn't change noticeably]

   and its density

   [increases] [decrease] [doesn't change noticeably].

   When a gas is under pressure, its volume

   [increases] [decreases] [doesn't change noticeably]

   and its density

   [increases] [decreases] [doesn't change noticeably].

PHYSICS
SIGH

GROUND-LEVEL SIZE

2. The sketch above shows the launching of a weather balloon at sea level. Make a sketch of the same weather balloon when it is high in the atmosphere. In words, what is different about its size and why?

   ____

   ____

   ____

   ____

HIGH-ALTITUDE SIZE

3. A hydrogen-filled balloon that weighs 10 N must displace

   __________ N of air in order to float in air. If it displaces

   less than __________ N it will be buoyed up with

   less than __________ N and sink. If it displaces

   more than __________ N of air it will move upward.

4. Why is the cartoon more humorous to physics types than nonphysics types? What physics concept has occurred?

   ____

   ____

   ____

   ____

RATS TO YOU TOO, DANIEL BERNOULLI!

Name ______________________________ Date __________

CONCEPTUAL Physics FUNDAMENTALS PRACTICE PAGE

Mechanics Overview—Chapters 1 to 7

1. The sketch shows the elliptical path described by a satellite about Earth. In which of the labeled positions, A - D, (place an "S" for "same everywhere") does the satellite experience the maximum

   a. gravitational force? __________

   b. speed? __________

   c. momentum? __________

   d. kinetic energy? __________

   e. gravitational potential energy? __________

   f. total energy (KE + PE)? __________

   g. acceleration? __________

   h. angular momentum? __________

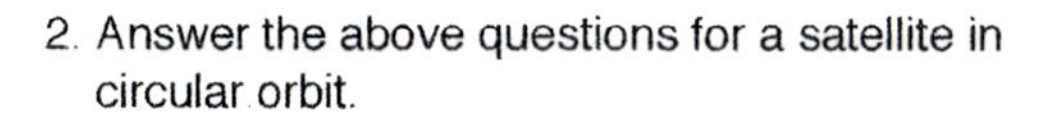

2. Answer the above questions for a satellite in circular orbit.

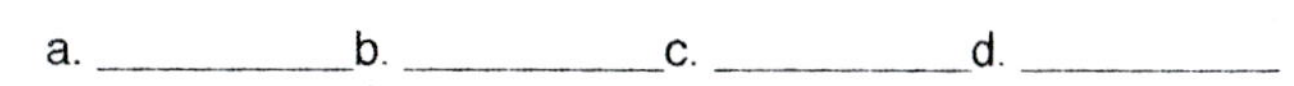

   a. __________ b. __________ c. __________ d. __________

   e. __________ f. __________ g. __________ h. __________

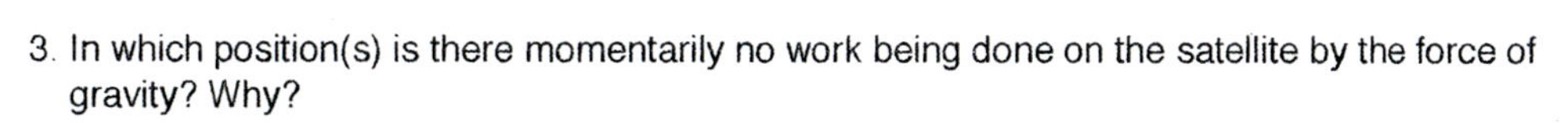

3. In which position(s) is there momentarily no work being done on the satellite by the force of gravity? Why?

   ____

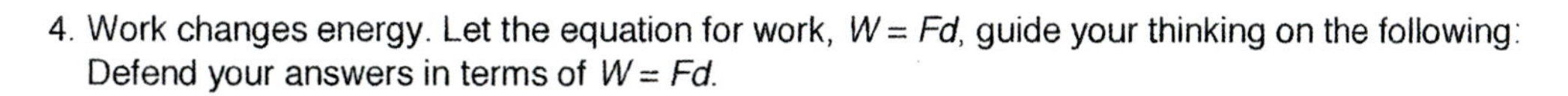

4. Work changes energy. Let the equation for work, $W = Fd$, guide your thinking on the following: Defend your answers in terms of $W = Fd$.

   a. In which position will a several-minutes thrust of rocket engines pushing the satellite forward do the most work on the satellite and give it the greatest change in kinetic energy? (Hint: Think about where the most distance will be traveled during the application of a several-minutes thrust?)

   ____

   b. In which position will a several-minutes thrust of rocket engines pushing the satellite forward do the least work on the satellite and give it the least boost in kinetic energy?

   ____

   c. In which position will a several-minutes thrust of a retro-rocket (pushing opposite to the satellite's direction of motion) do the most work on the satellite and change its kinetic energy the most?

   ____

Hewitt Drew it!

Name ______________________________ Date __________

CONCEPTUAL **Physics** FUNDAMENTALS PRACTICE PAGE

Chapter 8 Temperature, Heat, and Thermodynamics

Measuring Temperatures

1. Complete the table:

TEMPERATURE OF MELTING ICE	°C	32 °F	K
TEMPERATURE OF BOILING WATER	°C	212 °F	K

2. Suppose you apply a flame and warm one liter of water, raising its temperature 10°C. If you transfer the same heat energy to two liters, how much will the temperature rise? For three liters? Record your answers on the blanks in the drawing at the right.

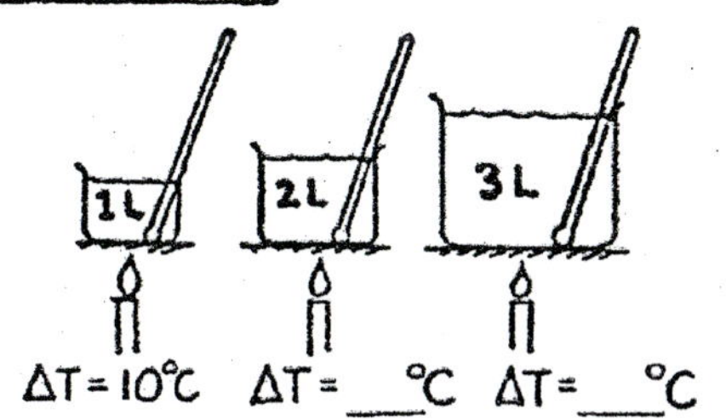

3. A thermometer is in a container half-filled with 20°C water.

   a. When an equal volume of 20°C water is added, the temperature of the mixture will be

   [10°C] [20°C] [40°C].

   b. When instead an equal volume of 40°C water is added, the temperature of the mixture will be

   [20°C] [30°C] [40°C].

   c. When instead a small amount of 40°C water is added, the temperature of the mixture will be

   [20°C] [between 20°C and 30°C] [30°C] [more than 30°C].

Circle one:

4. A small red-hot piece of iron is placed into a large bucket of cool water. (Ignore the heat transfer to the bucket.)

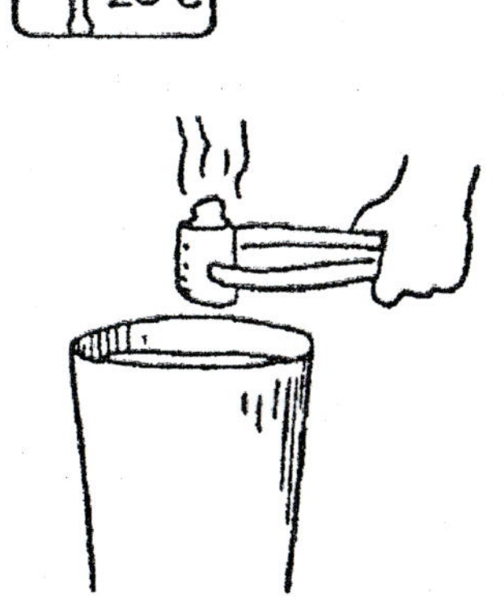

   a. [True] [False] The decrease in iron temperature equals the increase in the water temperature.

   b. [True] [False] The quantity of heat lost by the iron equals the quantity of heat gained by the water.

   c. [True] [False] The iron and water both will eventually reach the same temperature.

   d. [True] [False] The final temperature of the iron and water is halfway between the initial temperatures of each.

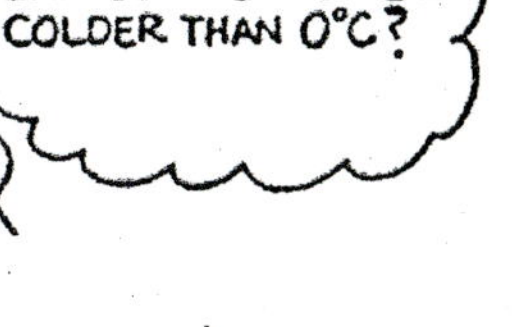

Chapter 8 Temperature, Heat, and Thermodynamics

Thermal Expansion

1. The weight hangs above the floor from the copper wire. When a candle is moved along the wire and warms it, what happens to the height of the weight above the floor? Why?

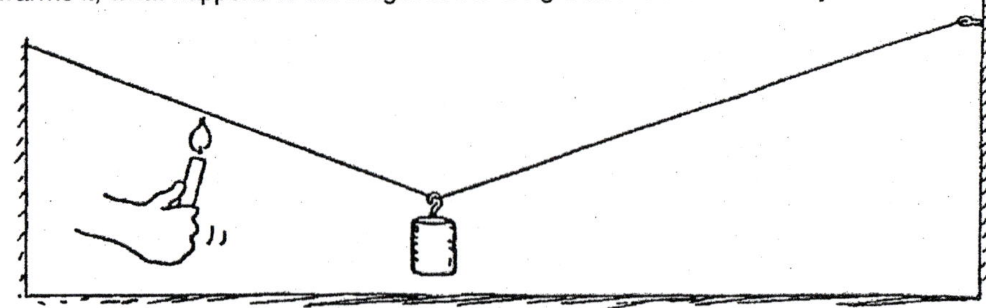

2. The levels of water at 0°C and 1°C are shown below in the first two flasks. At these temperatures there is microscopic slush in the water. There is slightly more slush at 0°C than at 1°C. As the water is warmed, some of the slush collapses as it melts, and the level of the water falls in the tube. That's why the level of water is slightly lower in the 1°C-tube. Make rough estimates and sketch in the appropriate levels of water at the other temperatures shown. What is important about the level when the water reaches 4°C?

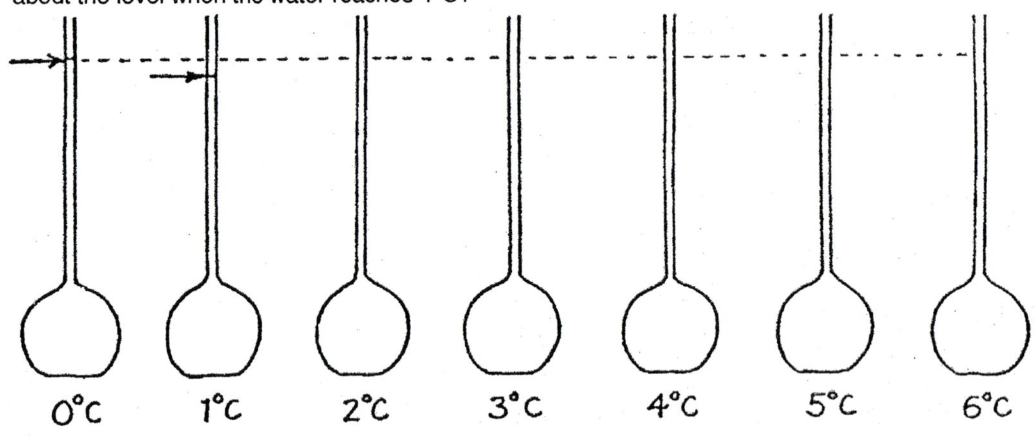

3. The diagram to the left shows an ice-covered pond. Fill in the blanks for likely temperatures of the water at the top and bottom of the pond.

Name ______________________ Date __________

CONCEPTUAL **Physics** FUNDAMENTALS PRACTICE PAGE

Chapter 8 Temperature, Heat, and Thermodynamics

Absolute Zero

A mass of air is contained so that the volume can change but the pressure remains constant. Table I shows air volumes at various temperatures when the air is warmed slowly.

T

W

AIR

HEAT

1. Plot the data in Table I on the graph and connect the points.

TABLE I

TEMP. (°C)	VOLUME (mL)
0	50
25	55
50	60
75	65
100	70

VOLUME (mL)
70
60
50
40
30
20
10
-200 -100 0 50 100
TEMPERATURE (°C)

2. The graph shows how the volume of air varies with temperature at constant pressure. The straightness of the line means that the air expands uniformly with temperature. From your graph, you can predict what will happen to the volume of air when it is cooled.

   Extrapolate (extend) the straight line of your graph to find the temperature at which the volume of the air would become zero. Mark this point on your graph. Estimate this temperature: ________

3. Although air would liquefy before cooling to this temperature, the procedure suggests that there is a lower limit to how cold something can be. This is the absolute zero of temperature.

   Careful experiments show that absolute zero is ________ °C.

4. Scientists measure temperature in *kelvins* instead of degrees Celsius, where the absolute zero of temperature is 0 kelvins. If you relabeled the temperature axis on the graph in Question 1 so that it shows temperature in kelvins, would your graph look like the one below?

   ____

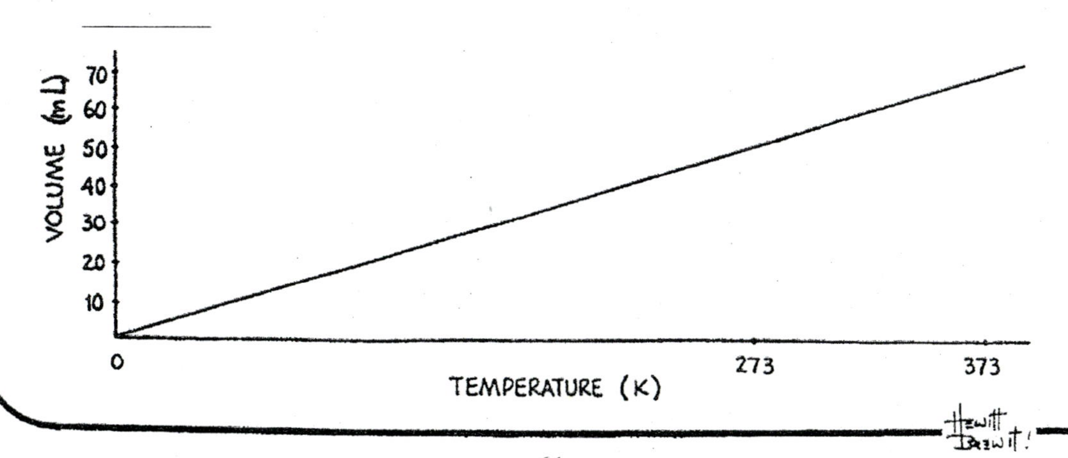

Hewitt Drew it!

Name ______________________________ Date __________

CONCEPTUAL Physics FUNDAMENTALS PRACTICE PAGE

Chapter 9 Heat Transfer and Change of Phase

Transmission of Heat

Circle one:

1. The tips of both brass rods are held in the gas flame.

   a. [True] [False] Heat is conducted only along Rod A.

   b. [True] [False] Heat is conducted only along Rod B.

   c. [True] [False] Heat is conducted equally along both Rod A and Rod B.

   d. [True] [False] The idea that "heat rises" applies to heat transfer by *convection*, not by *conduction*.

2. Why does a bird fluff its feathers to keep warm on a cold day?

   ____

3. Why does a down-filled sleeping bag keep you warm on a cold night? Why is it useless if the down is wet?

   ____

4. What does *convection* have to do with the holes in the shade of the desk lamp?

   ____

5. The warmth of equatorial regions and coldness of polar regions on Earth can be understood by considering light from a flashlight striking a surface. If it strikes perpendicularly, light energy is more concentrated as it covers a smaller area; if it strikes at an angle, the energy spreads over a larger area. So the energy per unit area is less.

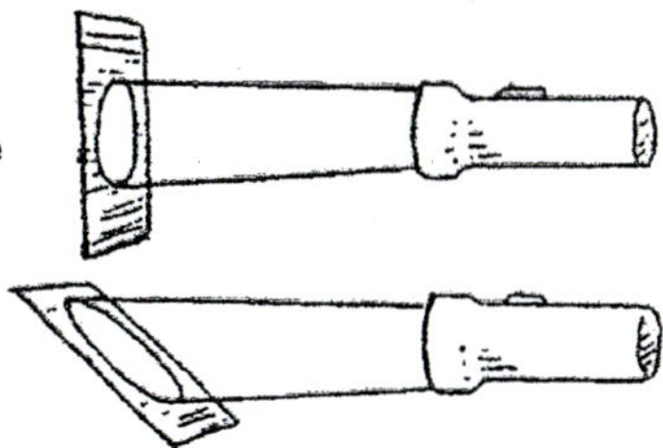

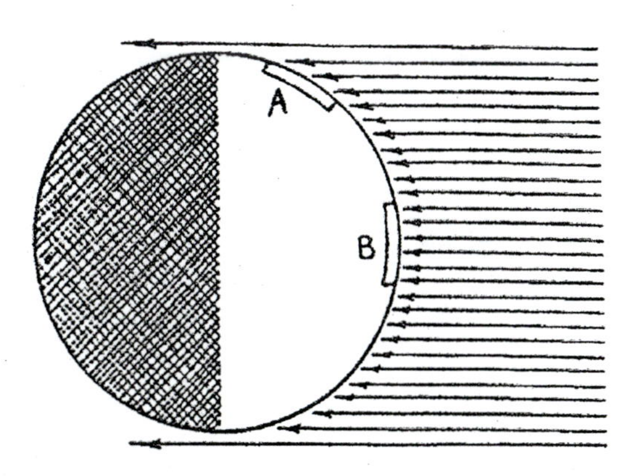

The arrows represent rays of light from the distant Sun incident upon Earth. Two areas of equal size are shown, Area A near the North Pole and Area B near the equator. Count the rays that reach each area, and explain why B is warmer than A.

____

____

____

Chapter 9 Heat Transfer and Change of Phase

***Transmission of Heat**—continued*

6. The Earth's seasons arise from the 23.5-degree tilt of Earth's daily spin axis as it orbits the Sun. When Earth is at the position shown on the right in the sketch below (not to scale), the Northern Hemisphere tilts toward the Sun, and sunlight striking it is strong (more rays per area). Sunlight striking Southern Hemisphere is weak (fewer rays per area). Days in the north are warmer, and daylight is longer. You can see this by imagining Earth making its complete daily 24-hour spin.

   Do two things on the sketch:
   (i) Shade the part of Earth in nighttime darkness for all positions, as is already done in the left position.
   (ii) Label each position with the proper month—March, June, September, or December.

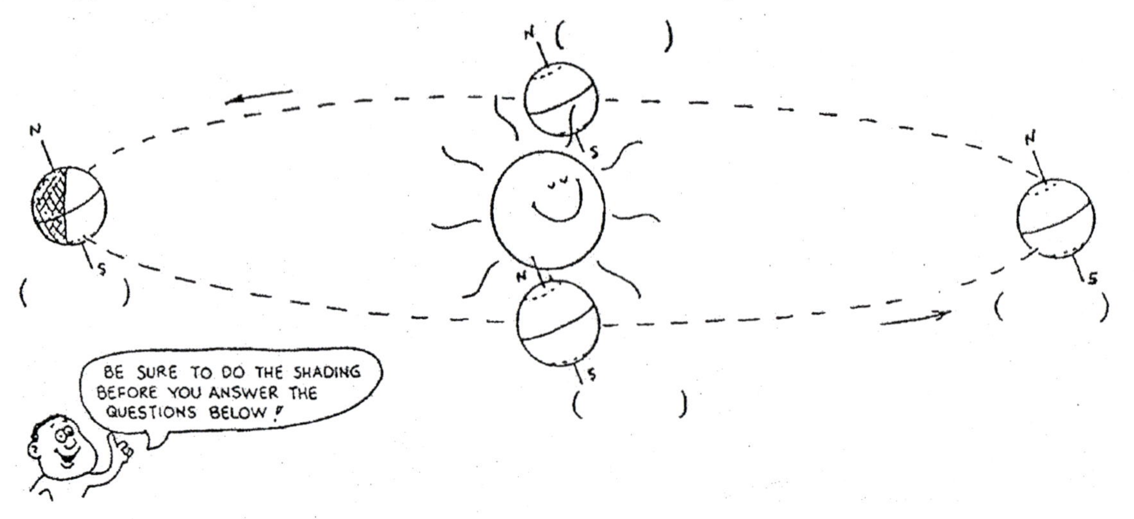

N
S
BE SURE TO DO THE SHADING BEFORE YOU ANSWER THE QUESTIONS BELOW!

a. When Earth is in any of the four positions shown, during one 24-hour spin, a location at the equator receives sunlight half the time and is in darkness the other half the time. This means that regions at the equator always receive about __________ hours of sunlight and __________ hours of darkness.

b. Can you see that in the June position regions farther north have longer daylight hours and shorter nights? Locations north of the Arctic Circle (dotted line in Northern Hemisphere) are continually in view of the Sun as Earth spins, so they get daylight __________ hours a day.

c. How many hours of light and darkness are there in June at regions south of the Antarctic Circle (dotted line in Southern Hemisphere)?

____

d. Six months later, when Earth is at the December position, is the situation in the Antarctic Circle the same or is it the reverse?

____

e. Why do South America and Australia enjoy warm weather in December instead of June?

____

Hewitt Drew it!

Name ______________________ Date __________

CONCEPTUAL *Physics* FUNDAMENTALS PRACTICE PAGE

Chapter 9 Heat Transfer and Change of Phase

Ice, Water, and Steam

All matter can exist in the solid, liquid, or gaseous phases. The solid phase normally exists at relatively low temperatures, the liquid phase at higher temperatures, and the gaseous phase at still higher temperatures. Water is the most common example, not only because of its abundance but also because the temperatures for all three phases are common. Study "Energy and Changes of Phase" in your textbook and then answer the following:

1. How many calories are needed to change 1 gram of 0°C ice to water?

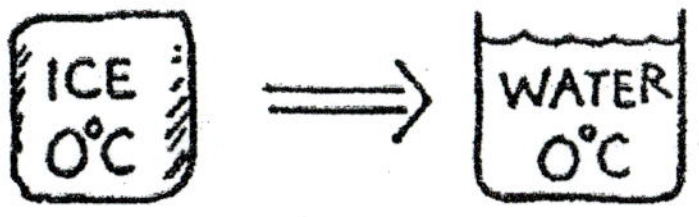

____

2. How many calories are needed to change the temperature of 1 gram of water by 1°C?

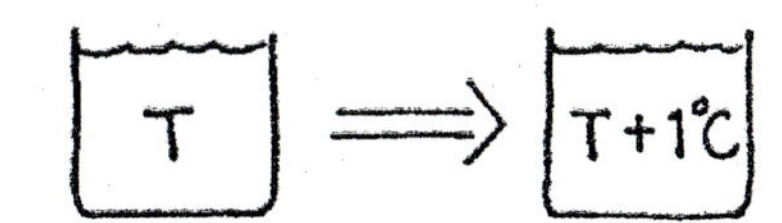

____

3. How many calories are needed to melt 1 gram of 0°C ice and turn it to water at a room temperature of 23°C?

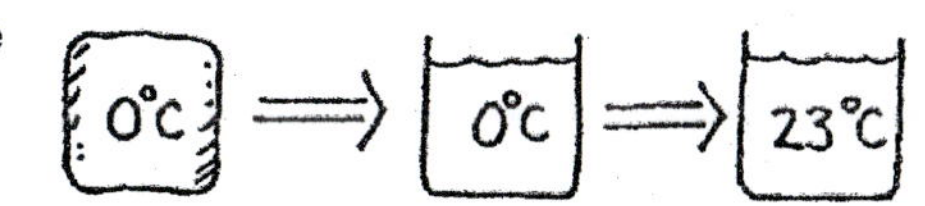

____

4. A 50-gram sample of ice at 0°C is placed in a glass beaker that contains 200 g of water at 20°C.

   a. How much heat is needed to melt the ice? __________

   b. By how much would the temperature of the water change if it released this much heat to the ice? __________

   c. What will be the final temperature of the mixture? (Disregard any heat absorbed by the glass or given off by the surrounding air.)

   ____

5. How many calories are needed to change 1 gram of 100°C boiling water to 100°C steam?

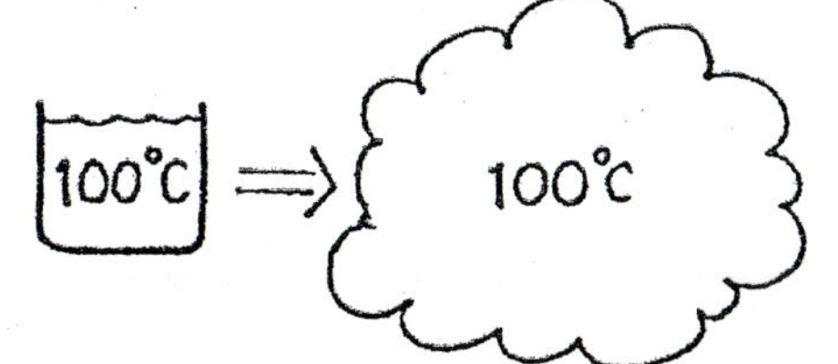

____

6. Fill in the number of calories at each step below for changing the phase of 1 gram of 0°C ice to 100°C steam.

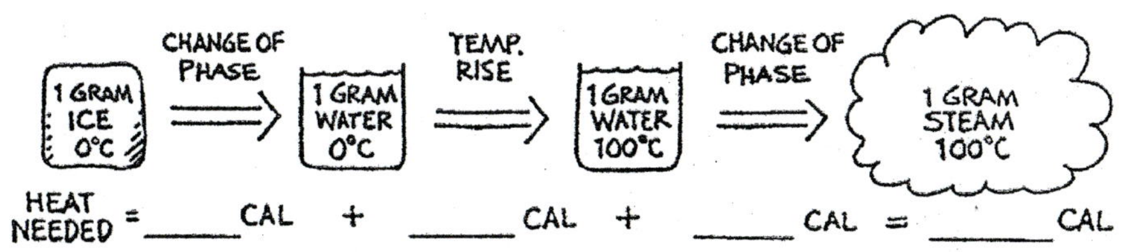

HEAT NEEDED = ______ CAL + ______ CAL + ______ CAL = ______ CAL

Hewitt Drew it!

Chapter 9 Heat Transfer and Change of Phase

Ice, Water, and Steam—continued

7. One gram of steam at 100°C condenses, and the water cools to 22°C.

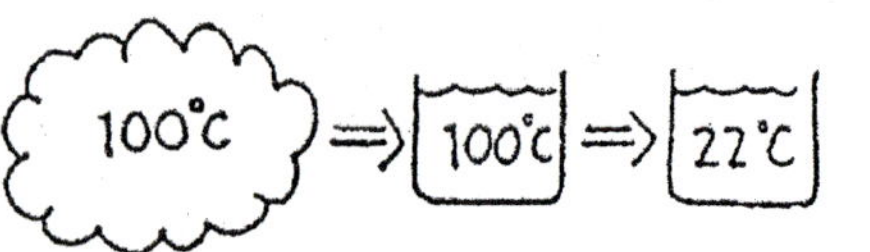

a. How much heat is released when the steam condenses? ________

b. How much heat is released when the water cools from 100°C to 22°C? ________

c. How much heat is released altogether? ________

8. In a household radiator 1000 g of steam at 100°C condenses, and the water cools to 90°C.

a. How much heat is released when the steam condenses? ________

b. How much heat is released when the water cools from 100°C to 90°C?

____

c. How much heat is released altogether? ________

9. Why is it difficult to brew tea on the top of a high mountain?

____

10. How many calories are given up by 1 gram of 100°C steam that condenses to 100°C water?

____

11. How many calories are given up by 1 gram of 100°C steam that condenses and drops in temperature to 22°C water?

____

12. How many calories are given to a household radiator when 1000 grams of 100°C steam condenses, and drops in temperature to 90°C water?

____

13. To get water from the ground, even in the hot desert, dig a hole about a half meter wide and a half meter deep. Place a cup at the bottom. Spread a sheet of plastic wrap over the hole and place stones along the edge to hold it secure. Weight the center of the plastic with a stone so it forms a cone shape. Why will water collect in the cup? (Physics can save your life if you're ever stranded in a desert!)

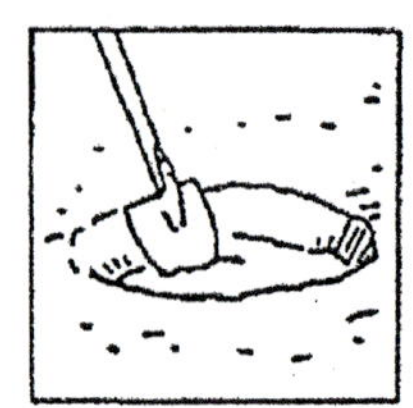

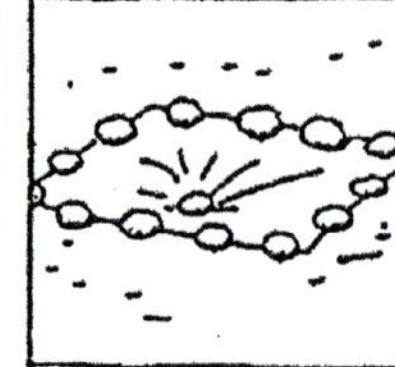

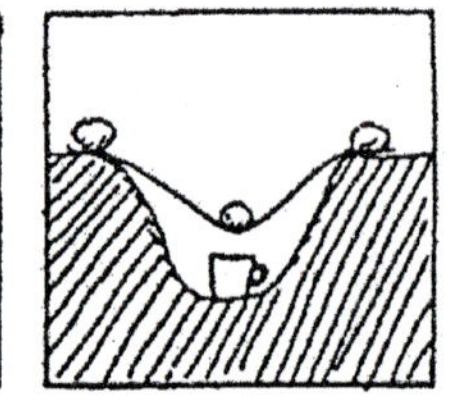

____

____

____

Hewitt Drew it!

Name ______________________ Date __________

CONCEPTUAL **Physics** FUNDAMENTALS PRACTICE PAGE

Chapter 9 Heat Transfer and Change of Phase

Evaporation

1. Why do you feel colder when you swim in a pool on a windy day?

____

2. Why does your skin feel cold when a little rubbing alcohol is applied to it?

____

3. Briefly explain from a molecular point of view why evaporation is a cooling process.

____

____

4. When hot water rapidly evaporates, the result can be dramatic. Consider 4 g of boiling water spread over a large surface so that 1 g rapidly evaporates. Suppose further that the surface and surroundings are very cold so that all 540 calories for evaporation come from the remaining 3 g of water.

   a. How many calories are taken from each gram of water that remains?

   ____

   b. How many calories are released when 1 g of 100°C water cools to 0°C?

   ____

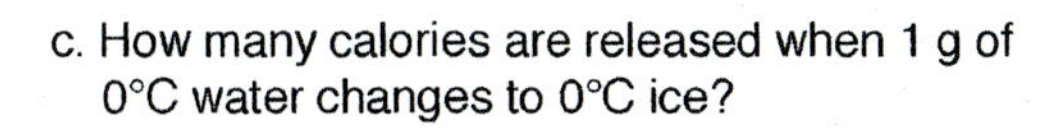

   c. How many calories are released when 1 g of 0°C water changes to 0°C ice?

   ____

   d. What happens in this case to the remaining 3 g of boiling water when 1 g rapidly evaporates?

____

____

____

Chapter 9 Heat Transfer and Change of Phase

Our Earth's Hot Interior

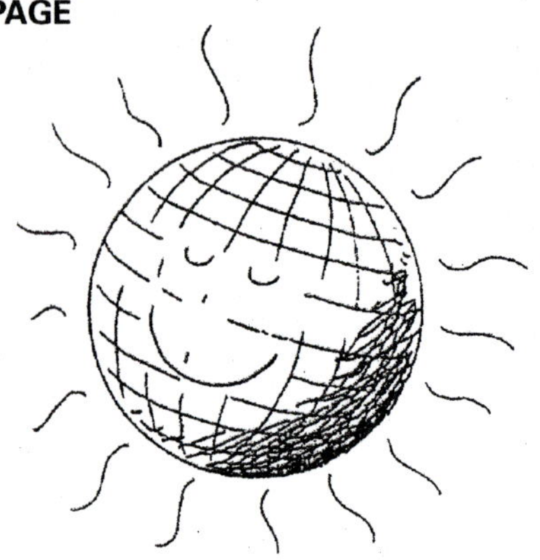

A major puzzle faced scientists in the 19th century. Volcanoes showed that Earth is molten beneath its crust. Penetration into the crust by bore holes and mines showed that Earth's temperature increases with depth. Scientists found that heat flows from the interior to the surface. They assumed that the source of Earth's internal heat was primordial, the afterglow of its fiery birth. Measurements of cooling rates indicated a relatively young Earth—some 25 to 30 millions years in age. But geological evidence indicated an older Earth. This puzzle wasn't solved until the discovery of radioactivity. Then it was learned that the interior is kept hot by the energy of radioactive decay. We now know the age of Earth is some 4.5 billions years—a much older Earth.

All rock contains trace amounts of radioactive minerals. Those in common granite release energy at the rate 0.03 Joule/kilogram•year. Granite at Earth's surface transfers this energy to the surroundings as fast as it is generated, so we don't find granite warm to the touch. But what if a sample of granite were thermally insulated? That is, suppose the increase of internal energy due to radioactivity were contained. Then it would get hotter. How much? Let's figure it out, using 790 joule/kilogram kelvin as the specific heat of granite.

Calculations to make:

1. How many joules are required to increase the temperature of 1 kg of granite by 1000 K?

   ____

2. How many years would it take radioactive decay in a kilogram of granite to produce this many joules?

   ____

Questions to answer:

1. How many years would it take a thermally insulated 1-kg chunk of granite to undergo a 1000 K increase in temperature?

   ____

2. How many years would it take a thermally insulated one-million-kilogram chunk of granite to undergo a 1000 K increase in temperature?

   ____

3. Why are your answers to the above the same (or different)?

   ____

An electric toaster stays hot while electric energy is supplied, and doesn't cool until switched off. Similarly, do you think the energy source now keeping the Earth hot will one day suddenly switch off like a disconnected toaster — or gradually decrease over a long time?

Circle one:

4. [True] [False] The energy produced by Earth radioactivity ultimately becomes terrestrial radiation.

Name _______________________________________ Date ______________

Chapter 10 Static and Current Electricity

Static Charge

1. Consider the diagram below.
   a. A pair of insulated metal spheres, A and B, touch each other, so in effect they form a single uncharged conductor.
   b. A positively charged rod is brought near A, but not touching, and electrons in the metal sphere are attracted toward the rod. Charges in the spheres have redistributed, and the negative charge is labeled. Draw the appropriate + signs that are repelled to the far side of B.
   c. Draw the signs of charge when the spheres are separated while the rod is still present, and
   d. after the rod has been removed. Your completed work should be similar to Figure 22.7 in the textbook. The spheres have been charged by *induction*.

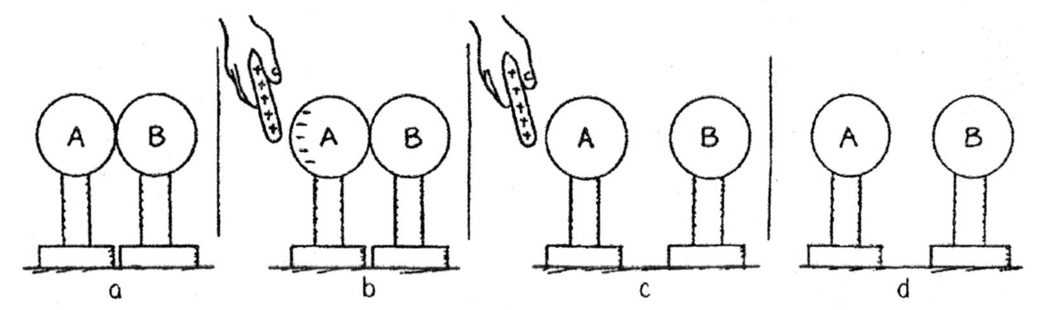

2. Consider below a single metal insulated sphere, (*a*) initially uncharged. When a negatively charged rod is nearby, (*b*), charges in the metal are separated. Electrons are repelled to the far side. When the sphere is touched with your finger, (*c*), electrons flow out of the sphere to Earth through your hand. The sphere is "grounded." Note the positive charge remaining (*d*) while the rod is still present and your finger removed, and (*e*) when the rod is removed. This is an example of *charge induction by grounding*. In this procedure the negative rod "gives" a positive charge to the sphere.

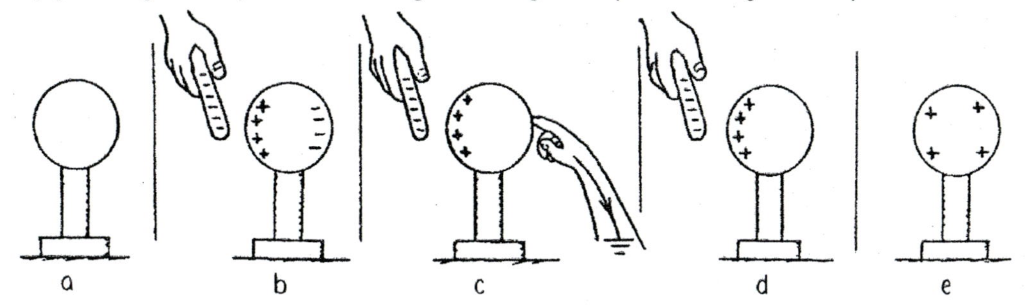

The diagrams below show a similar procedure with a positive rod. Draw the correct charges for *a* through *e*.

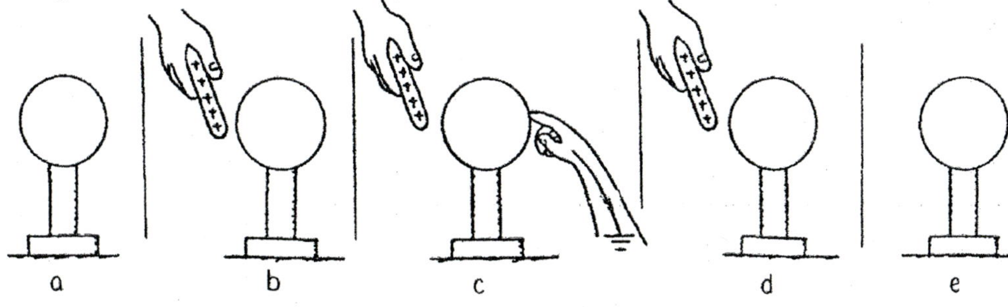

Chapter 10 Static and Current Electricity

Electric Potential

1. Just as PE (potential energy) transforms to KE (kinetic energy) for a mass lifted against the gravitational field (left), the electric PE of an electric charge transforms to other forms of energy when it changes location in an electric field (right). When released, how does the KE acquired by each compare to the decrease in PE?

   ____

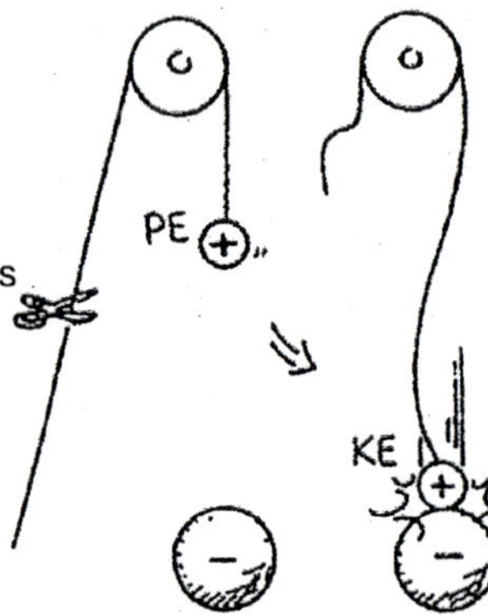

Complete the statements:

2. A force compresses the spring. The work done in compression is the product of the average force and the distance moved. $W = Fd$. This work increases the PE of the spring.

   Similarly, a force pushes the charge (call it a test charge) closer to the charged sphere. The work done in moving the test charge

   is the product of the average __________ and the __________

   moved. $W =$ __________ . This work __________ the PE of the test charge.

   At any point, a greater quantity of test charge means a greater amount of PE, but not a greater amount of PE *per quantity* of charge. The quantities PE (measured in joules) and PE/charge (measured in volts) are different concepts.

   By definition: **Electric Potential** = $\dfrac{\textbf{PE}}{\textbf{charge}}$. 1 volt = 1 joule/coulomb.

3. Complete the statements:

ELECTRIC PE/CHARGE HAS THE SPECIAL NAME ELECTRIC __________
SINCE IT IS MEASURED IN VOLTS IT IS COMMONLY CALLED __________

4. If a conductor connected to the terminal of a battery has a potential of 12 volts, then each coulomb

   of charge on the conductor has a PE of __________ J.

5. Some people are confused between force and pressure. Recall that pressure is force *per area*. Similarly, some people get mixed up between electric PE and voltage. According to this chapter,

   voltage is electric PE per __________ .

Hewitt Drew it!

Name ________________________________ Date ____________

CONCEPTUAL *Physics* FUNDAMENTALS PRACTICE PAGE

Chapter 10 Static and Current Electricity

Flow of Charge

1. Water doesn't flow in the pipe when both ends (*a*) are at the same level. Another way of saying this is that water will not flow in the pipe when both ends have the same potential energy (PE). Similarly, charge will not flow in a conductor if both ends of the conductor are the same electric potential. But tip the water pipe, as in (*b*), and water will flow. Similarly, charge will flow when you increase the electric potential of an electric conductor so there is a potential difference across the ends.

   a. The unit of electric potential difference is

      [volt] [ampere] [ohm] [watt].

   b. It is common to call electric potential difference

      [voltage] [amperage] [wattage].

   c. The flow of electric charge is called electric

      [voltage] [current] [power]

      and is measured in

      [volts] [amperes] [ohms] [watts].

A VOLT IS A UNIT OF __________
AND AN AMPERE IS A UNIT OF __________

DOES VOLTAGE CAUSE CURRENT, OR DOES CURRENT CAUSE VOLTAGE? WHICH IS THE CAUSE AND WHICH IS THE EFFECT?

Complete the statements:

2. a. A current of 1 ampere is a flow of charge at the rate of ________ coulomb per second.

   b. When a charge of 15 C flows through any area in a circuit each second, the current is _______ A.

   c. One volt is the potential difference between two points if 1 joule of energy is needed to move ________ coulomb of charge between the two points.

   d. When a lamp is plugged into a 120-V socket, each coulomb of charge that flows in the circuit is raised to a potential energy of ________ joules.

   e. Which offers more resistance to water flow, a wide pipe or a narrow pipe? ______________

      Similarly, which offers more resistance to the flow of charge, a thick wire or a thin wire?

      ____

Chapter 10 Static and Current Electricity

Ohm's Law

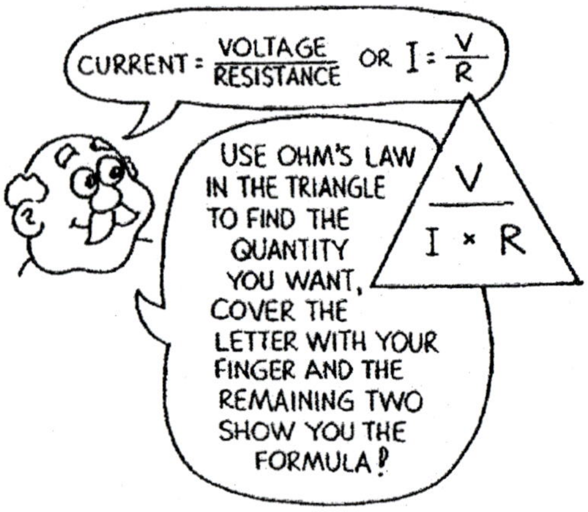

CURRENT = $\frac{\text{VOLTAGE}}{\text{RESISTANCE}}$ OR $I = \frac{V}{R}$

USE OHM'S LAW IN THE TRIANGLE TO FIND THE QUANTITY YOU WANT, COVER THE LETTER WITH YOUR FINGER AND THE REMAINING TWO SHOW YOU THE FORMULA!

V
I × R

CONDUCTORS AND RESISTORS HAVE RESISTANCE TO THE CURRENT IN THEM.

1. How much current flows in a 1000-ohm resistor when 1.5 volts are impressed across it?

   ____

2. If the filament resistance in an automobile headlamp is 3 ohms, how many amps does it draw when connected to a 12-volt battery?

   ____

3. The resistance of the side lights on an automobile are 10 ohms. How much current flows in them when connected to 12 volts?

   ____

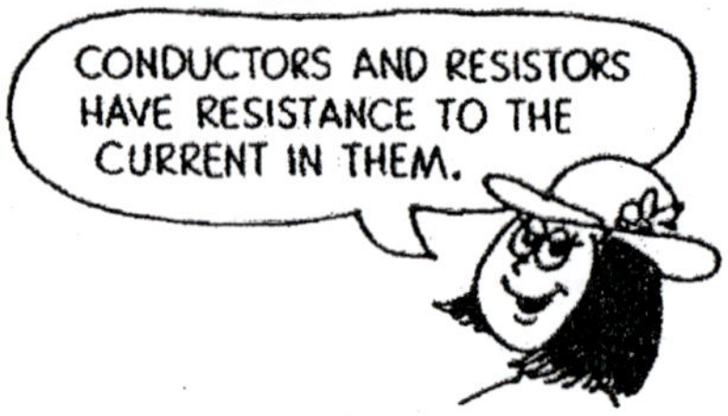

4. What is the current in the 30-ohm heating coil of a coffee maker that operates on a 120-volt circuit?

   ____

5. During a lie detector test, a voltage of 6 V is impressed across two fingers. When a certain question is asked, the resistance between the fingers drops from 400,000 ohms to 200,000 ohms.

   a. What is the current initially through the fingers? ________

   b. What is the current through the fingers when the resistance between them drops? ________

6. How much resistance allows an impressed voltage of 6 V to produce a current of 0.006 A?

   ____

7. What is the resistance of a clothes iron that draws a current of 12 A at 120 V?

   ____

8. What is the voltage across a 100-ohm circuit element that draws a current of 1 A?

   ____

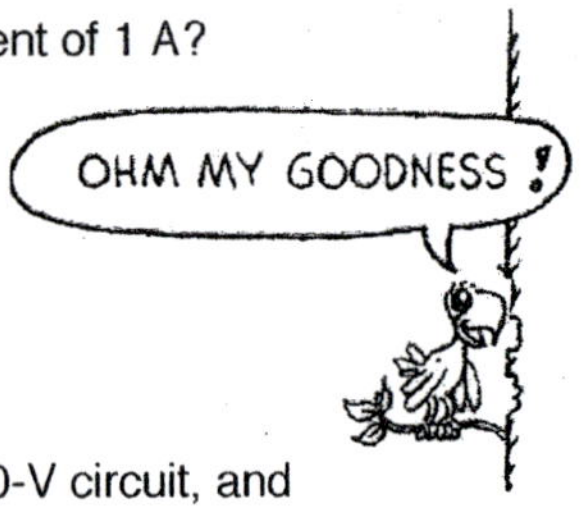

OHM MY GOODNESS!

9. What voltage will produce 3 A through a 15-ohm resistor?

   ____

10. The current in an incandescent lamp is 0.5 A when connected to a 120-V circuit, and 0.2 A when connected to a 10-V source. Does the resistance of the lamp change in these cases? Explain your answer and defend it with numerical values.

    ____

    ____

Hewitt Drew it!

Name ________________________________ Date ____________

CONCEPTUAL Physics FUNDAMENTALS PRACTICE PAGE

Chapter 10 Static and Current Electricity

Electric Power

Recall that the rate at which energy is converted from one form to another is *power*.

$$\text{Power} = \frac{\text{energy converted}}{\text{time}} = \frac{\text{voltage} \times \text{charge}}{\text{time}} = \text{voltage} \times \frac{\text{charge}}{\text{time}} = \text{voltage} \times \text{current}$$

The unit of power is the *watt* (or *kilowatt*), so in units form,

Electric power (*watts*) = current (*amperes*) × voltage (*volts*), where 1 *watt* = 1 *ampere* × 1 *volt*.

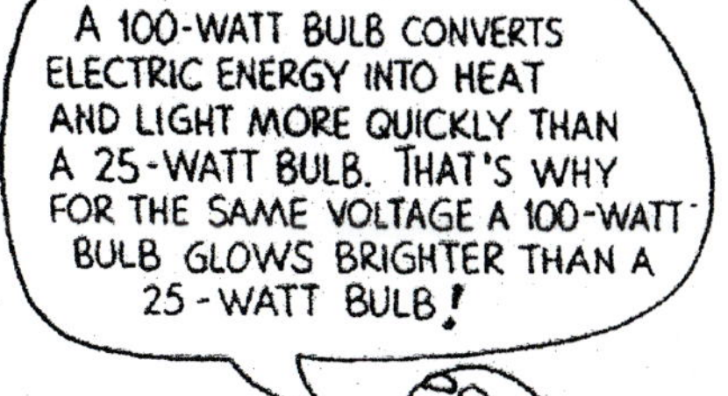

AND CHARGE/TIME = CURRENT ; NEAT ;

1. What is the power when a voltage of 120 V drives a 2-A current through a device?

   ____

2. What is the current when a 60-W lamp is connected to 120 V?

   ____

3. How much current does a 100-W lamp draw when connected to 120 V?

   ____

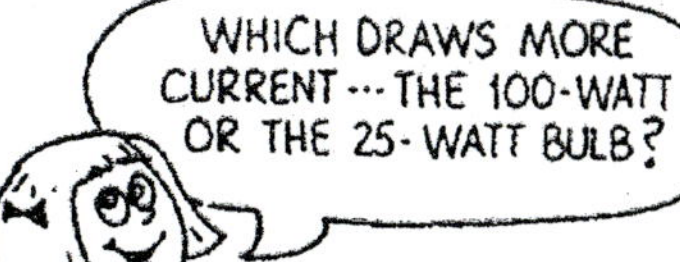

WHICH DRAWS MORE CURRENT --- THE 100-WATT OR THE 25-WATT BULB?

4. If part of an electric circuit dissipates energy at 6 W when it draws a current of 3 A, what voltage is impressed across it?

   ____

5. The equation

$$\text{power} = \frac{\text{energy converted}}{\text{time}}$$

rearranged gives energy converted = ____________________

WATT'S HAPPENING ?

6. Explain the difference between a kilowatt and a kilowatt-hour.

   ____

7. One deterrent to burglary is to leave your front porch light constantly on. If your fixture contains a 60-W bulb at 120 V, and your local power utility sells energy at 10 cents per kilowatt-hour, how much will it cost to leave the light on for the entire month? Show your work on the other side of this page.

   ____

   ____

Hewitt Drew it!

Name ______________________ Date __________

CONCEPTUAL Physics FUNDAMENTALS PRACTICE PAGE

Chapter 10 Static and Current Electricity

Series Circuits

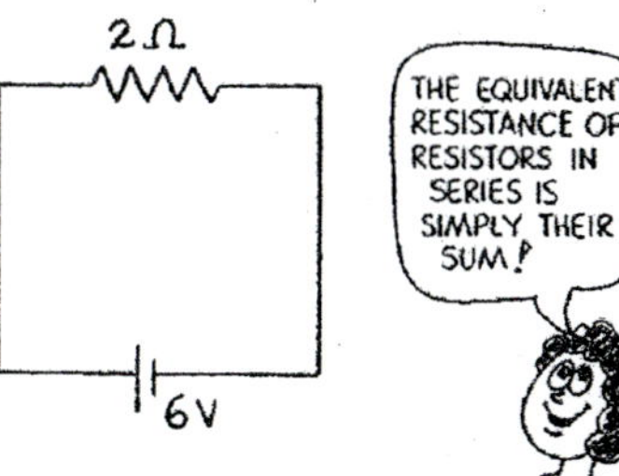

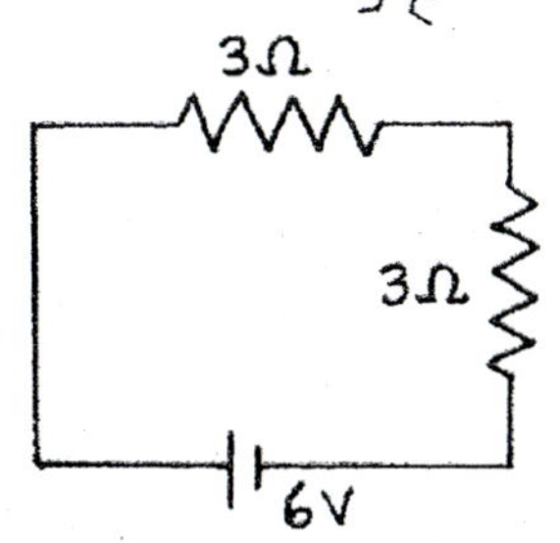

1. In the circuit shown at the right, a voltage of 6 V pushes charge through a single resistor of 2 Ω. According to Ohm's law, the current in the resistor (and therefore in the whole circuit) is

   __________ A.

2. Two 3-Ω resistors and a 6-V battery comprise the circuit on

   the right. The total resistance of the circuit is __________ Ω.

   The current in the circuit is then __________ A.

3. The equivalent resistance of three 4-Ω resistors in series would be

   __________ Ω.

4. Does current flow *through* a resistor, or *across* a resistor? ______________________

   Is voltage established *through* a resistor, or *across* a resistor? ______________________

5. Does current in the lamps of a circuit occur simultaneously, or does charge flow first through one lamp, then the other, and finally the last in turn?

   ____

6. Circuits *a* and *b* below are identical with all bulbs rated at equal wattage (therefore equal resistance). The only difference between the circuits is that Bulb 5 has a short circuit, as shown.

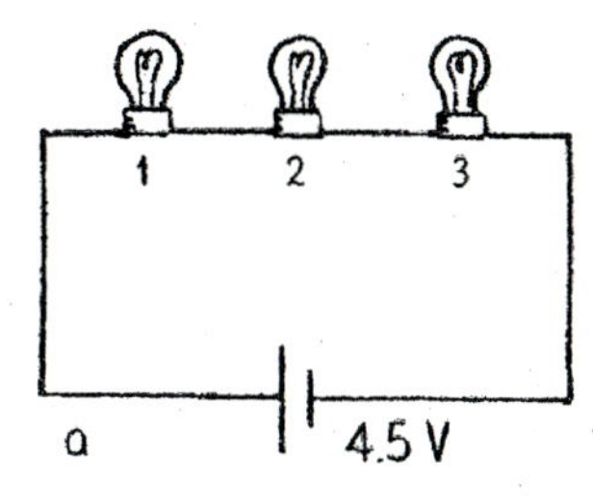

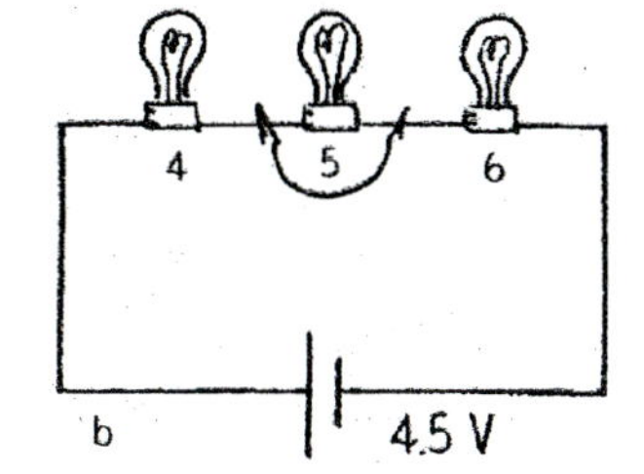

a. In which circuit is the current greater? __________

b. In which circuit are all three bulbs equally bright? __________

c. Which bulbs are the brightest? __________

d. Which bulb is the dimmest? __________

e. Which bulbs have the largest voltage drops across them? __________

f. Which circuit dissipates more power? __________

g. Which circuit produces more light? __________

Hewitt Drew it!

Chapter 10 Static and Current Electricity

Parallel Circuits

THE SUM OF THE CURRENTS IN THE TWO BRANCH PATHS EQUALS THE CURRENT BEFORE IT DIVIDES.

WATER FLOW

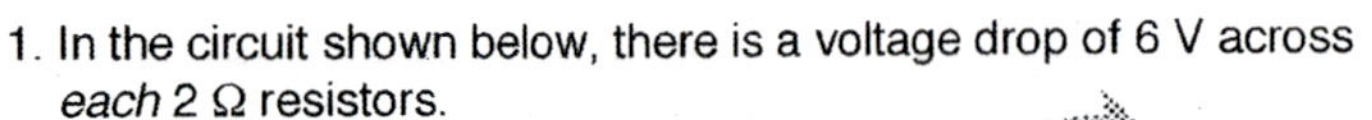

1. In the circuit shown below, there is a voltage drop of 6 V across *each* 2 Ω resistors.

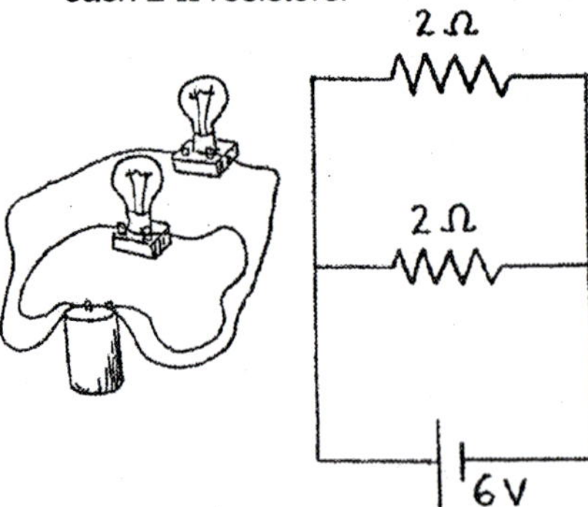

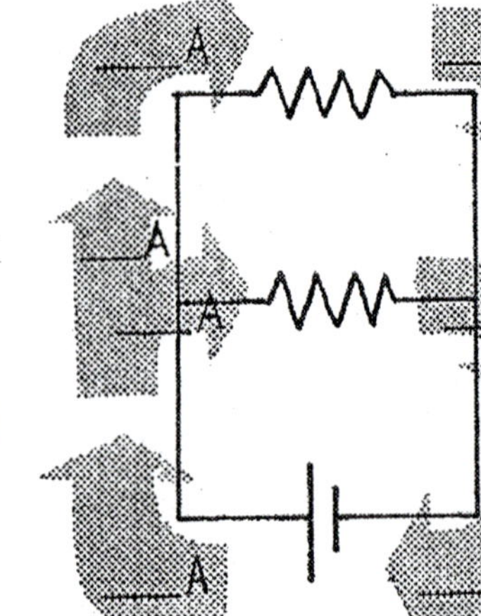

a. By Ohm's law, the current in *each* resistor is __________ A.

b. The current through the battery is the sum of the currents in the resistors, __________ A.

c. Fill in the current in the eight blank spaces in the diagram above of the same circuit.

2. Cross out the circuit below that is *not* equivalent to the circuit above.

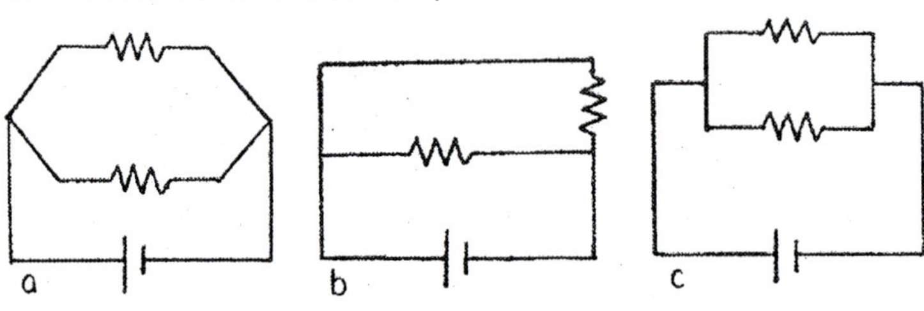

3. Consider the parallel circuit at the right.

a. The voltage drop across each resistor is

__________ V.

b. The current in each branch is:

2-Ω resistor __________ A.

2-Ω resistor __________ A.

1-Ω resistor __________ A.

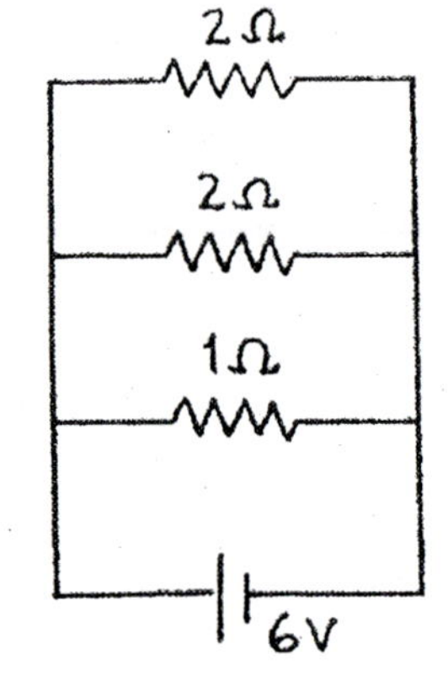

THE EQUIVALENT RESISTANCE OF A PAIR OF RESISTORS IN PARALLEL IS THEIR PRODUCT DIVIDED BY THEIR SUM!

c. The current through the battery equals the sum of the currents which

equals __________ A.

d. The equivalent resistance of the circuit equals __________ Ω.

Hewitt DREW it!

Name ______________________________ Date ____________

CONCEPTUAL PHYSICS FUNDAMENTALS

Figure what the resistances are, then show their values in the blanks to the left of each lamp.

Chapter 10 Static and Current Electricity

Circuit Resistance

All circuits below have the same lamp A with resistance of 6 Ω, and the same 12-volt battery with negligible resistance. The unknown resistances of lamps B through L are such that the current in lamp A remains 1 ampere. *Fill in the blanks*:

a.

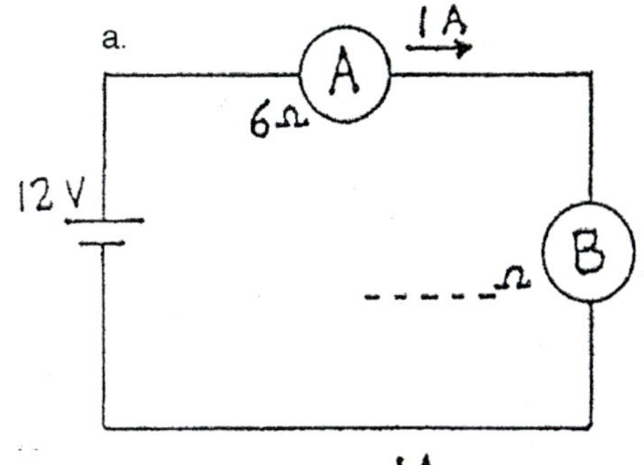

b.

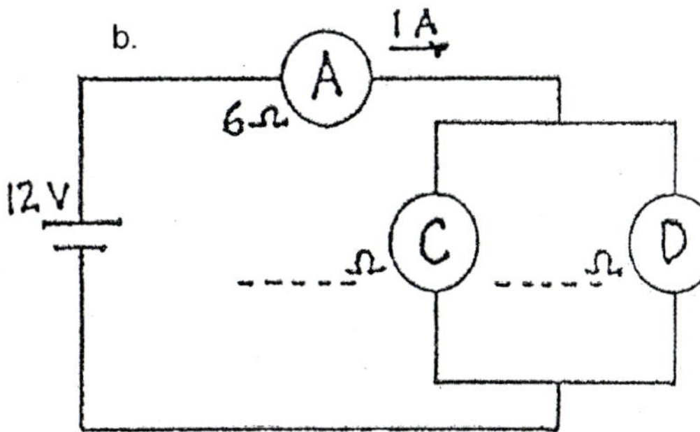

c.

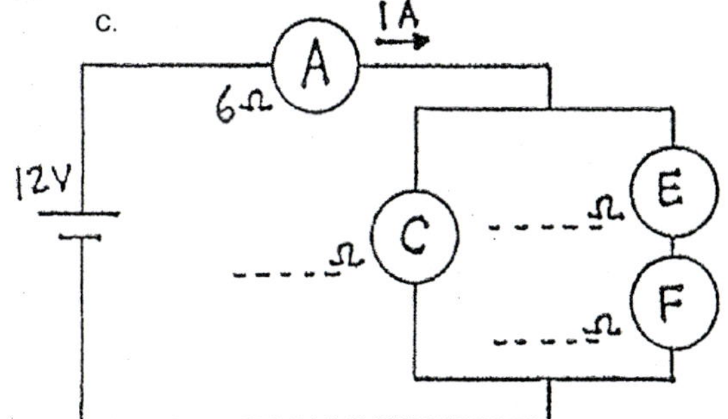

d.

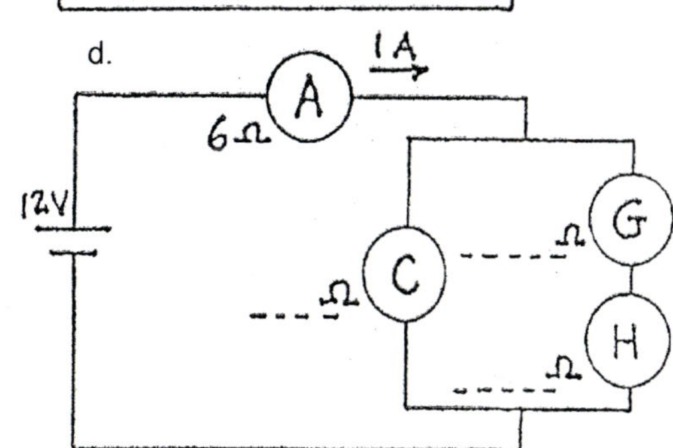

Circuit a: How much current flows through the battery? __________ A.

Circuit b: Assume lamps C and D are identical. Current through lamp D is __________ A.

Circuit c: Identical lamps E and F replace replace lamp D. Current through lamp C is __________ A.

Circuit d: Lamps G and H replace lamps E and F, and the resistance of lamp G is twice that of lamp H. Current through lamp H is __________ A.

e.

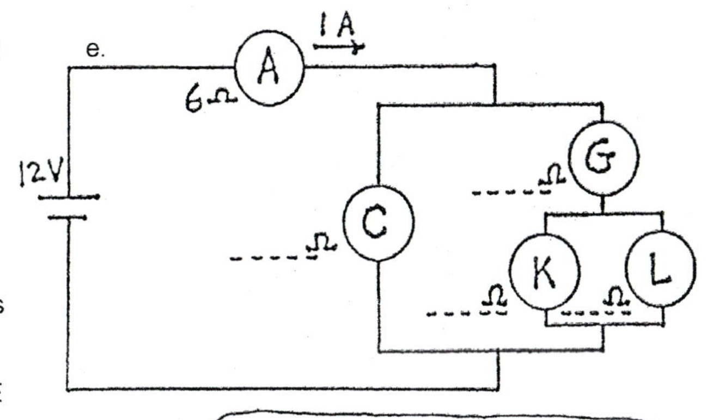

Handy rule: For a pair of resistors in parallel:

$$\text{Equivalent resistance} = \frac{\text{product of resistances}}{\text{sum of resistances}}$$

Circuit e: Identical lamps K and L replace lamp H. Current through lamp L is __________ A.

The equivalent resistance of a circuit is the value of a single resistor that will replace all the resistors of the circuit to produce the same load on the battery. How do the equivalent resistances of the circuits *a* through *e* compare?

____

Hewitt Drew it!

Chapter 10 Static and Current Electricity

Electric Power in Circuits

The table beside circuit *a* below shows the current through each resistor, the voltage across each resistor, and the power dissipated as heat in each resistor. Find the similar correct values for circuits *b* through *d*, and *write* your answers in the tables shown.

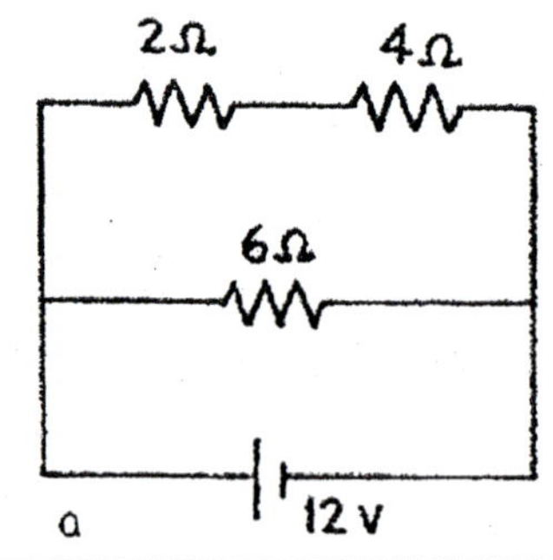

RESISTANCE	CURRENT ×	VOLTAGE =	POWER
2 Ω	2 A	4 V	8 W
4 Ω	2 A	8 V	16 W
6 Ω	2 A	12 V	24 W

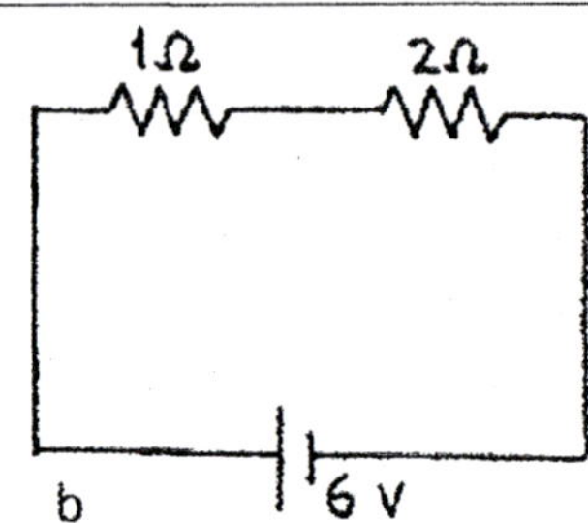

RESISTANCE	CURRENT ×	VOLTAGE =	POWER
1 Ω			
2 Ω			

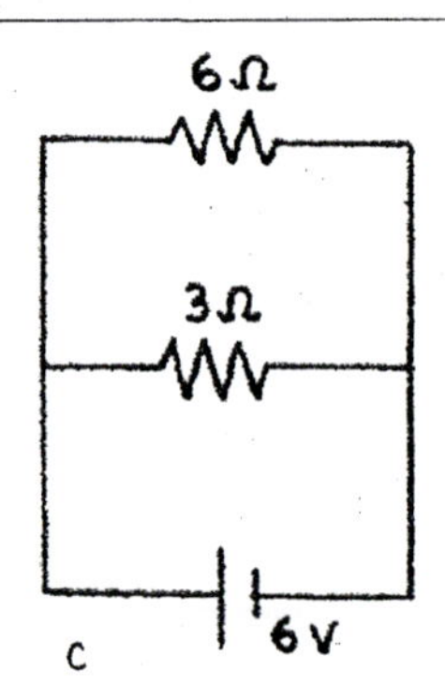

RESISTANCE	CURRENT ×	VOLTAGE =	POWER
6 Ω			
3 Ω			

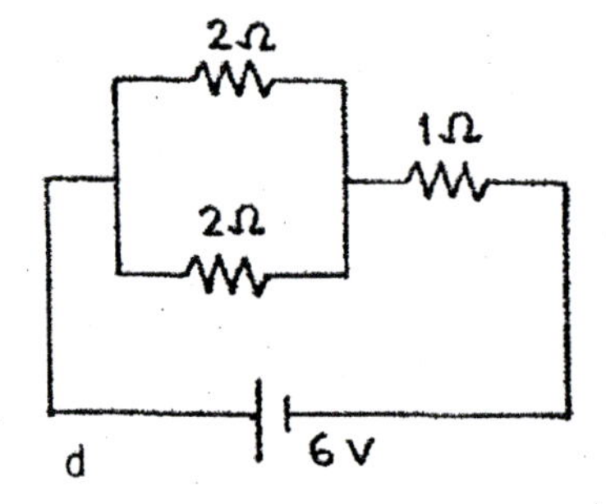

RESISTANCE	CURRENT ×	VOLTAGE =	POWER
2 Ω			
2 Ω			
1 Ω			

Hewitt Drew it!

Name ______________________ Date __________

CONCEPTUAL **Physics** FUNDAMENTALS PRACTICE PAGE

Chapter 11 Magnetism and Electromagnetic Induction

Magnetic Fundamentals

Fill in the blanks:

1. Attraction or repulsion of charges depends on their *signs*, positives or negatives. Attraction or repulsion of magnets depends on their magnetic ______________: ______________ or ______________.

2. Opposite poles attract; like poles ______________.

3. A magnetic field is produced by the ______________ of electric charge.

4. Clusters of magnetically aligned atoms are magnetic ______________.

5. A magnetic ______________ surrounds a current-carrying wire.

6. When a current-carrying wire is made to form a coil around a piece of iron, the result is an ______________________________.

7. A charged particle moving in a magnetic field experiences a deflecting ______________ that is maximum when the charge moves ______________ to the field.

8. A current-carrying wire experiences a deflecting __________ that is maximum when the wire and magnetic field are ______________ to one another.

9. A simple instrument designed to detect electric current is the ______________; when calibrated to measure current, it is an ______________; when calibrated to measure voltage, it is a ______________

10. The largest size magnet in the world is the ______________ itself.

Chapter 11 Magnetism and Electromagnetic Induction

***Magnetic Fundamentals**—continued*

11. The illustration below is similar to Figure 24.2 in your textbook. Iron filings trace out patterns of magnetic field lines about a bar magnet. In the field are some magnetic compasses. The compass needle in only one compass is shown. Draw in the needles with proper orientation in the other compasses.

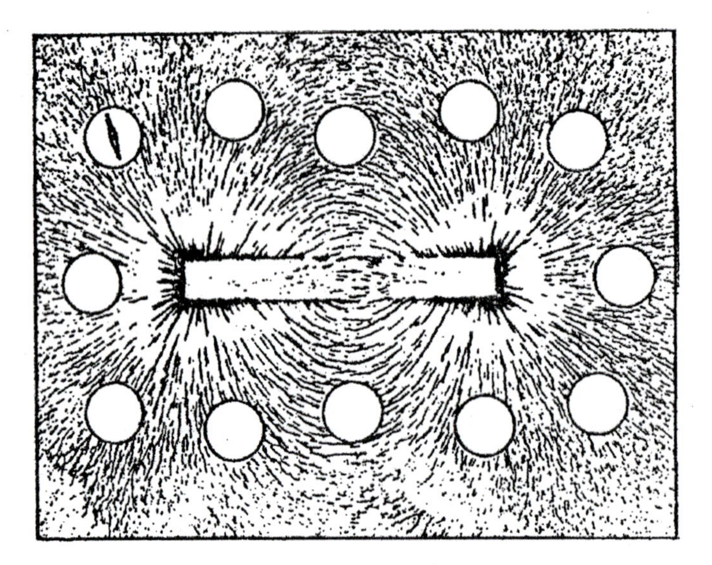

12. The illustration below is similar to Figure 24.10b in your textbook. Iron filings trace out magnetic field pattern about the loop of current-carrying wire. Draw in the compass needle orientations for all the compasses.

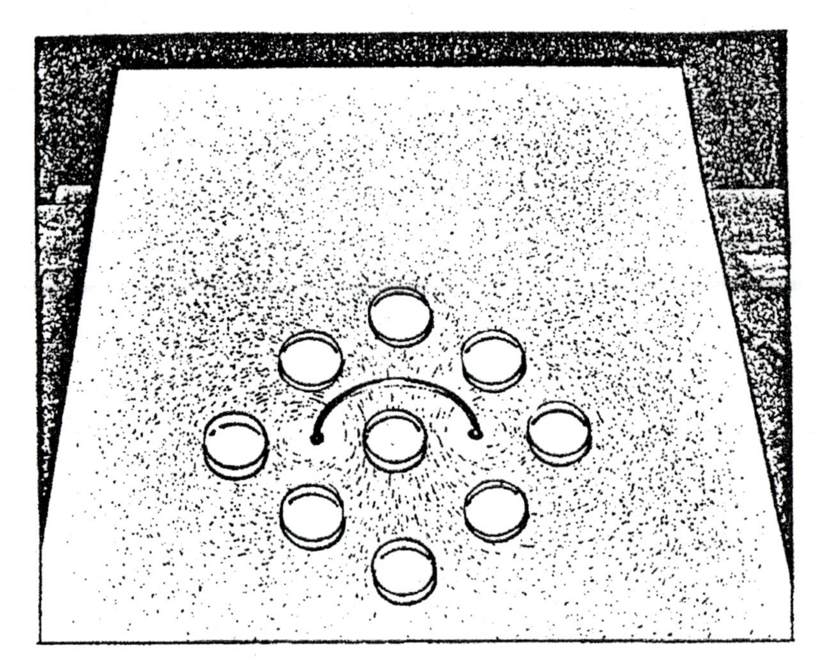

Hewitt Drew it!

Name ______________________ Date ________

CONCEPTUAL **Physics** FUNDAMENTAL$ PRACTICE PAGE

Chapter 11 Magnetism and Electromagnetic Induction

Faraday's Law

Circle the correct answers:

1. Hans Christian Oersted discovered that magnetism and and electricity are

   [related] [independent of each other].

   Magnetism is produced by

   [batteries] [motion of electric charges].

   Faraday and Henry discovered that electric current can be produced by

   [batteries] [motion of a magnet].

   More specifically, voltage is induced in a loop of wire if there is a change in

   [batteries] [magnetic field in the loop].

   This phenomenon is called

   [electromagnetism] [electromagnetic induction].

2. When a magnet is plunged in and out of a coil of wire, voltage is induced in the coil. If the rate of the in-and-out motion of the magnet is doubled, the induced voltage

   [doubles] [halves] [remains the same].

   If instead the number of loops in the coil is doubled, the induced voltage

   [doubles] [halves] [remains the same].

3. A rapidly changing magnetic field in any region of space induces a rapidly changing

   [electric field] [magnetic field] [gravitational field]

   which in turn induces a rapidly changing

   [magnetic field] [electric field] [baseball field].

   This generation and regeneration of electric and magnetic fields make up

   [electromagnetic waves] [sound waves] [both of these].

Hewitt Drew it!

Chapter 11 Magnetism and Electromagnetic Induction

Transformers

Consider a simple transformer that has a 100-turn primary coil and a 1000-turn secondary coil. The primary is connected to a 120-V AC source and the secondary is connected to an electrical device with a resistance of 1000 ohms.

1. What will be the voltage output of the secondary?

   __________ V.

2. What current flows in the secondary circuit?

   __________ A.

3. Now that you know the voltage and the current, what is the power in the secondary coil?

   __________ W.

4. Neglecting small heating losses, and knowing that energy is conserved, what is the power in the primary coil?

   __________ W.

5. Now that you know the power and the voltage across the primary coil, what is the current drawn by the primary coil?

   __________ A.

Circle the answers:

6. The results show voltage is stepped [up] [down] from primary to secondary, and that current is correspondingly stepped [up] [down].

7. For a step-up transformer, there are [more] [fewer] turns in the secondary coil than in the primary. For such a transformer, there is [more] [less] current in the secondary than in the primary.

8. A transformer can step up [voltage] [energy and power] , but in no way can it step up [voltage] [energy and power].

9. If 120 V is used to power a toy electric train that operates on 6 V, then a [step up] [step down] transformer should be used that has a primary to secondary turns ratio of [1/20] [20/1].

10. A transformer operates on [dc] [ac] because the magnetic field within the iron core must [continually change] [remain steady].

Electricty and magnetism connect to become light!

Name ______________________________ Date __________

CONCEPTUAL **Physics** FUNDAMENTALS PRACTICE PAGE

Chapter 12 Waves and Sound

Vibration and Wave Fundamentals

1. A sine curve that represents a transverse wave is drawn below. With a ruler, measure the wavelength and amplitude of the wave.

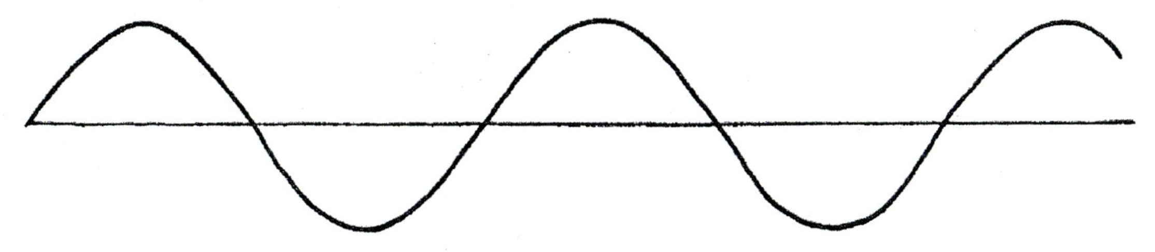

a. Wavelength = __________

b. Amplitude = __________

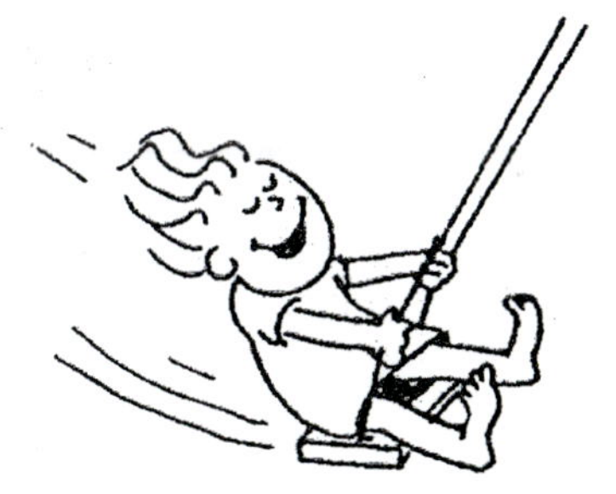

2. A kid on a playground swing makes a complete to-and-fro swing each 2 seconds. The frequency of swing is

[0.5 hertz] [1 hertz] [2 hertz]

and the period is

[0.5 seconds] [1 second] [2 seconds].

3. *Complete the statements*:

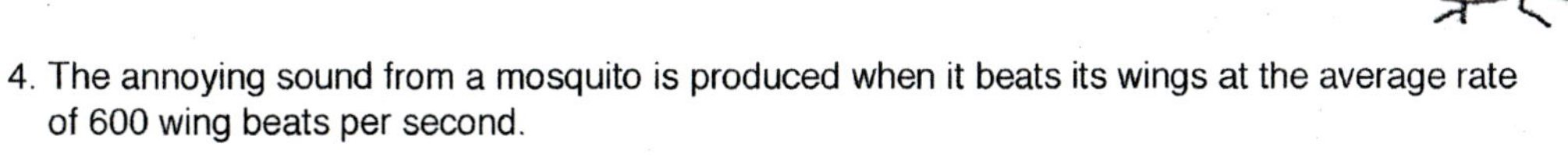

THE PERIOD OF A 440-HERTZ SOUND WAVE IS ______ SECOND.
A MARINE WEATHER STATION REPORTS WAVES ALONG THE SHORE THAT ARE 8 SECONDS APART. THE FREQUENCY OF THE WAVES IS THEREFORE ______ HERTZ.

4. The annoying sound from a mosquito is produced when it beats its wings at the average rate of 600 wing beats per second.

a. What is the frequency of the sound waves?

____

b. What is the wavelength?
(Assume the speed of sound is 340 m/s.)

____

Hewitt Drew it!

Chapter 12 Waves and Sound

***Vibration and Wave Fundamentals**—continued*

5. A machine gun fires 10 rounds per second. The speed of the bullets is 300 m/s.

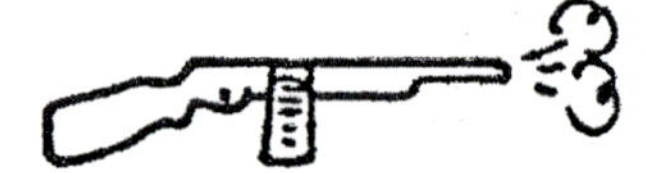

   a. What is the distance in the air between the flying bullets? __________

   b. What happens to the distance between the bullets if the rate of fire is increased?

   ____

6. Consider a wave generator that produces 10 pulses per second. The speed of the waves is 300 cm/s.

   a. What is the wavelength of the waves? __________

   b. What happens to the wavelength if the frequency of pulses is increased?

   ____

7. The bird at the right watches the waves. If the portion of a wave between 2 crests passes the pole each second,

   a. what is the speed of the waves? __________

   b. what is the period of wave motion? __________

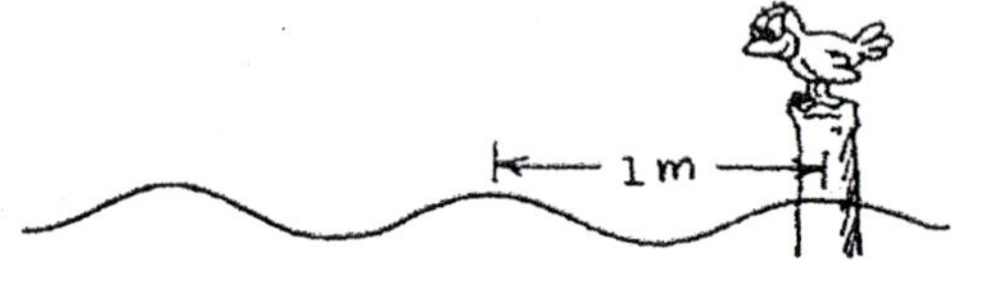

   c. If the distance between crests were 1.5 meters apart, and 2 crests pass the pole each second, what would be the speed of the wave?

   ____

   d. What would the period of wave motion be for 7.c ?

   ____

8. When an automobile moves toward a listener, the sound of its horn seems relatively

   [low pitched] [high pitched] [normal]

   and when moving away from the listener, its horn seems

   [low pitched] [high pitched] [normal].

9. The changed pitch of the Doppler effect is due to changes in wave

   [speed] [frequency] [both].

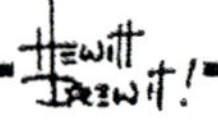

Name ______________________ Date __________

CONCEPTUAL **Physics** FUNDAMENTALS PRACTICE PAGE

Chapter 12 Waves and Sound

Shock Waves

The cone-shaped shock wave produced by a supersonic aircraft is actually the result of overlapping spherical waves of sound, as indicated by the overlapping circles in Figure 19.19 in your textbook. Sketches *a* through *e* below show the "animated" growth of only one of the many spherical sound waves (shown as an expanding circle in the two-dimensional drawing).

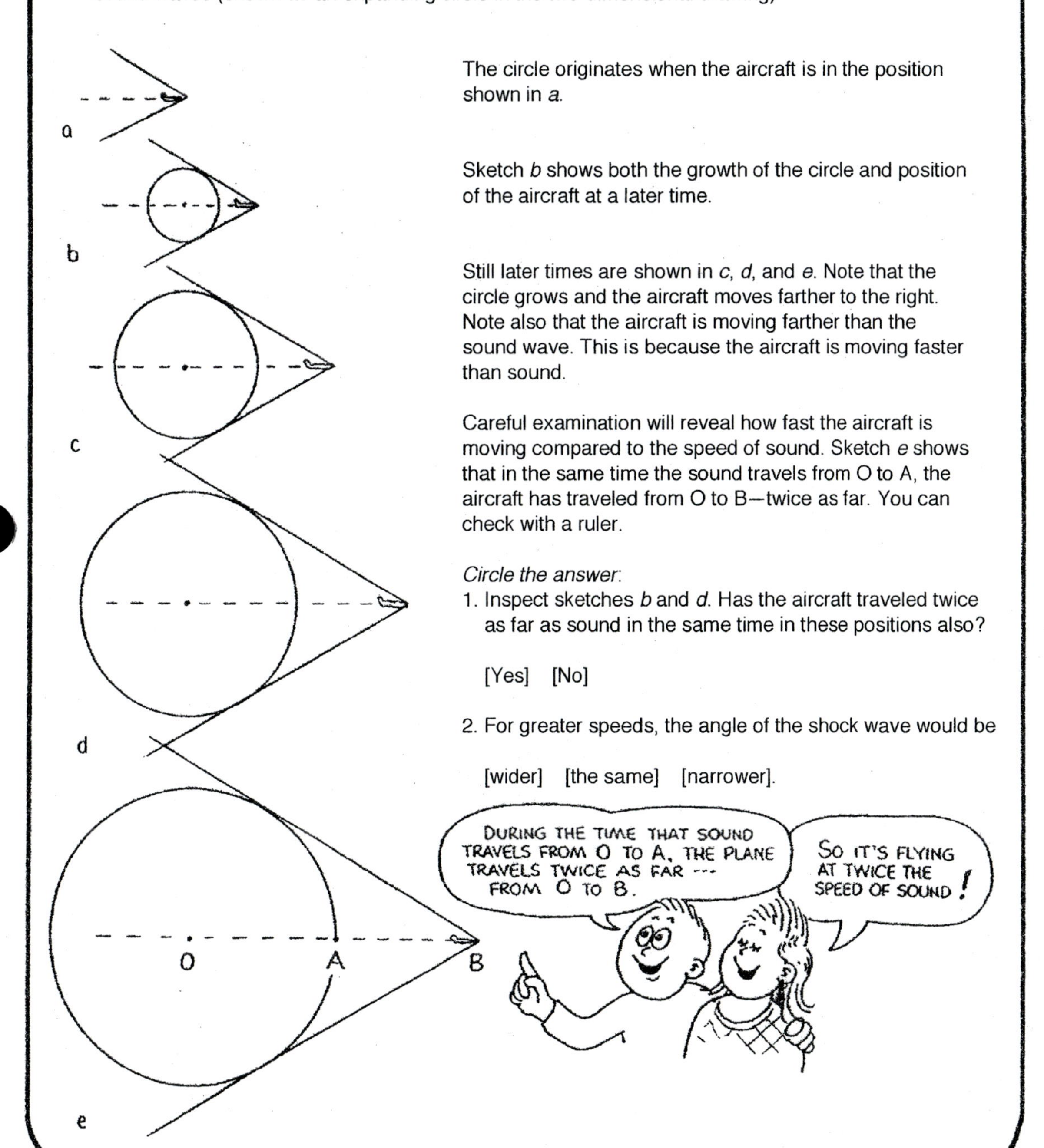

a
b
c
d
e
O
A
B
DURING THE TIME THAT SOUND TRAVELS FROM O TO A, THE PLANE TRAVELS TWICE AS FAR --- FROM O TO B.
SO IT'S FLYING AT TWICE THE SPEED OF SOUND!

The circle originates when the aircraft is in the position shown in *a*.

Sketch *b* shows both the growth of the circle and position of the aircraft at a later time.

Still later times are shown in *c*, *d*, and *e*. Note that the circle grows and the aircraft moves farther to the right. Note also that the aircraft is moving farther than the sound wave. This is because the aircraft is moving faster than sound.

Careful examination will reveal how fast the aircraft is moving compared to the speed of sound. Sketch *e* shows that in the same time the sound travels from O to A, the aircraft has traveled from O to B—twice as far. You can check with a ruler.

Circle the answer:

1. Inspect sketches *b* and *d*. Has the aircraft traveled twice as far as sound in the same time in these positions also?

   [Yes] [No]

2. For greater speeds, the angle of the shock wave would be

   [wider] [the same] [narrower].

Chapter 12 Waves and Sound

Shock Waves*—continued*

3. Use a ruler to estimate the speeds of the aircraft that produce the shock waves in the two sketches below.

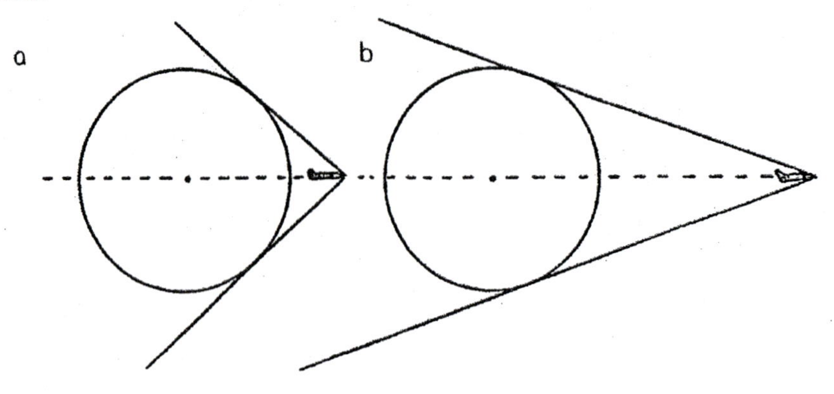

Aircraft *a* is traveling about __________ times the speed of sound.

Aircraft *b* is traveling about __________ times the speed of sound.

4. Draw your own circle (anywhere) and estimate the speed of the aircraft to produce the shock wave shown below:

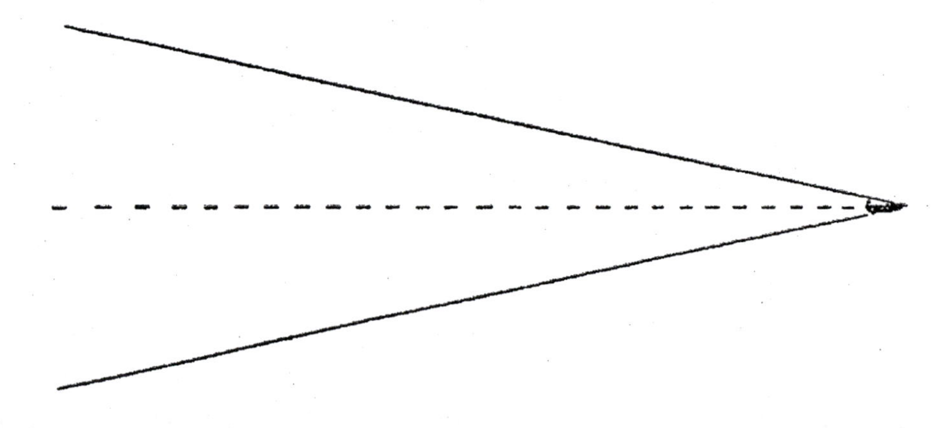

The speed is about __________ times the speed of sound.

5. In the space below, draw the shock wave made by a supersonic missile that travels at four times the speed of sound.

Hewitt Drew it!

Name ______________________ Date ________

CONCEPTUAL **Physics** FUNDAMENTALS PRACTICE PAGE

Chapter 12 Waves and Sound

Wave Superposition

A pair of pulses travel toward each at equal speeds. The composite waveforms, as they pass through each other and interfere, are shown at 1-second intervals. In the left column note how the pulses interfere to produce the composite waveform (solid line). Make a similar construction for the two wave pulses in the right column. Like the pulses in the first column, they each travel at 1 space per second.

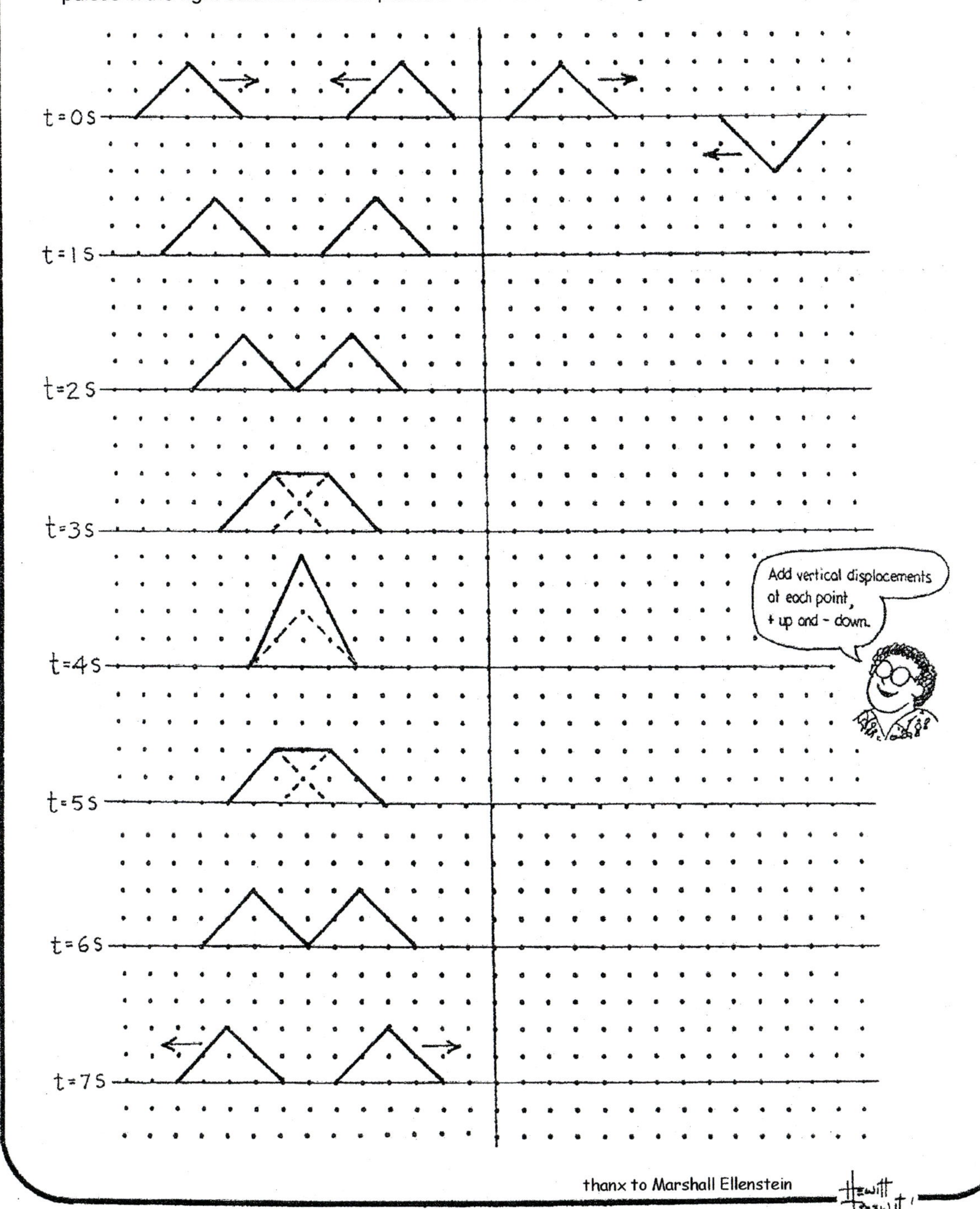

thanx to Marshall Ellenstein

Chapter 12 Waves and Sound

Wave Superposition—continued

Construct the composite waveforms at 1-second intervals for the two waves traveling toward each other at equal speed.

t=0s

t=1s

t=2s

t=3s

t=4s

t=5s

t=6s

t=7s

t=8s

Hewitt Drew it!

Name ________________________________ Date ____________

CONCEPTUAL *Physics* FUNDAMENTALS PRACTICE PAGE

Chapter 13 Light Waves

Speed, Wavelength, and Frequency

1. The first investigation that led to a determination of the speed of light was performed in about 1675 by the Danish astronomer Olaus Roemer. He made careful measurements of the period of Io, a moon about the planet Jupiter, and was surprised to find an irregularity in Io's observed period. While Earth was moving away from Jupiter, the measured periods were slightly longer than average. While Earth approached Jupiter, they were shorter than average. Roemer estimated that the cumulative discrepancy amounted to about 16.5 minutes. Later interpretations showed that what occurs is that light takes about 16.5 minutes to travel the extra 300,000,000-km distance across Earth's orbit. Aha! We have enough information to calculate the speed of light!

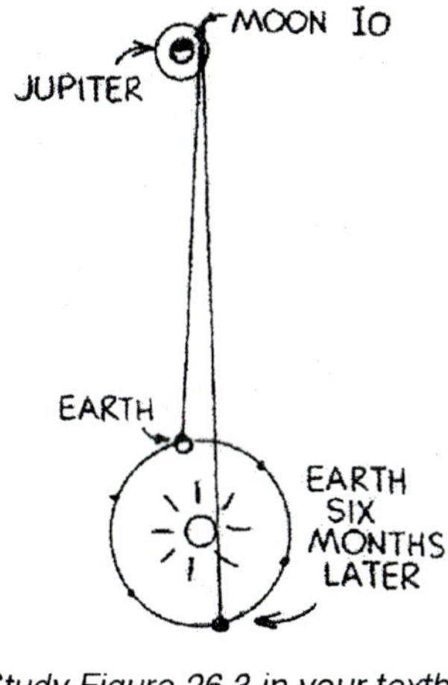

a. Write a formula for speed in terms of the distance traveled and the time spent traveling that distance.

____

b. Using Roemer's data, and changing 16.5 minutes to seconds, calculate the speed of light.

____

Study Figure 26.3 in your textbook and answer the following:

2. a. Which has the longer *wavelengths*? [radio waves] [light waves].

   b. Which has the longer *wavelengths*? [light waves] [gamma waves].

   c. Which has the higher *frequencies*? [ultraviolet waves] [infrared waves].

   d. Which has the higher *frequencies*? [ultraviolet waves] [gamma rays].

LIGHT IS THE *ONLY* THING WE SEE!

Carefully study the section "Transparent Materials" in your textbook and answer the following:

3. a. Exactly what do vibrating electrons emit?

____

   b. When ultraviolet light shines on glass, what does it do to electrons in the glass structure?

____

   c. When energetic electrons in the glass structure vibrate against neighboring atoms, what happens to the energy of vibration?

____

   d. What happens to the energy of a vibrating electron that does not collide with neighboring atoms?

____

Hewitt Drew it!

Chapter 13 Light Waves

***Speed, Wavelength, and Frequency**—continued*

e. Light in which range of frequencies is absorbed in glass? [visible] [ultraviolet].

f. Light in which range of frequencies is transmitted through glass? [visible] [ultraviolet].

g. How is the speed of light in glass affected by the succession of time delays that accompany the absorption and re-emission of light from atom to atom in the glass?

____

h. How does the speed of light compare in water, glass, and diamond?

____

4. The Sun normally shines on both Earth and Moon. Both cast shadows. Sometimes the Moon's shadow falls on Earth, and at other times Earth's shadow falls on the Moon.

a. The sketch shows the Sun and Earth. Draw the Moon at a position for a solar eclipse.

b. This sketch also shows the Sun and Earth. Draw the Moon at a position for a lunar eclipse.

5. The diagram shows the limits of light rays when a large lamp makes a shadow of a small object on a screen. Make a sketch of the shadow on the screen, shading the umbra darker than the penumbra. In what part of the shadow could an ant on the screen see part of the lamp?

____

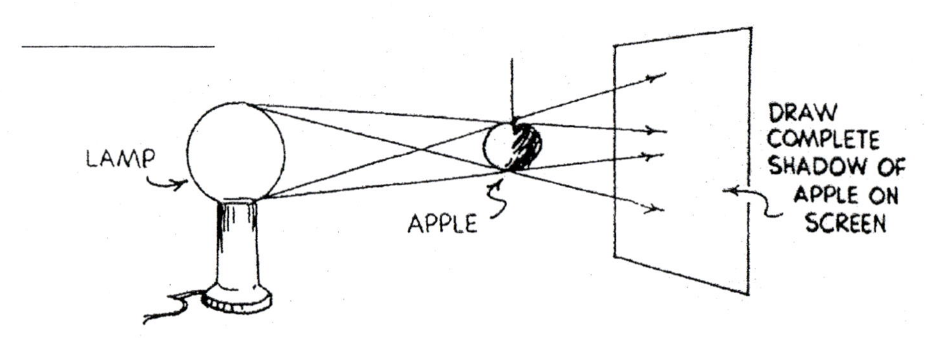

Hewitt Drew it!

Name ______________________ Date __________

CONCEPTUAL **Physics** FUNDAMENTALS PRACTICE PAGE

Chapter 13 Light Waves

Color Addition

The sketch to the right shows the shadow of an instructor in front of a white screen in a dark room. The light source is red, so the screen looks red and the shadow looks black. Color the sketch, or label the colors with pen or pencil.

A green lamp is added and makes a second shadow. The shadow cast by the red lamp is no longer black, but is illuminated by green light. So it is green. Color or mark it green. The shadow cast by the green lamp is not black because it is illuminated by the red lamp. Indicate its color. Do the same for the background, which receives a mixture of red and green light.

A blue lamp is added and three shadows appear. Indicate the appropriate colors of the shadows and the background.

The lamps are placed closer together so the shadows overlap. Indicate the colors of all screen areas.

Chapter 13 Light Waves

***Color Addition**—continued*

If you have colored markers or pencils, have a try at these.

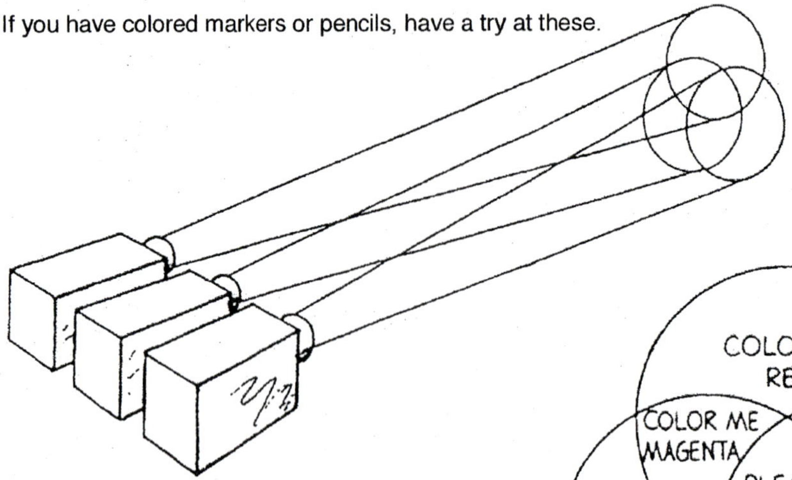

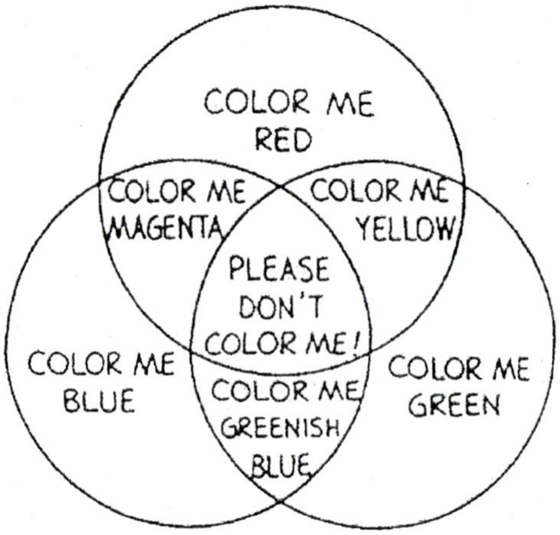

Name ______________________________ Date __________

CONCEPTUAL *Physics* FUNDAMENTALS PRACTICE PAGE

Chapter 13 Light Waves

Diffraction and Interference

1. Shown are concentric solid and dashed circles, each different in radius by 1 cm. Consider the circular pattern a top view of water waves, where the solid circles are crests and the dashed circles are troughs.

   a. Draw another set of the same concentric circles with a compass. Choose any part of the paper for your center (except the present central point). Let the circles run off the edge of the paper.

   b. Find where a dashed line crosses a solid line and draw a large dot at the intersection. Do this for ALL places where a solid and dashed line intersect.

   c. With a wide felt marker, connect the dots with the solid lines. These *nodal lines* lie in regions where the waves have cancelled—where the crest of one wave overlaps the trough of another (see Figures 29.15 and 29.16 in your textbook).

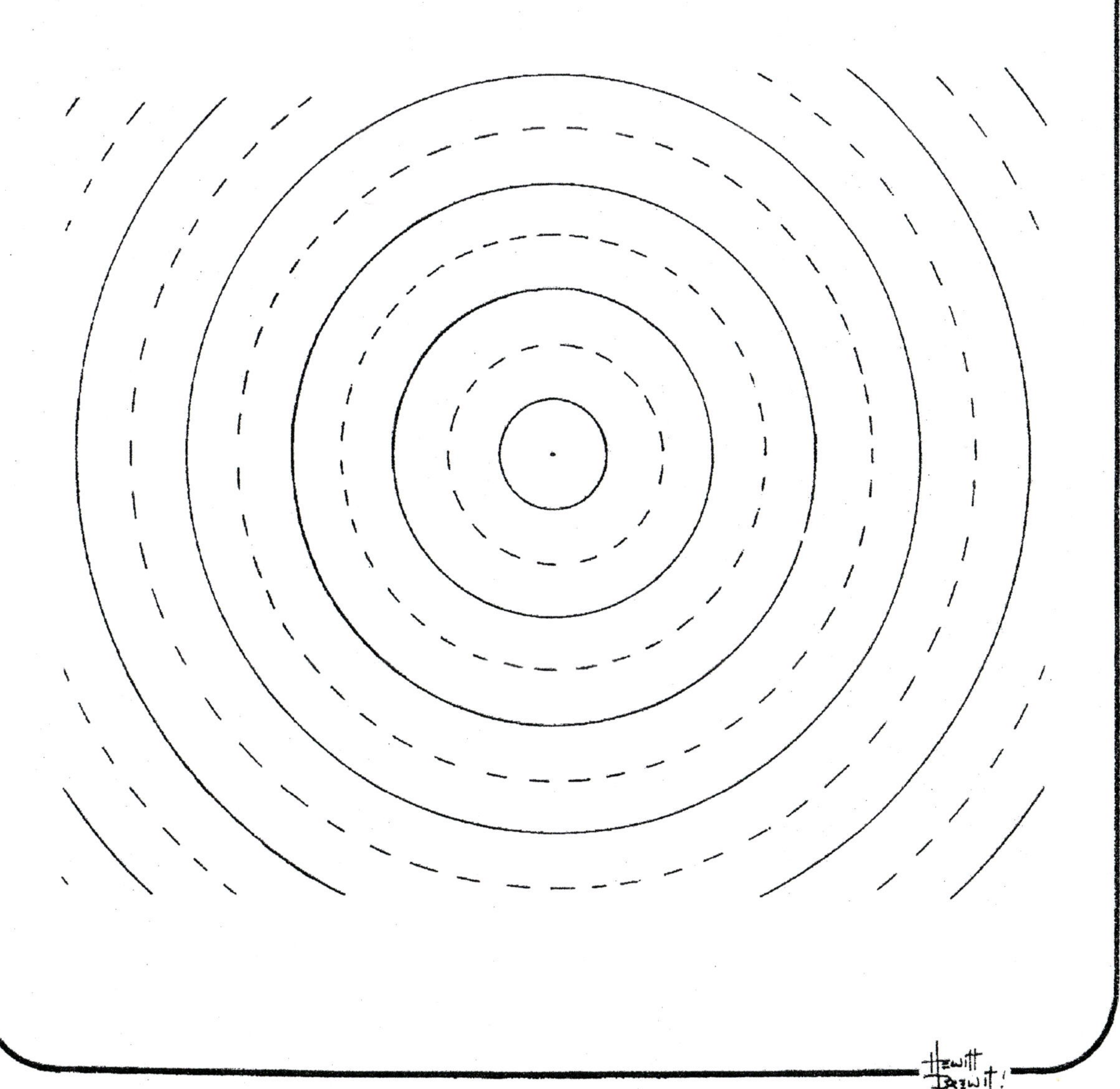

Hewitt Drewit!

Chapter 13 Light Waves

Diffraction and Interference*—continued*

2. Look at the construction of overlapping circles on your classmates' papers. Some will have more nodal lines than others, due to different starting points. How does the number of nodal lines in a pattern relate to the distance between centers of circles, (or sources of waves)?

____

____

3. Figure 29.19 from your textbook is repeated below. Carefully count the number of wavelengths (same as the number of wave crests) along the following paths between the slits and the screen.

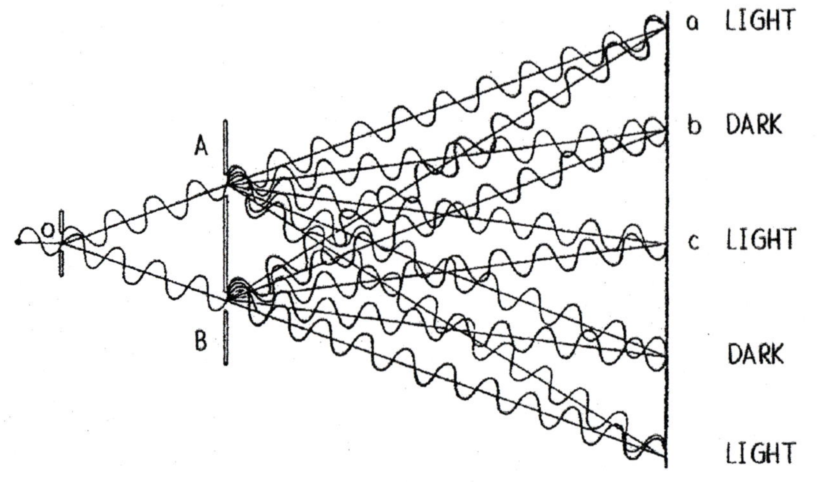

a. Number of wavelengths between slit A and point a is __________.

b. Number of wavelengths between slit B and point a is __________.

c. Number of wavelengths between slit A and point b is __________.

d. Number of wavelengths between slit B and point b is __________.

e. Number of wavelengths between slit A and point c is __________.

f. Number of wave crests between slit B and point c is __________.

4. When the number of wavelengths along each path is the same or differs by one or more whole wavelengths, interference is

   [constructive] [destructive]

   and when the number of wavelengths differ by a half-wavelength (or odd multiples of a half-wavelength), interference is

   [constructive] [destructive].

It's nice how knowing some physics really changes the way we see things!
Hewitt Drew it!

Name ______________________________ Date __________

CONCEPTUAL **Physics** FUNDAMENTALS PRACTICE PAGE

Chapter 14 Properties of Light

Pool Room Optics

The law of reflection for optics is useful in playing pool. A ball bouncing off the bank of a pool table behaves like a photon reflecting off a mirror. As the sketch shows, angles become straight lines with the help of mirrors. The diagram shows a top view of this, with a flattened "mirrored" region. Note that the angled path on the table appears as a straight line (dashed) in the mirrored region.

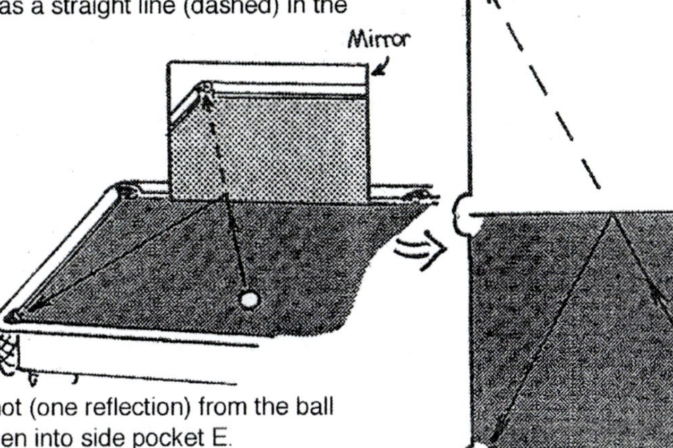

1. Consider a one-bank shot (one reflection) from the ball to the north bank and then into side pocket E.

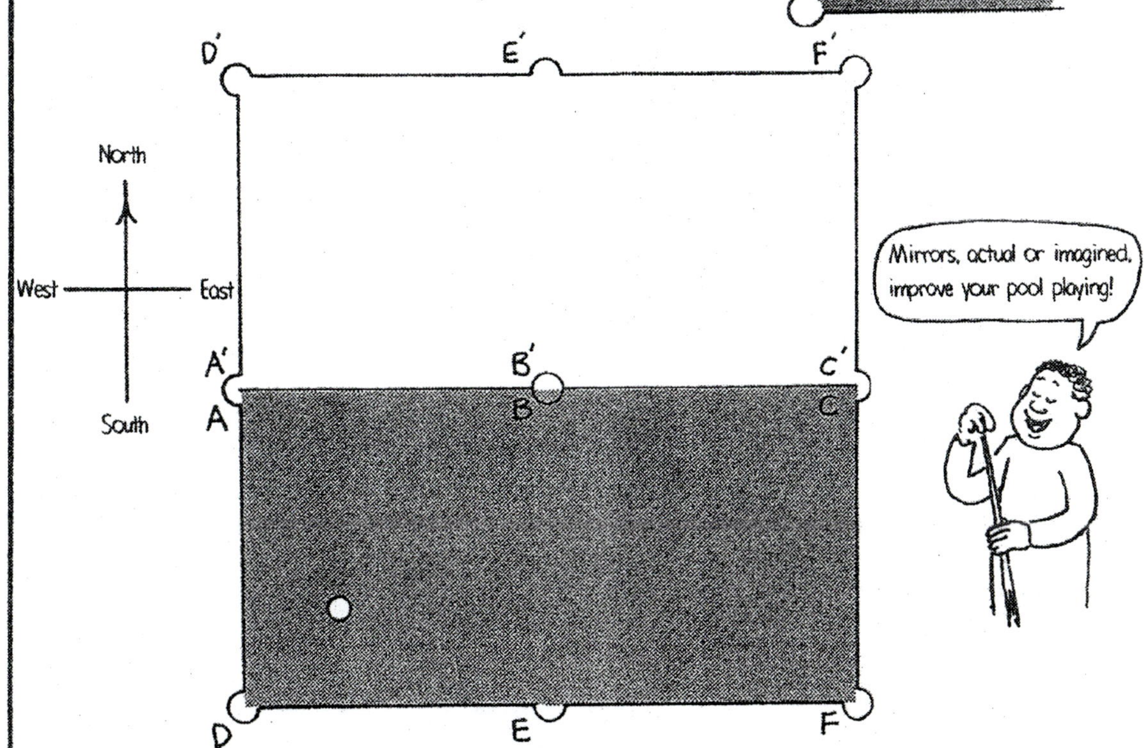

a. Use the mirror method to construct a straight line path to mirrored E'. Then construct the actual path to E.

b. Without using off-center strokes or other tricks, can a one-bank shot off the north bank put the ball in corner pocket F? __________ Show why or why not using the diagram.

Chapter 14 Properties of Light

Pool Room Optics—continued

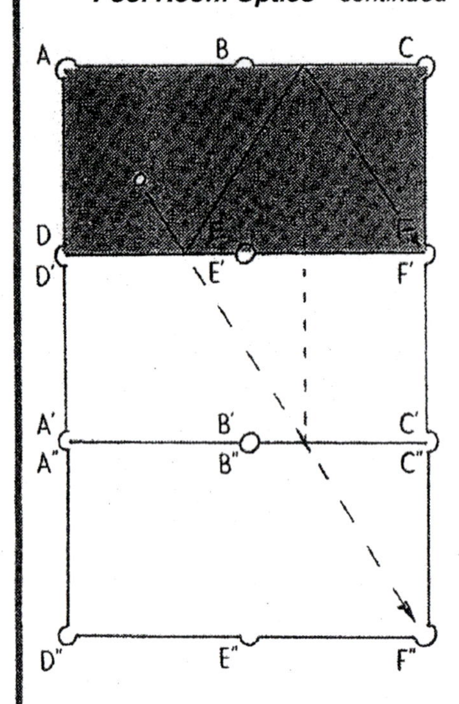

2. Consider the left diagram, a two-bank shot (2 reflections) into corner pocket F. Here we use 2 mirrored regions. Note the straight line of sight to F", and how the north-bank impact point matches the intersection between B' and C'.

   a. On the same diagram to the left, construct a similar path for a two-bank shot to get the ball in the side pocket E.

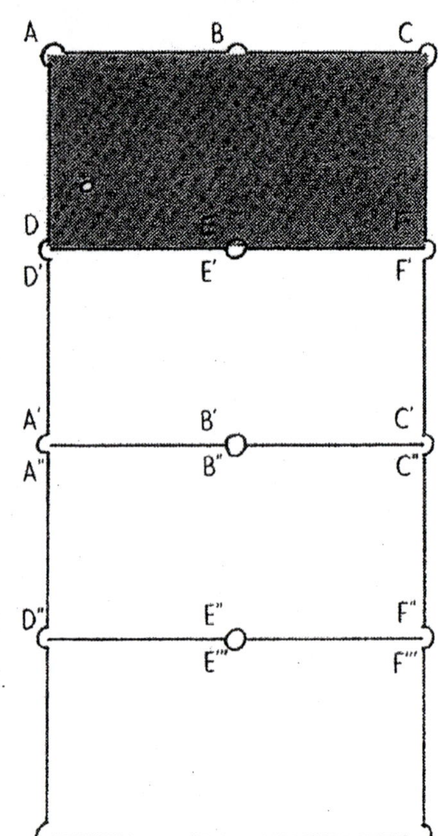

3. Consider above right, a three bank-shot into corner pocket C, first bouncing against the south bank, then to the north, again to the south, and into pocket C.

   a. Construct the path. (First construct the single dashed line to C'''.)

   b. Construct the path to make a three-bank shot into side pocket B.

4. Let's try banking from adjacent banks of the table. Consider a two-bank shot to corner pocket F (first off the west bank, then to and off the north bank, then into F). Note how our two mirrored regions permit a straight-line path from the ball to F".

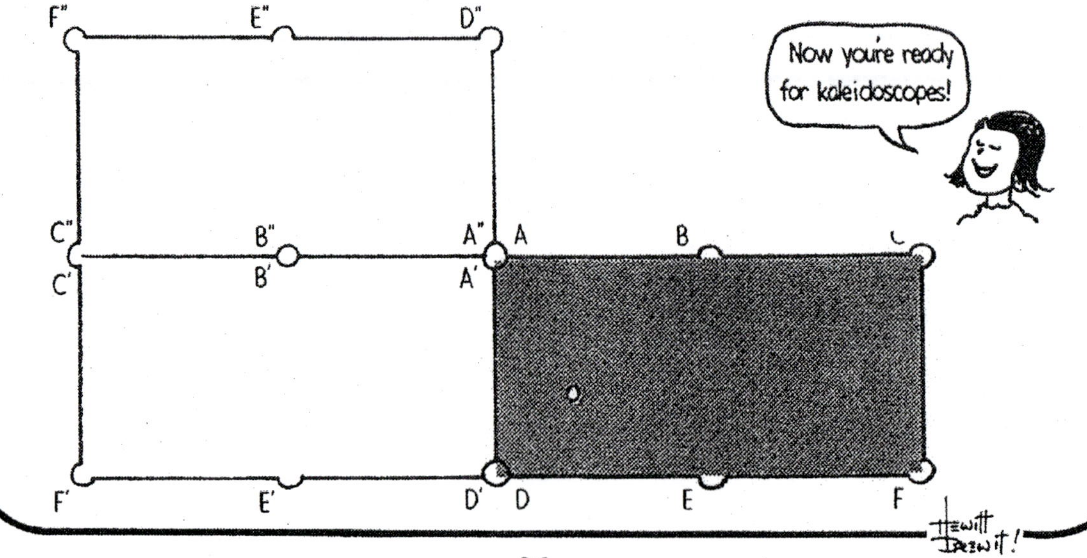

Name ______________________ Date ____________

CONCEPTUAL Physics FUNDAMENTALS PRACTICE PAGE

Chapter 14 Properties of Light

Reflection

Abe and Bev both look in a plane mirror directly in front of Abe (left view). Abe can see himself while Bev cannot see herself—but can Abe see Bev, and can Bev see Abe?

←MIRROR

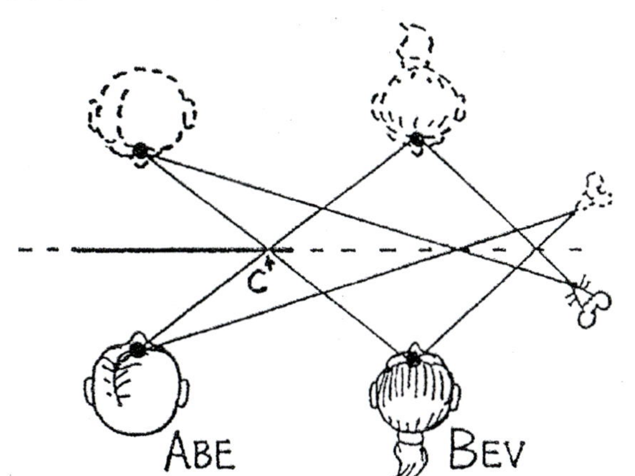

To find the answer, we construct their artificial locations "through" the mirror, the same distance behind as Abe and Bev are in front (right view). If straight-line connections intersect the mirror, as at point C, then each sees the other. The mouse, for example, cannot see or be seen by Abe and Bev (because there's no mirror in its line of sight).

Here we have eight students in front of a small plane mirror. Their positions are shown in the diagram below. Make appropriate straight-line constructions to answer the following:

←MIRROR

ABE BEV CIS DON EVA FLO GUY HAN

Abe can see ____________ Abe cannot see ____________
Bev can see ____________ Bev cannot see ____________
Cis can see ____________ Cis cannot see ____________
Don can see ____________ Don cannot see ____________
Eva can see ____________ Eva cannot see ____________
Flo can see ____________ Flo cannot see ____________
Guy can see ____________ Guy cannot see ____________
Han can see ____________ Han cannot see ____________

thanx to Marshall Ellenstein

Hewitt Drew it!

Chapter 14 Properties of Light

***Reflection**—continued*

Six of our group are now arranged differently in front of the same plane mirror. Their positions are shown below. Make appropriate constructions for this interesting arrangement, and answer the questions provided below:

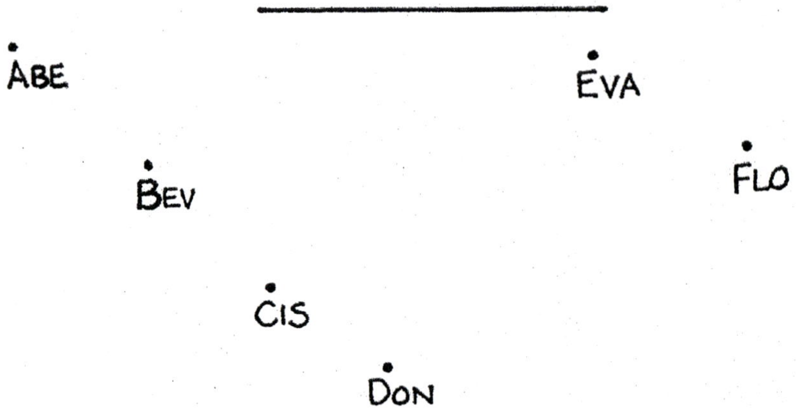

ABE

EVA

BEV

FLO

CIS

DON

Who can Abe see? ____________ Who can Abe not see? ____________

Who can Bev see? ____________ Who can Bev not see? ____________

Who can Cis see? ____________ Who can Cis not see? ____________

Who can Don see? ____________ Who can Don not see? ____________

Who can Eva see? ____________ Who can Eva not see? ____________

Who can Flo see? ____________ Who can Flo not see? ____________

Harry Hostshot views himself in a full-length mirror (right). Construct straight lines from Harry's eyes to the image of his feet, and to the top of his head. Mark the mirror to indicate the minimum area Harry uses to see a full view of himself.

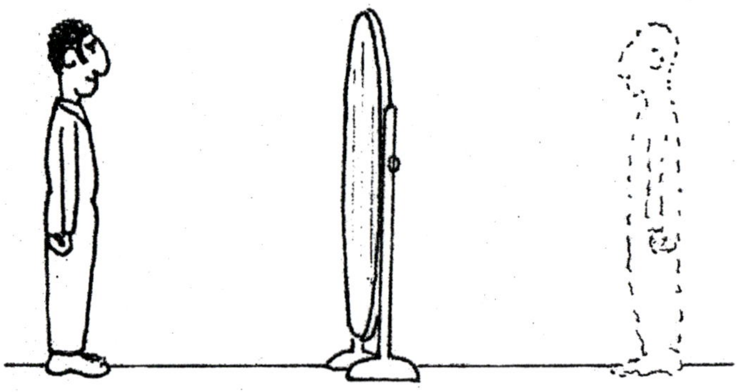

Does this region of the mirror depend on Harry's distance from the mirror? ____________

Hewitt Drew it!

Chapter 14 Properties of Light

Reflected Views

1. The ray diagram below shows the extension of one of the reflected rays from the plane mirror.

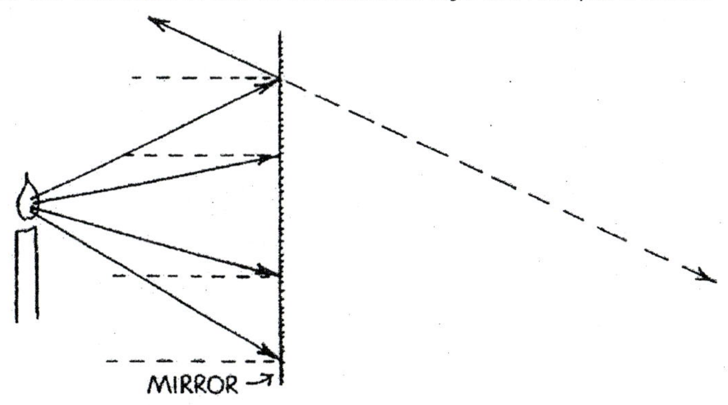

Complete the above diagram:

a. Carefully draw the three other reflected rays.

b. Extend your drawn rays behind the mirror to locate the image of the flame. (Assume the candle and image are viewed by an observer on the left.)

2. A girl takes a photograph of the bridge as shown. Which of the two sketches below correctly shows the reflected view of the bridge? Defend your answer.

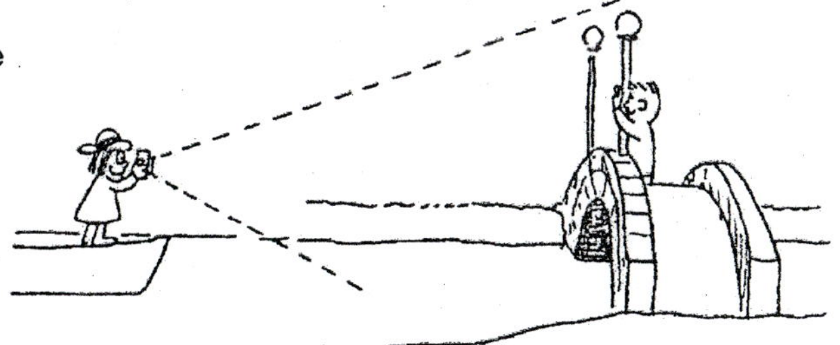

____

____

Hewitt Drew it!

Chapter 14 Properties of Light

More Reflection

1. Light from a flashlight shines on a mirror and illuminates one of the cards. Draw the reflected beam to indicate the illuminated card.

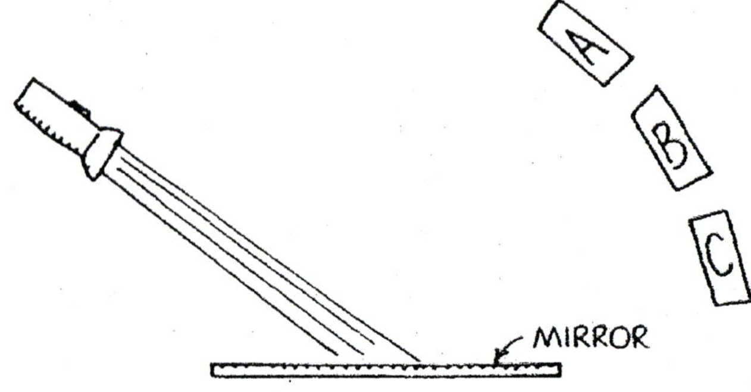

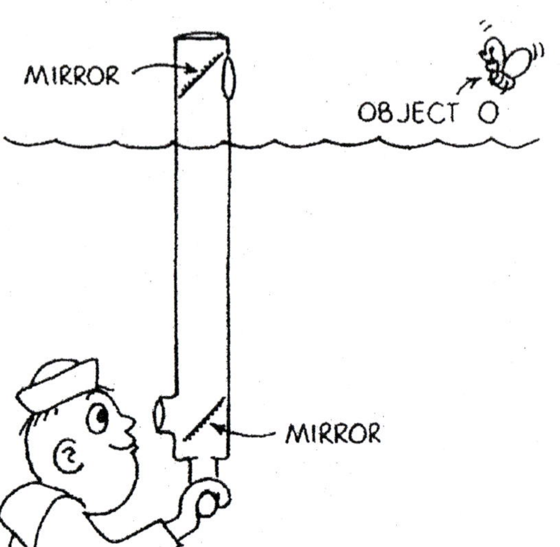

2. A periscope has a pair of mirrors in it. Draw the light path from the object "O" to the eye of the observer.

3. The ray diagram below shows the reflection of one of the rays that strikes the parabolic mirror. Notice that the law of reflection is obeyed, and the angle of incidence (from the normal, the dashed line) equals the angle of reflection (from the normal). Complete the diagram by drawing the reflected rays of the other three rays that are shown. (Do you see why parabolic mirrors are used in automobile headlights?)

MIRROR

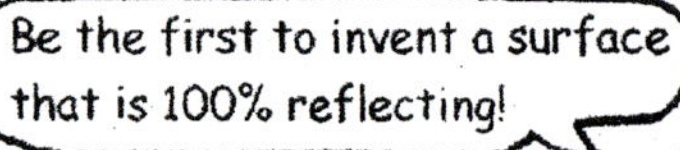

Be the first to invent a surface that is 100% reflecting!

Name ____ Date ____________

CONCEPTUAL **Physics** FUNDAMENTALS PRACTICE PAGE

Chapter 14 Properties of Light

Refraction

1. A pair of toy cart wheels are rolled obliquely from a smooth surface onto two plots of grass—a rectangular plot on the left, and a triangular plot on the right. The ground is on a slight incline so that after slowing down in the grass, the wheels speed up again when emerging on the smooth surface. Finish each sketch and show some positions of the wheels inside the plots and on the other side. Clearly indicate their paths and directions of travel.

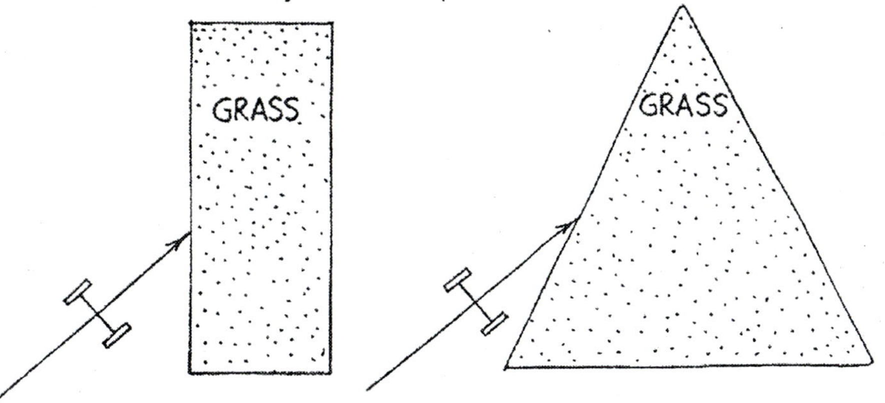

2. Red, green, and blue rays of light are incident upon a glass prism as shown below. The average speed of red light in the glass is less than in air, so the red ray is refracted. When it emerges into the air it regains its original speed and travels in the direction shown. Green light takes longer to get through the glass. Because of its slower speed it is refracted as shown. Blue light travels even slower in glass. Complete the diagram by estimating the path of the blue ray.

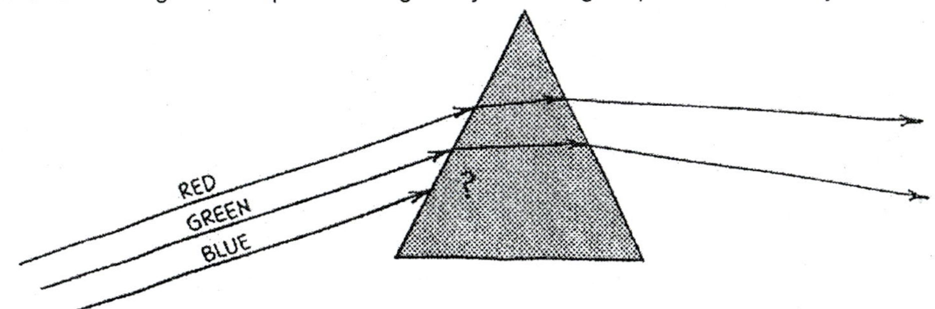

3. Below we consider a prism-shaped hole in a piece of glass—that is, an "air prism." Complete the diagram, showing likely paths of the beams of red, green, and blue light as they pass through this "prism" and then into glass.

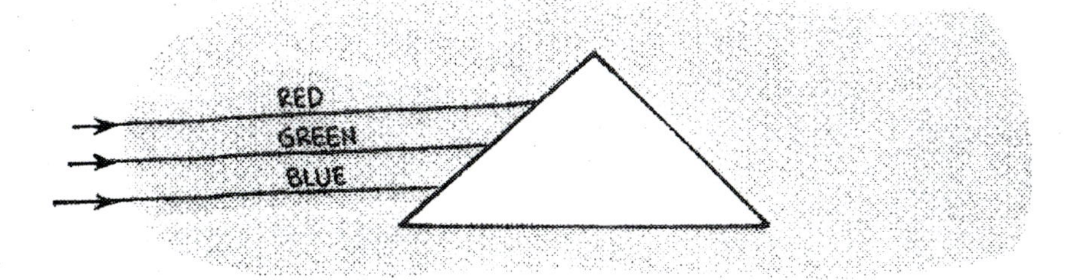

Hewitt Drew it!

Chapter 14 Properties of Light

***Refraction**—continued*

4. Light of different colors diverges when emerging from a prism. Newton showed that with a second prism he could make the diverging beams become parallel again. Which placement of the second prism will do this?

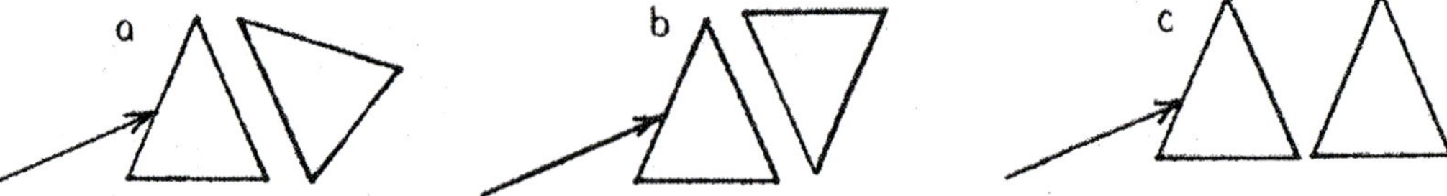

5. The sketch shows that due to refraction, the man sees the fish closer to the water surface than it actually is.

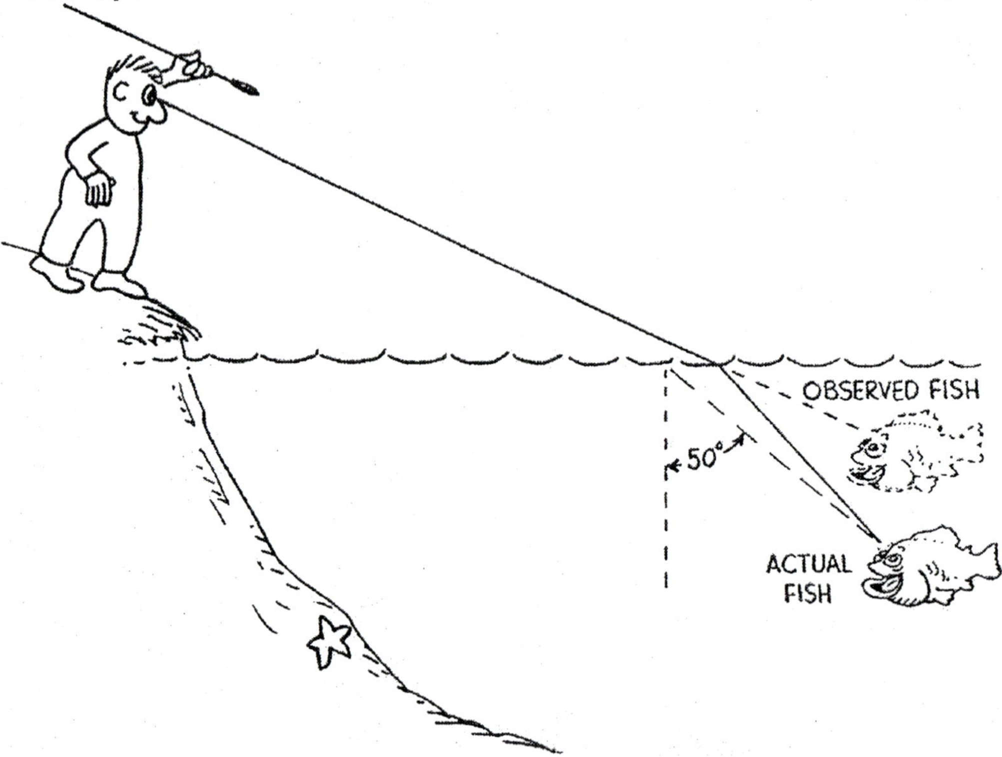

a. Draw a ray beginning at the fish's eye to show the line of sight of the fish when it looks upward at 50° to the normal at the water surface. Draw the direction of the ray after it meets the surface of water.

b. At the 50° angle, does the fish see the man, or does it see the reflected view of the starfish at the bottom of the pond? Explain.

____

c. To see the man, should the fish look higher or lower than the 50° path? ______________

d. If the fish's eye were barely above the water surface, it would see the world above in a 180° view, horizon to horizon. The fisheye view of the world above as seen beneath the water, however, is very different. Due to the 48° critical angle of water, the fish sees a normally 180° horizon-to-horizon view compressed within an angle of ______________.

Hewitt Drew it!

Name ______________________ Date __________

CONCEPTUAL **Physics** FUNDAMENTALS PRACTICE PAGE

Chapter 14 Properties of Light

More Refraction

1. The sketch to the right shows a light ray moving from air into water, at 45° to the normal. Which of the three rays indicated with capital letters is most likely the light ray that continues inside the water?

   ____

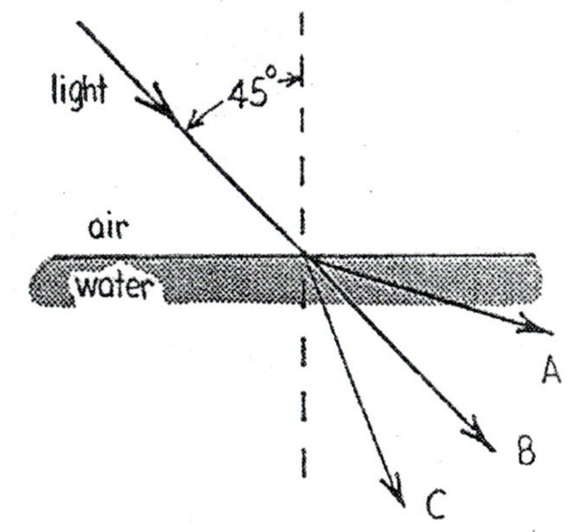

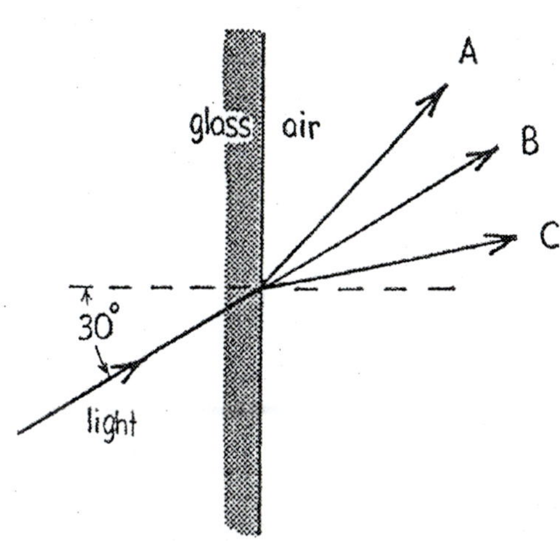

2. The sketch on the left shows a light ray moving from glass into air, at 30° to the normal. Which of the three is most likely the light ray that continues in the air?

   ____

3. To the right, a light ray is shown moving from air into a glass block, at 40° to the normal. Which of the three rays is most likely the light ray that travels in the air after emerging from the opposite side of the block? (Sketch the path the light would take inside the glass.)

   ____

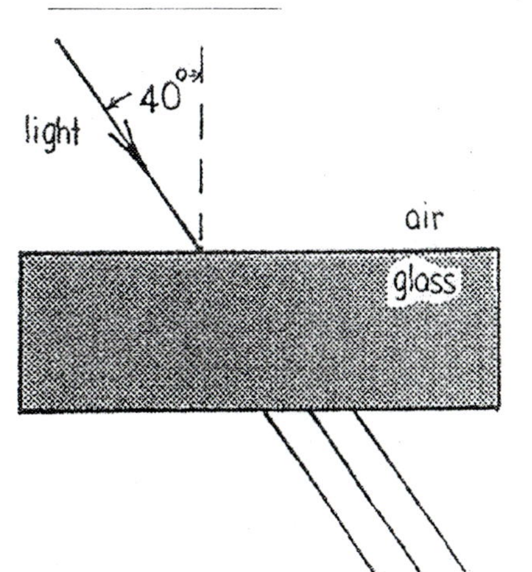

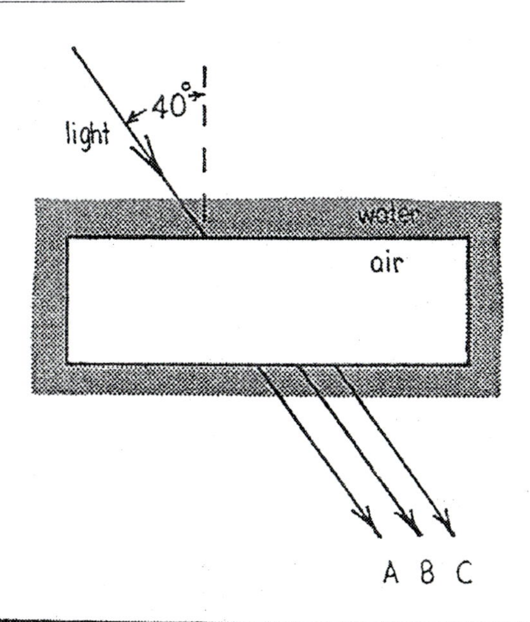

4. To the left, a light ray is shown moving from water into a rectangular block of air (inside a thin-walled plastic box), at 40° to the normal. Which of the rays is most likely the light ray that continues into the water on the opposite side of the block?

   ____

   Sketch the path the light would take inside the air.

thanx to Clarence Bakken

Hewitt Drew it!

Chapter 14 Properties of Light

More Refraction—continued

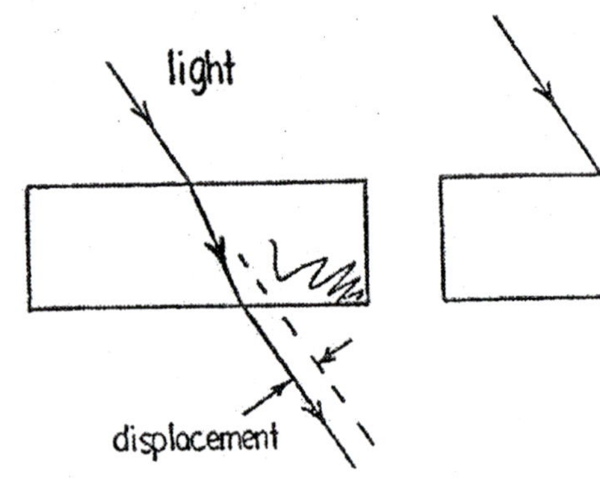

5. The two transparent blocks (right) are made of different materials. The speed of light in the left block is greater than the speed of light in the right block. Draw an appropriate light path through and beyond the right block. Is the light that emerges displaced more or less than light emerging from the left block?

____

6. Light from the air passes through plates of glass and plastic below. The speeds of light in the different materials are shown to the right (these different speeds are often implied by the "index of refraction" of the material). Construct a rough sketch showing an appropriate path through the system of four plates.

Compared to the 50° incident ray at the top, what can you say about the angles of the ray in the air between and below the block pairs?

____

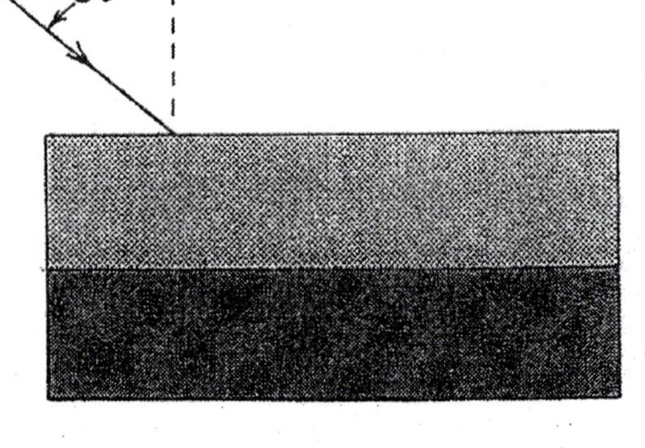

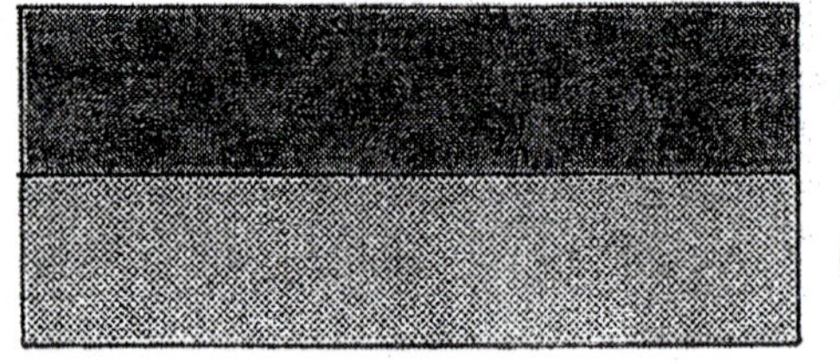

7. Parallel rays of light are refracted as they change speed in passing from air into the eye (left below). Construct a rough sketch showing appropriate light paths when parallel light under water meets the same eye (right below).

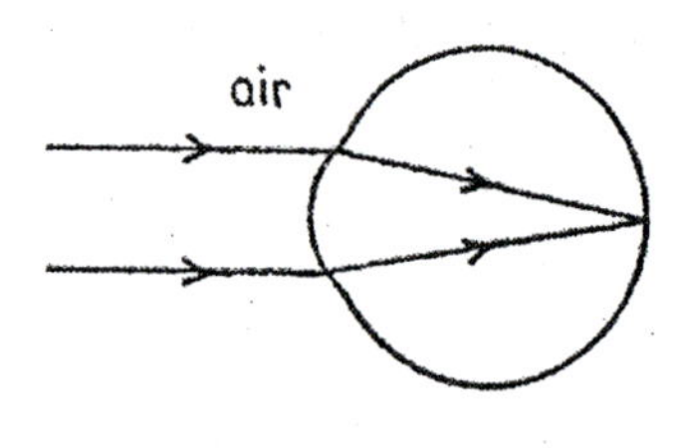

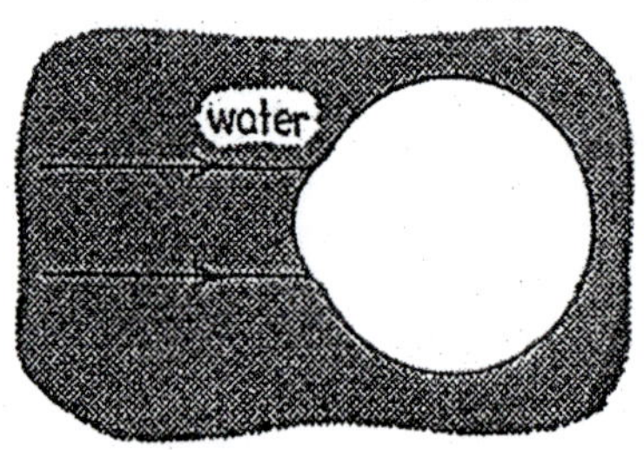

If a fish out of water wishes to clearly view objects in air, should it wear goggles filled with water or with air

8. Why do we need to wear a face mask or goggles to see clearly when under water?

____

Hewitt Drew it!

Name ________________________________ Date __________

CONCEPTUAL Physics FUNDAMENTALS PRACTICE PAGE

Chapter 14 Properties of Light

Lenses

Rays of light bend as shown when passing through the glass blocks.

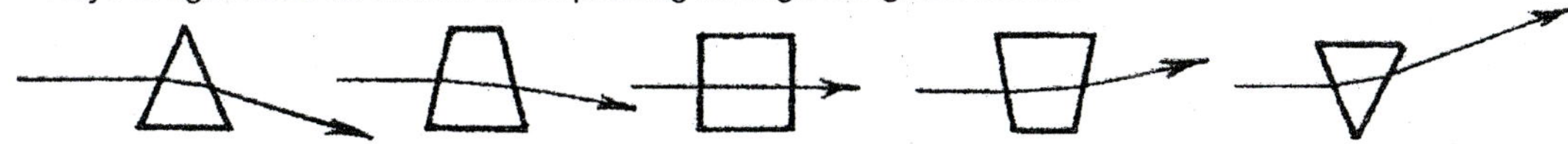

1. Show how light rays bend when they pass through the arrangement of glass blocks below.

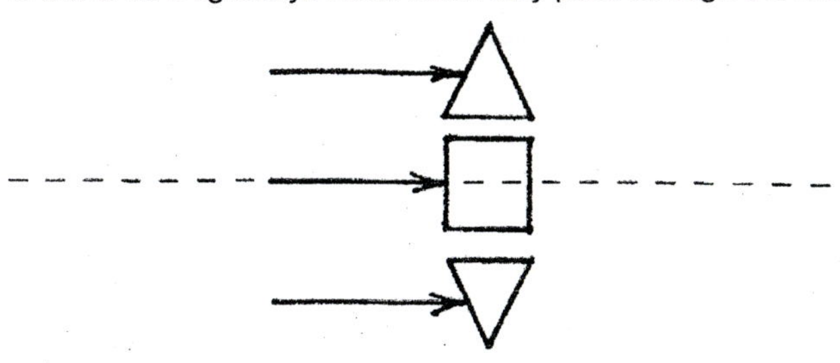

2. Show how light rays bend when they pass through the lens below. Is the lens a converging or a diverging lens? What is your evidence?

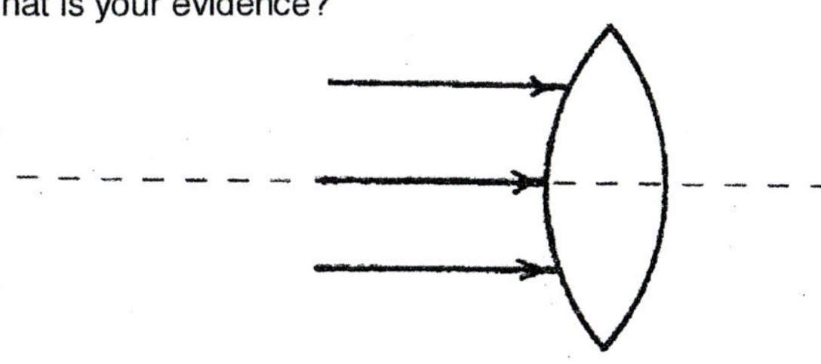

____

3. Show how light rays bend when they pass through the arrangement of glass blocks below.

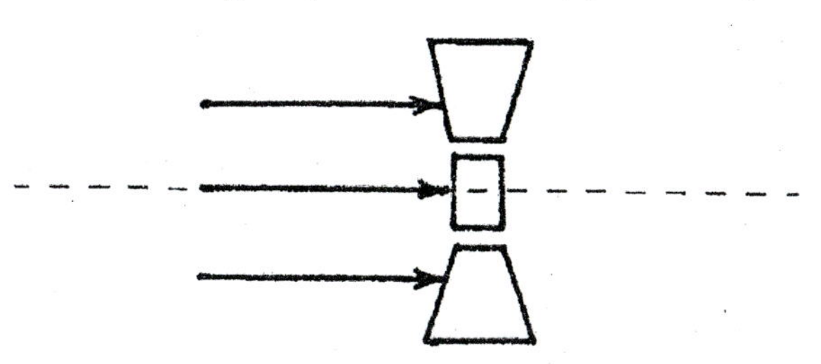

4. Show how light rays bend when they pass through the lens shown below. Is the lens a converging or diverging lens? What is your evidence?

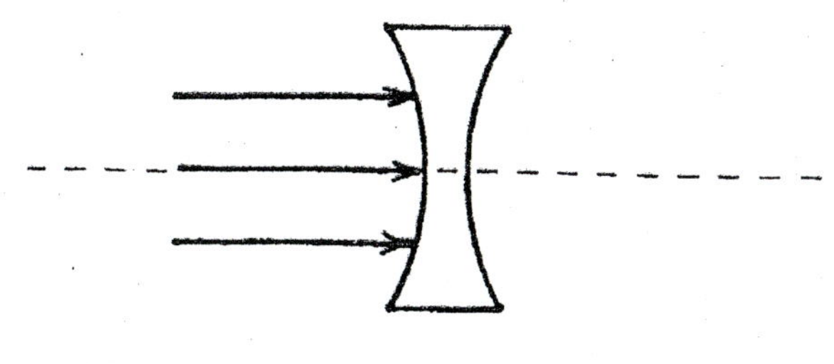

____

Hewitt Drew it!

Chapter 14 Properties of Light

***Lenses**—continued*

5. Which type of lens is used to corrected farsightedness? ________________
   Nearsightedness? ________________

6. Construct rays to find the location and relative size of the arrow's image for each of the lenses. Rays that pass through the middle of a lens continue undeviated. In a converging lens, rays from the tip of the arrow that are parallel to the optic axis extend through the far focal point after going through the lens. Rays that go through the near focal point travel parallel to the axis after going through the lens. In a diverging lens, rays parallel to the axis diverge and appear to originate from the near focal point after passing through the lens. Have fun!

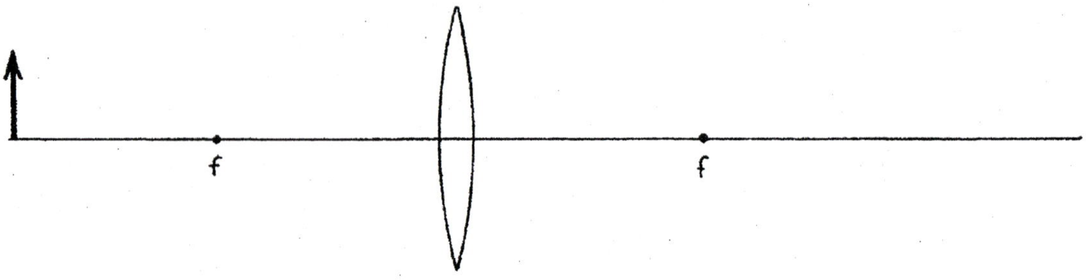

f f

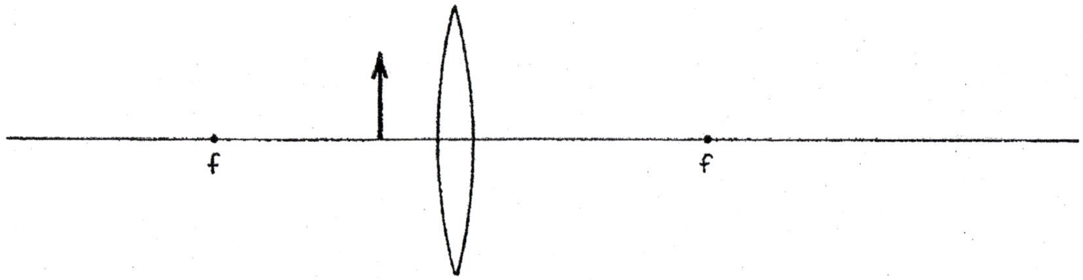

f f

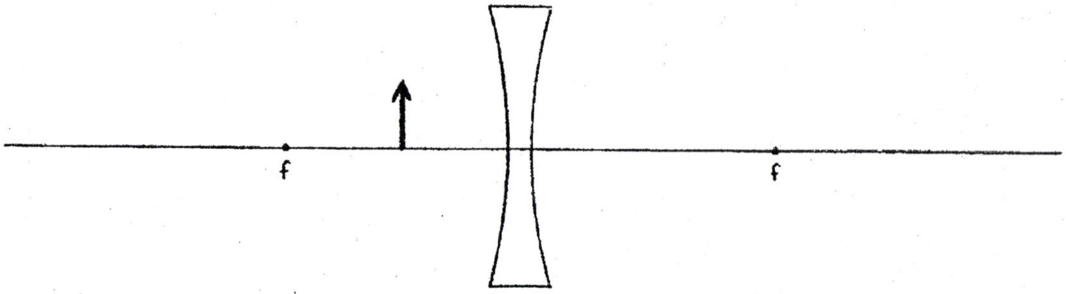

f f

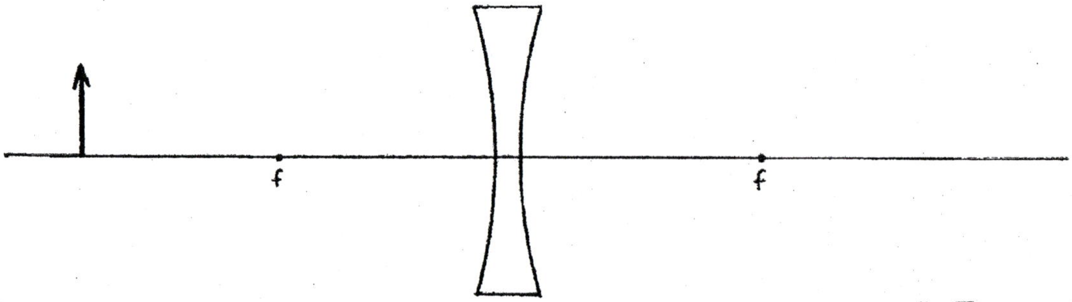

f f

Hewitt Drew it!

Name ____________________ Date ________

CONCEPTUAL **Physics** FUNDAMENTALS PRACTICE PAGE

Chapter 14 Properties of Light

Polarization

The amplitude of a light wave has magnitude and direction, and can be represented by a vector. Polarized light that vibrates in a single direction is represented by a single vector. To the left the single vector represents vertically polarized light. The pair of perpendicular vectors to the right represents nonpolarized light. The vibrations of nonpolarized light are equal in all directions, with as many vertical components as horizontal components.

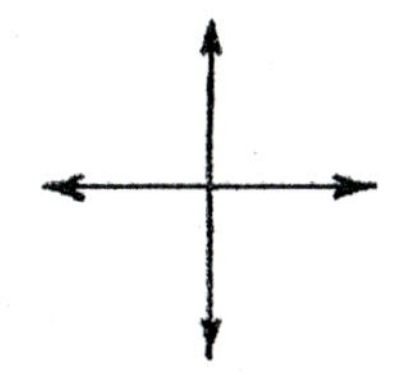

1. In the sketch below, nonpolarized light from a flashlight strikes a pair of Polaroid filters.

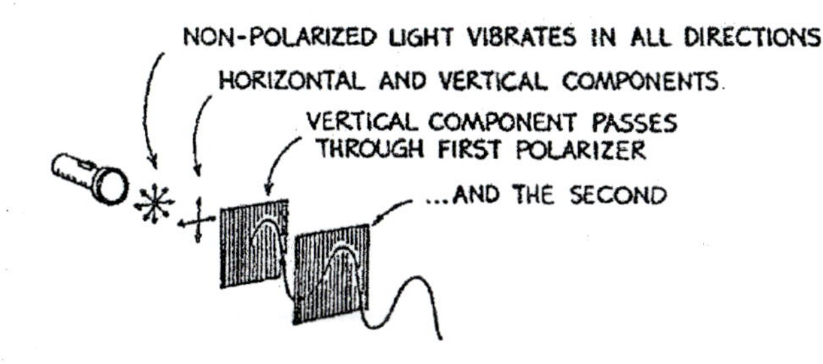
NON-POLARIZED LIGHT VIBRATES IN ALL DIRECTIONS
HORIZONTAL AND VERTICAL COMPONENTS.
VERTICAL COMPONENT PASSES THROUGH FIRST POLARIZER
...AND THE SECOND

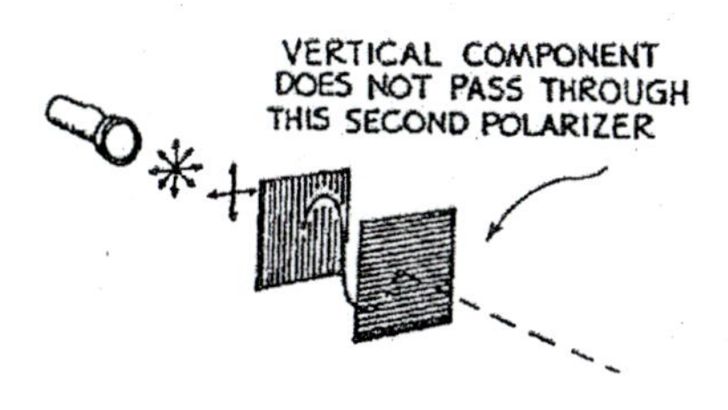
VERTICAL COMPONENT DOES NOT PASS THROUGH THIS SECOND POLARIZER

a. Light is transmitted by a pair of Polaroids when their axes are [aligned] [crossed at right angles]

and light is blocked when their axes are [aligned] [crossed at right angles].

b. Transmitted light is polarized in a direction [the same as] [different than] the polarization axis of the filter.

2. Consider the transmission of light through a pair of Polaroids with polarization axes at 45° to each other. Although in practice the Polaroids are one atop the other, we show them spread out side by side below. From left to right:
(a) Nonpolarized light is represented by its horizontal and vertical components.
(b) These components strike filter A.
(c) The vertical component is transmitted, and
(d) falls upon filter B. This vertical component is not aligned with the polarization axis of filter B, but it has a component that is aligned—component *t*,
(e) which is transmitted.

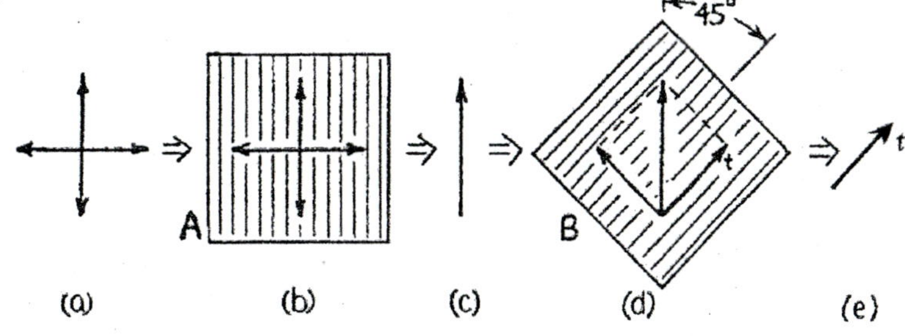
45°
A
B
t
t
(a)
(b)
(c)
(d)
(e)

a. The amount of light that gets through Filter B, compared to the amount that gets through

Filter A is [more] [less] [the same].

b. The component perpendicular to *t* that falls on Filter B is [also transmitted] [absorbed].

Hewitt Drew it!

Chapter 14 Properties of Light

***Polarization**—continued*

3. Below are a pair of Polaroids with polarization axes at 30° to each other. Carefully draw vectors and appropriate components (as in Question 2) to show the vector that emerges at *e*.

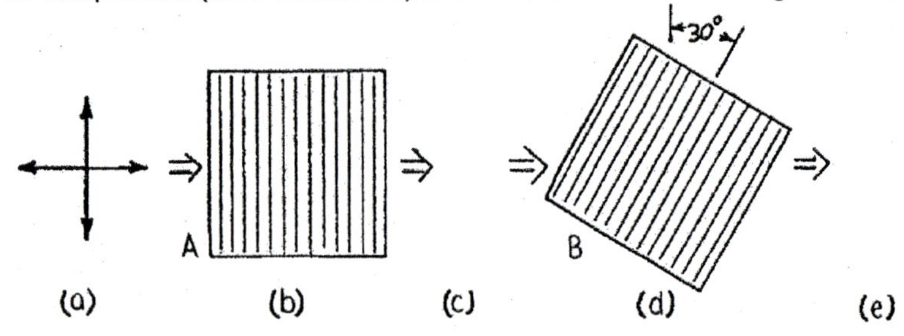

a. The amount of light that gets through the Polaroids at 30°, compared to the amount that gets through the 45° Polaroids is [less] [more] [the same].

4. Figure 29.35 in your textbook shows the smile of Ludmila Hewitt emerging through three Polaroids. Use vector diagrams to complete steps *b* through *g* below to show how light gets through the three-Polaroid system.

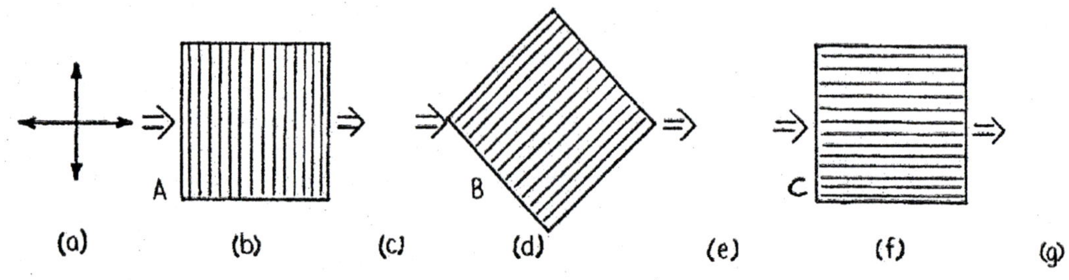

5. A novel use of polarization is shown below. How do the polarized side windows in these next-to-each-other houses provide privacy for the occupants? (Who can see what?)

____

____

Hewitt Drew it!

Name ________________________________ Date __________

CONCEPTUAL **Physics** FUNDAMENTALS PRACTICE PAGE

Chapter 15 Quantum Theory

Light Quanta

1. To say that light is quantized means that light is made up of

   [elemental units] [waves].

2. Compared to photons of low-frequency light, photons of higher-frequency light have more

   [energy] [speed] [quanta].

3. The photoelectric effect supports the

   [wave model of light] [particle model of light].

4. The photoelectric effect is evident when light shone on certain

   photosensitive materials ejects [photon] [electrons].

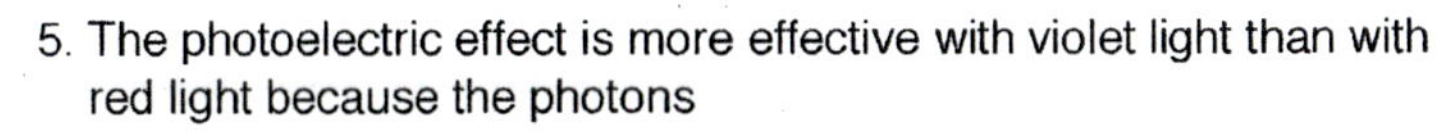

5. The photoelectric effect is more effective with violet light than with red light because the photons

   [resonate with the atoms in the material]

   [deliver more energy to the material]

   [are more numerous].

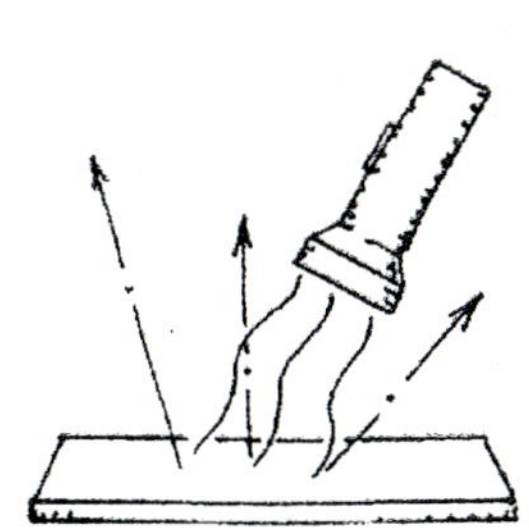

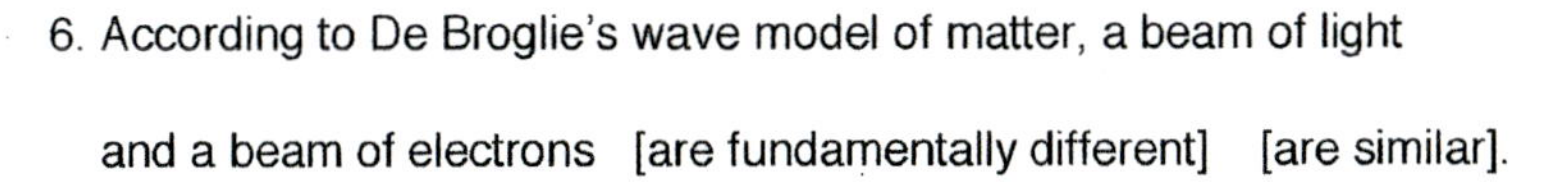

6. According to De Broglie's wave model of matter, a beam of light

   and a beam of electrons [are fundamentally different] [are similar].

7. According to De Broglie, the greater the speed of an electron beam, the

   [longer is its wavelength] [shorter is its wavelength].

8. The discreteness of the energy levels of electrons about the atomic nucleus is best understood

   by considering the electron to be a [wave] [particle].

9. Heavier atoms are not appreciably larger in size than lighter atoms. The main reason for this is that the greater nuclear charge

   [pulls surrounding electrons into tighter orbits]

   [holds more electrons about the atomic nucleus]

   [produces a denser atomic structure].

A QUANTUM MECHANIC!

10. Whereas in the everyday macroworld the study of motion is called *mechanics* in the microworld the study of quanta is called

    [Newtonian mechanics] [quantum mechanics].

Hewitt Drew it!

Name ______________________________ Date ____________

CONCEPTUAL **Physics** FUNDAMENTALS PRACTICE PAGE

Chapter 16 The Atomic Nucleus and Radioactivity

Radioactivity

Complete the following statements:

1.a. A lone neutron spontaneously decays into a proton plus an ______________ .

b. Alpha and beta rays are made of streams of particles, whereas gamma rays are streams of

______________ .

c. An electrically charged atom is called an ______________ .

PHYSICS SIGH

d. Different ______________ of an element are chemically identical but differ in the number of neutrons in the nucleus.

e. Transuranic elements are those beyond atomic number __________ .

f. If the amount of a certain radioactive sample decreases by half in four weeks,

in four more weeks the amount remaining should be ______________ the original amount.

g. Water from a natural hot spring is warmed by ______________ inside Earth.

2. The gas in the little girl's balloon is made up of former alpha and beta particles produced by radioactive decay.

a. If the mixture is electrically neutral, how many more beta particles than alpha particles are in the balloon?

____

____

b. Why is your answer to the above not the "same"?

____

____

c. Why are the alpha and beta particles no longer harmful to the child?

____

____

____

d. What element does this mixture make?

____

Hewitt Drew it!

Chapter 16 The Atomic Nucleus and Radioactivity

Nuclear Reactions

Complete these nuclear reactions:

1. $^{238}_{92}\text{U} \rightarrow {}^{234}_{90}\text{Th} + {}^{4}_{2}$ ____

THORIUM LATE,
I OVERTHLEPT!

2. $^{234}_{90}\text{Th} \rightarrow {}^{234}_{91}\text{Pa} + {}^{0}_{-1}$ ____

3. $^{234}_{91}\text{Pa} \rightarrow$ ____ $+ {}^{4}_{2}\text{He}$

4. $^{220}_{86}\text{Rn} \rightarrow$ ____ $+ {}^{4}_{2}\text{He}$

5. $^{216}_{84}\text{Po} \rightarrow$ ____ $+ {}^{0}_{-1}\text{e}$

6. $^{216}_{84}\text{Po} \rightarrow$ ____ $+ {}^{4}_{2}\text{He}$

7. $^{210}_{83}\text{Bi} \rightarrow$ ____ $+ {}^{0}_{-1}\text{e}$

NUCLEAR PHYSICS --- IT'S THE SAME TO ME WITH THE FIRST TWO LETTERS INTERCHANGED!

8. $^{1}_{0}\text{n} + {}^{10}_{5}\text{B} \rightarrow$ ____ $+ {}^{4}_{2}\text{He}$

Hewitt Drew it!

Name ______________________ Date __________

CONCEPTUAL **Physics** FUNDAMENTALS PRACTICE PAGE

Chapter 16 The Atomic Nucleus and Radioactivity

Natural Transmutation

Fill in the decay-scheme diagram below, similar to that shown in Figure 33.14 in your textbook, but beginning with U-235 and ending with an isotope of lead. Use the table at the left, and identify each element in the series with its chemical symbol.

Step	Particle emitted
1	Alpha
2	Beta
3	Alpha
4	Alpha
5	Beta
6	Alpha
7	Alpha
8	Alpha
9	Beta
10	Alpha
11	Beta
12	Stable

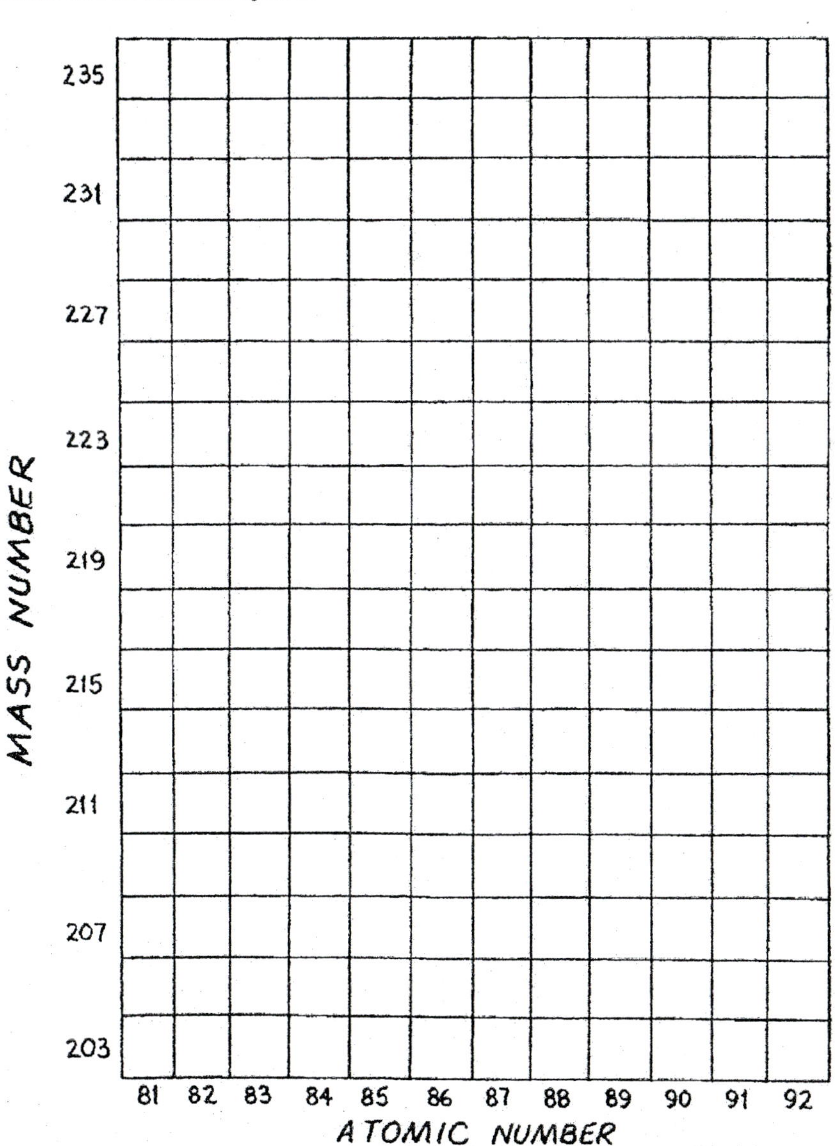

What is the final-product isotope?

____

Name ________________________________ Date ____________

CONCEPTUAL Physics FUNDAMENTALS PRACTICE PAGE

Chapter 16 The Atomic Nucleus and Radioactivity

Nuclear Reactions

1. Complete the table for a chain reaction in which two neutrons from each step individually cause a new reaction.

EVENT	1	2	3	4	5	6	7
NO. OF REACTIONS	1	2	4				

2. Complete the table for a chain reaction where three neutrons from each reaction cause a new reaction.

EVENT	1	2	3	4	5	6	7
NO. OF REACTIONS	1	3	9				

3. Complete these beta reactions, which occur in a breeder reactor.

$^{239}_{92}U \rightarrow \_\_\_\_ + ^{0}_{-1}e$

$^{239}_{93}Np \rightarrow \_\_\_\_ + ^{0}_{-1}e$

4. Complete the following fission reactions.

$^{1}_{0}n + ^{235}_{92}U \rightarrow ^{143}_{54}Xe + ^{90}_{38}Sr + \_\_\_(^{1}_{0}n)$

$^{1}_{0}n + ^{235}_{92}U \rightarrow ^{152}_{60}Nd + \_\_\_\_ + 4(^{1}_{0}n)$

$^{1}_{0}n + ^{239}_{94}Pu \rightarrow \_\_\_\_ + ^{97}_{40}Zr + 2(^{1}_{0}n)$

5. Complete the following fusion reactions.

$^{2}_{1}H + ^{2}_{1}H \rightarrow ^{3}_{2}He + \_\_\_\_$

$^{2}_{1}H + ^{3}_{1}H \rightarrow ^{4}_{2}He + \_\_\_\_$

Hewitt Drew it!

Name ______________________________ Date ____________

CONCEPTUAL **Physics** FUNDAMENTALS PRACTICE PAGE

Appendix B Linear and Rotational Motion

Torques

1. Apply what you know about torques by making a mobile. Shown below are five horizontal arms with fixed 1- and 2-kg masses attached, and four hangers with ends that fit in the loops of the arms, lettered *A* through *R*. You are to determine where the loops should be attached so that when the whole system is suspended from the spring scale at the top, it will hang as a proper mobile, with its arms suspended horizontally. This is best done by working from the bottom upward. Circle the loops where the hangers should be attached. When the mobile is complete, how many kilograms will be indicated on the scale? (Assume the horizontal struts and connecting hooks are practically massless compared with the 1- and 2-kg masses.) On a separate sheet of paper, make a sketch of your completed mobile.

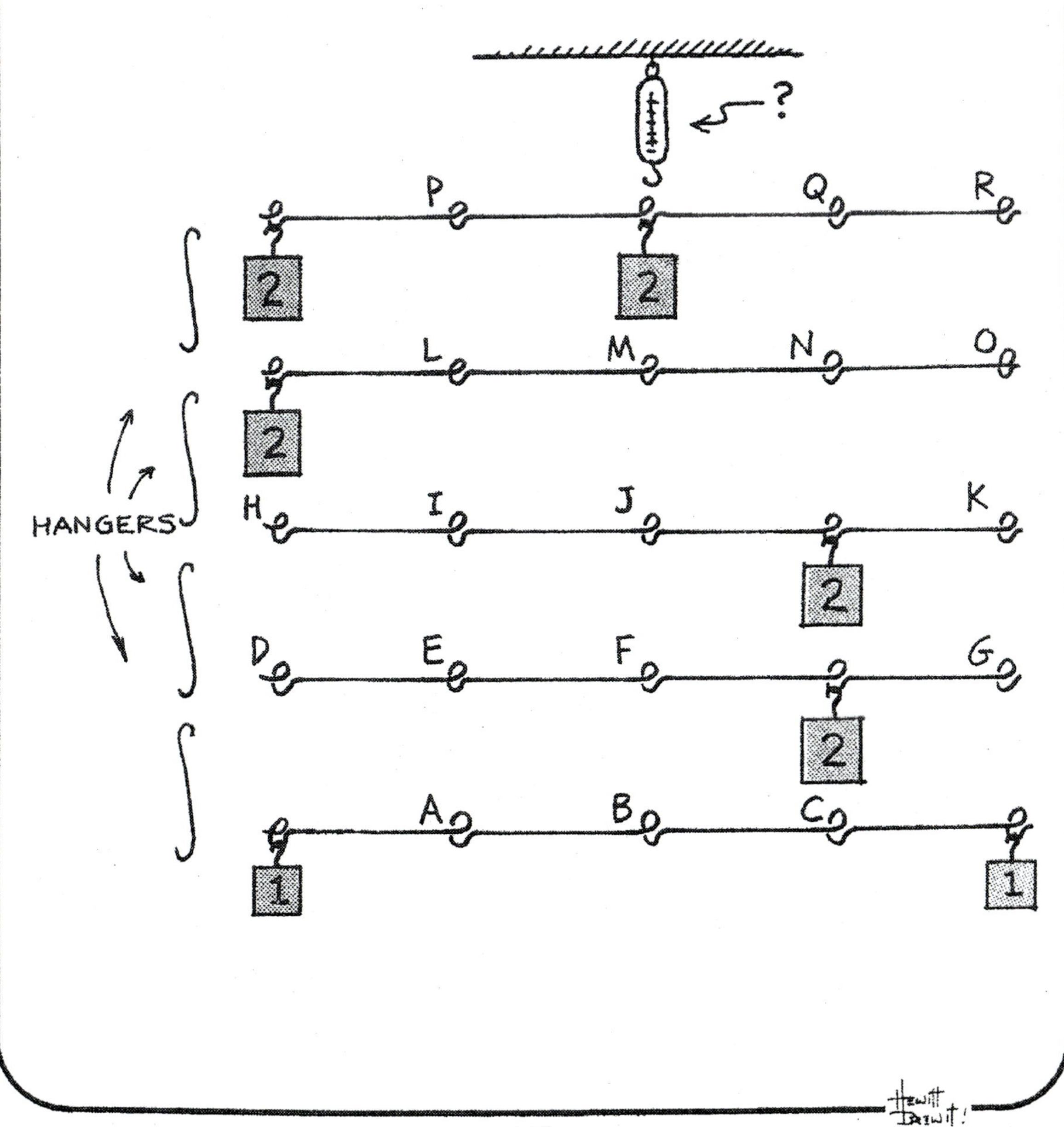

Appendix B Linear and Rotational Motion

Torques—continued

2. Complete the data for the three seesaws in equilibrium.

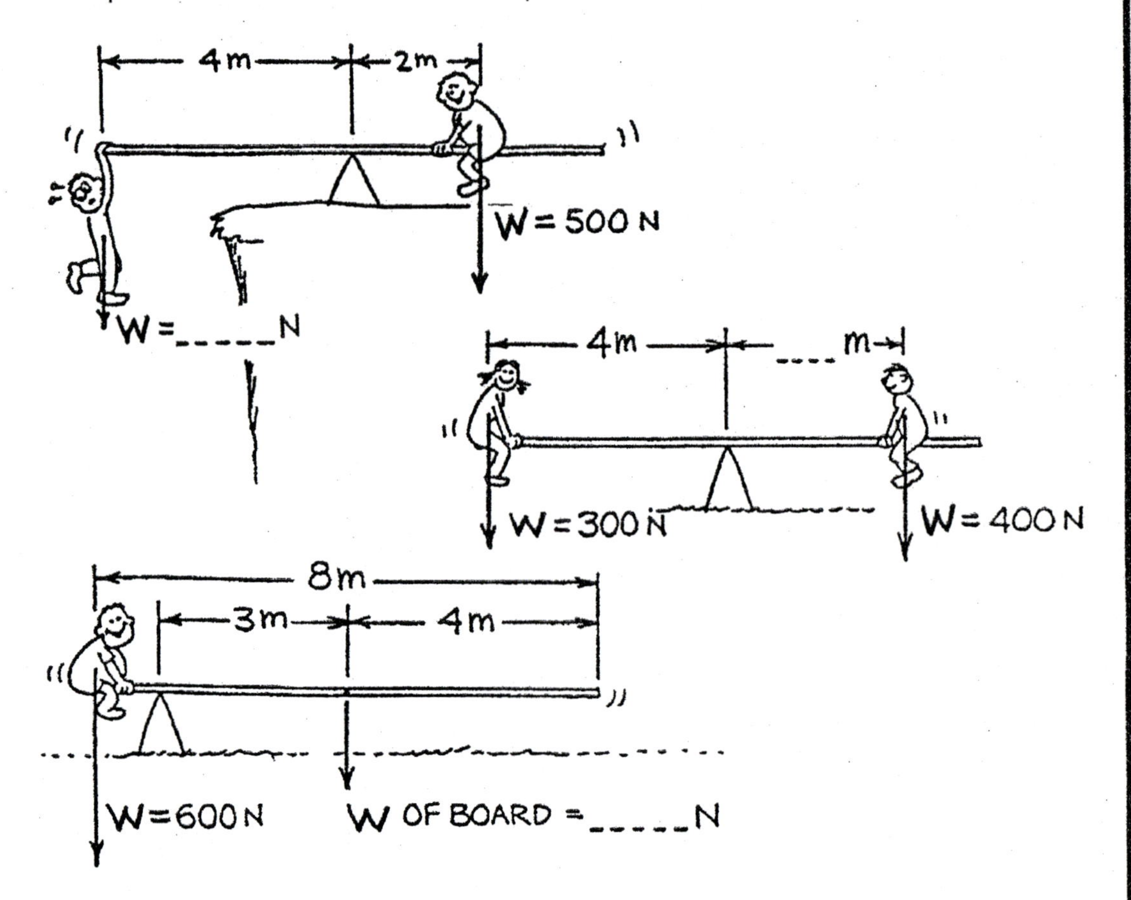

3. The broom balances at its CG. If you cut the broom in half at the CG and weigh each part of the broom, which end would weigh more?

____

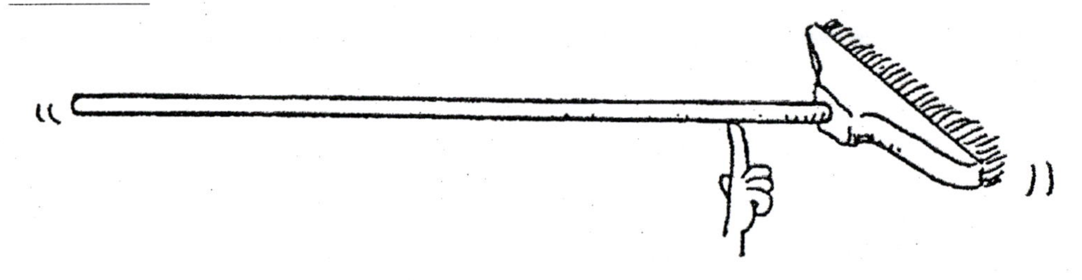

Explain why each end has or does not have the same weight? (Hint: Compare this to one of the seesaw systems above.)

____

____

Name ______________________________ Date __________

CONCEPTUAL **Physics** FUNDAMENTALS PRACTICE PAGE

Appendix B Linear and Rotational Motion

Torques and Rotation

1. Pull the string gently and the spool rolls. The direction of roll depends on the way the torque is applied.

   In (1) and (2) below, the force and lever arm are shown for the torque about the point where surface contact is made (shown by the triangular "fulcrum"). The lever arm is the heavy dashed line, which is different for each different pulling position.

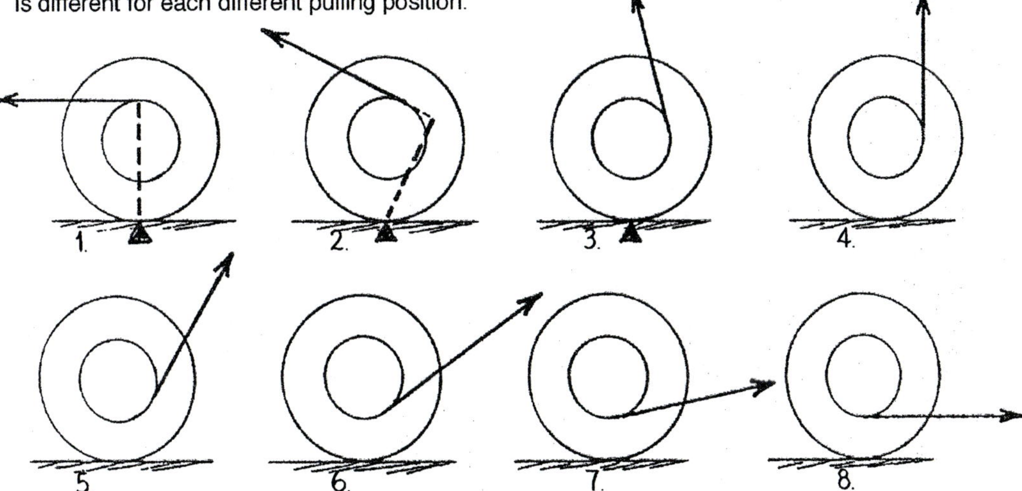

   a. Construct the lever arm for the other positions.

   b. Lever arm is longer when the string of the spool spindle is on the [top] [bottom].

   c. For a given pull, the torque is greater when the string is on the [top] [bottom].

   d. For the same pull, rotational acceleration is greater when the string is on the

   [top] [bottom] [makes no difference].

   e. At which position(s) does the spool roll to the left? __________

   f. At which position(s) does the spool roll to the right? __________

   g. At which position(s) does the spool not roll at all? __________

   h. Why does the spool slide rather than roll at this position?

Be sure your right angle is between the force's *line of action* and the lever arm.

___

2. Relatively few people know that the reason a ball picks up rotational speed rolling down an incline is because of a torque. In sketch A, we see the ingredients of the torque acting on the ball—the force due to gravity and the lever arm to the point where surface contact is made.

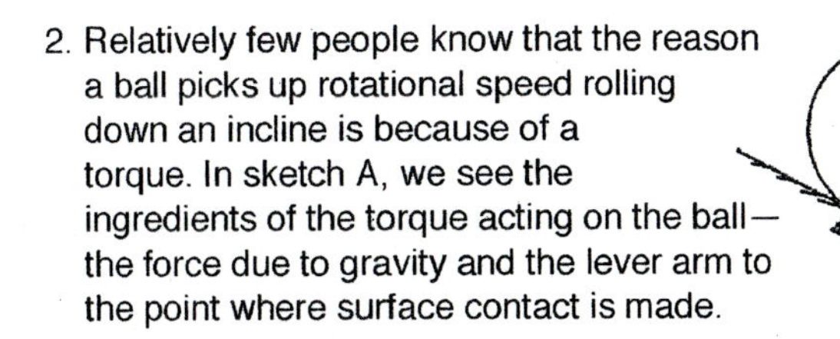

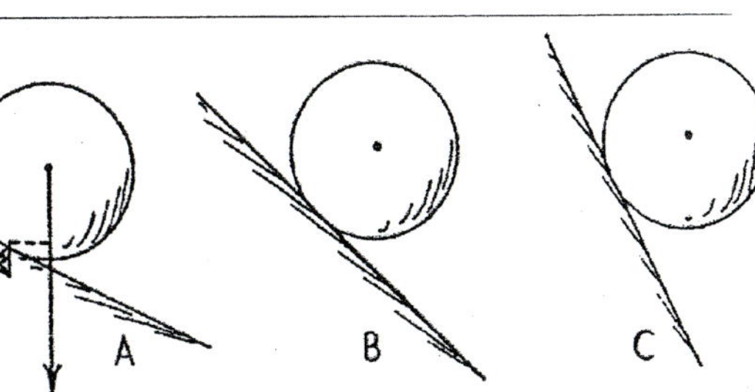

   a. Construct the lever arms for positions B and C.

   b. As the incline becomes steeper, the torque [increases] [decreases].

Appendix B Linear and Rotational Motion

Acceleration and Circular Motion

Newton's 2nd law, $a = F/m$, tells us that net force and its corresponding acceleration are always in the same direction. But force and acceleration vectors are not always in the direction of velocity (another vector).

1. You're in a car at a traffic light. The light turns green and the driver "steps on the gas." The sketch shows the top view of the car. Note the direction of the velocity and acceleration vectors.

   a. Your body tends to lurch [forward] [not at all] [backward].

   b. The car accelerates [forward] [not at all] [backward].

   c. The force on the car acts [forward] [not at all] [backward].

2. You're driving along and approach a stop sign. The driver steps on the brakes. The sketch shows the top view of the car. Draw vectors for velocity and acceleration.

   a. Your body tends to lurch [forward] [not at all] [backward].

   b. The car accelerates [forward] [not at all] [backward].

3. You continue driving, and round a sharp curve to the left at constant speed.

   a. Your body tends to lean [inward] [not at all] [outward].

   b. The direction of the car's acceleration is [inward] [not at all] [outward].

   c. The force on the car acts [inward] [not at all] [outward].

   Draw vectors for velocity and acceleration of the car.

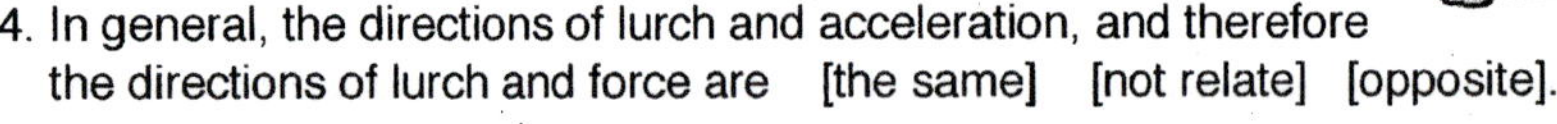
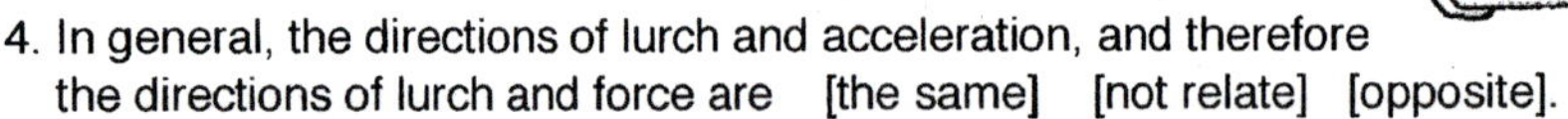

4. In general, the directions of lurch and acceleration, and therefore the directions of lurch and force are [the same] [not relate] [opposite].

5. The whirling stone's direction of motion keeps changing.

   a. If it moves faster, its direction changes [faster] [slower].

   b. This indicates that as speed increases, acceleration [increases] [decreases] [stays the same].

5. Like Question 4, consider whirling the stone on a shorter string—that is, of smaller radius.

   a. For a given speed, the rate that the stone changes direction is [less] [more] [the same].

   b. This indicates that as the radius decreases, acceleration [increases] [decreases] [stays the same].

thanx to Jim Harper

Hewitt Drew it!

Name ______________________________ Date ____________

CONCEPTUAL Physics FUNDAMENTALS PRACTICE PAGE

Appendix B Linear and Rotational Motion

The Flying Pig

The toy pig flies in a circle at constant speed. This arrangement is called a conical pendulum because the supporting string sweeps out a cone. Neglecting the action of its flapping wings, only two forces act on the pig—gravitational ***mg***, and string tension ***T***.

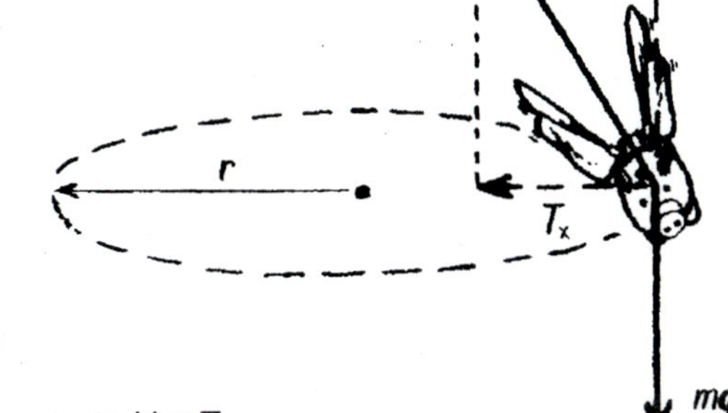

Vector Component Analysis:
Note that vector ***T*** can be resolved into two components—horizontal T_x, and vertical T_y. These vector components are dashed to distinguish them from the tension vector ***T***.

Circle the correct answers:

1. If ***T*** were somehow replaced with T_x and T_y the pig

   [would] [would not] behave identically to being supported by ***T***.

2. Since the pig doesn't accelerate vertically, compared with the magnitude of ***mg***, component T_y,

   must be [greater] [less] [equal and opposite].

3. The velocity of the pig at any instant is [along the radius of] [tangent to] its circular path.

4. Since the pig continues in circular motion, component T_x must be a

   [centripetal] [centrifugal] [nonexistent] force, which equals [zero] [mv^2/r].

   Furthermore, T_x is [along the radius] [tangent to] the circle swept out.

Vector Resultant Analysis:

5. Rather than resolving ***T*** into horizontal and vertical components, use your pencil to sketch the resultant of ***mg*** and ***T*** using the *parallelogram rule*.

6. The resultant lies in a [horizontal] [vertical]

   direction, and [toward] [away from] the

   center of the circular path. The resultant of ***mg***

   and ***T*** is a [centripetal] [centrifugal] force.

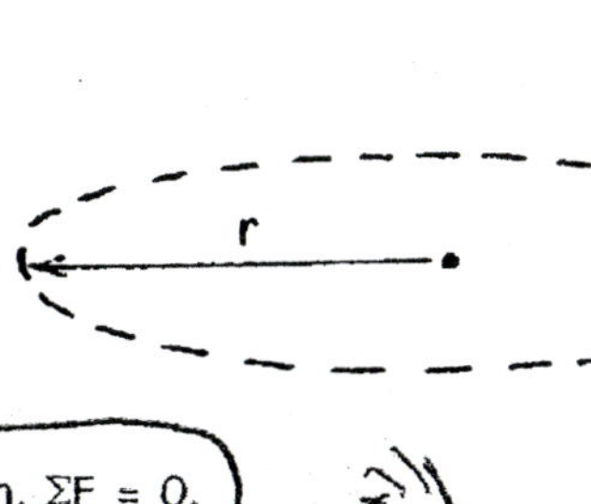

For straight-line motion with no acceleration, $\Sigma F = 0$.
But for uniform circular motion, $\Sigma F = mv^2/r$.

thanx to Pablo Robinson and Miss Piggy

Appendix B Linear and Rotational Motion

Banked Airplanes

An airplane banks as it turns along a horizontal circular path in the air. Except for the thrust of its engines and air resistance, the two significant forces on the plane are gravitational ***mg*** (vertical), and lift ***L*** (perpendicular to the wings).

Vector Component Analysis:
With a ruler and a pencil, resolve vector ***L*** into two perpendicular components, horizontal L_x, and vertical L_y. Make these vectors dashed to distinguish them from ***L***.

L

mg

Circle the correct answers:

1. The velocity of the airplane at any instant is

   [along the radius of]   [tangent to]   its circular path.

2. If ***L*** were somehow replaced with L_x and L_y,

   the airplane   [would]   [would not]   behave the

   same as being supported by ***L***.

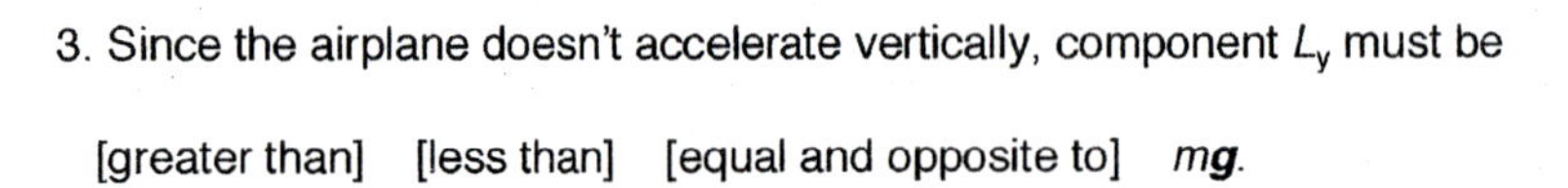

3. Since the airplane doesn't accelerate vertically, component L_y must be

   [greater than]   [less than]   [equal and opposite to]   ***mg***.

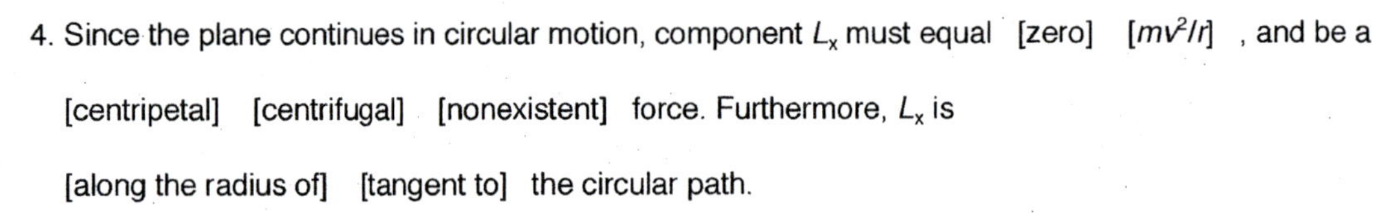

4. Since the plane continues in circular motion, component L_x must equal   [zero]   [mv^2/r] , and be a

   [centripetal]   [centrifugal]   [nonexistent]   force. Furthermore, L_x is

   [along the radius of]   [tangent to]   the circular path.

Vector Resultant Analysis:

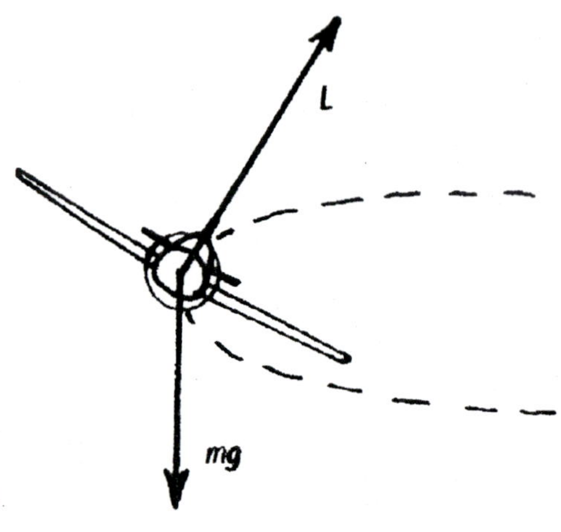

5. Rather than resolving ***L*** into horizontal and vertical components, use your pencil to sketch the resultant of ***mg*** and ***L*** using the *parallelogram rule.*

6. The resultant lies in a   [horizontal]   [vertical]

   direction, and   [toward]   [away from]   the center

   of the circular path. The resultant of ***mg*** and ***L*** is a

   [centripetal]   [centrifugal]   force.

7. The resultant of ***mg*** and ***L*** is the same as   [L_x]   [L_y].

Challenge: Explain in your own words why the resultant of two vectors can be the same as a single component of one of them.

Hewitt Drew it!

Appendix B Linear and Rotational Motion

Banked Track

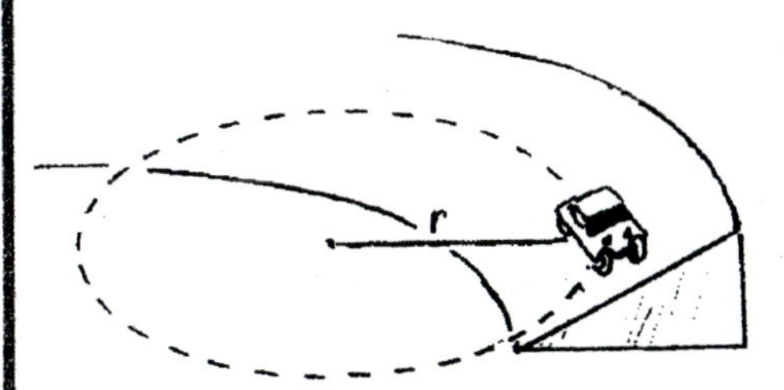

A car rounds a banked curve with just the right speed so that it has no tendency to slide down or up the banked road surface. Shown below are two main forces that act on the car perpendicular to its motion—gravitational ***mg*** and the normal force ***N*** (the support force of the surface).

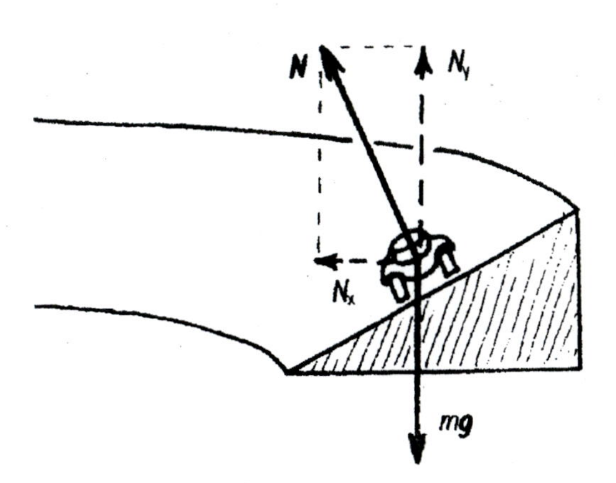

Vector Component Analysis:
Note that vector ***N*** is resolved into two perpendicular components, horizontal N_x, and vertical N_y. As usual, these vectors are dashed to distinguish them from ***N***.

Circle the correct answers:

1. If ***N*** were somehow replaced with N_x and N_y, the car [would] [would not] behave identically to being supported by ***N***.

2. Since the car doesn't accelerate vertically, component N_y must be [greater than] [equal and opposite to] [less than] ***mg***.

3. The velocity of the car at any instant is [along the radius of] [tangent to] its circular path.

4. Since the car continues in uniform circular motion, component N_x must equal [zero] [mv^2/r] and be a [centripetal] [centrifugal] [nonexistent] force. Furthermore, N_x [lies along the radius of] [is tangent to] the circular path.

Vector Resultant Analysis:

5. Rather than resolving ***N*** into horizontal and vertical components, use your pencil to sketch the resultant of ***mg*** and ***N*** using the *parallelogram rule*.

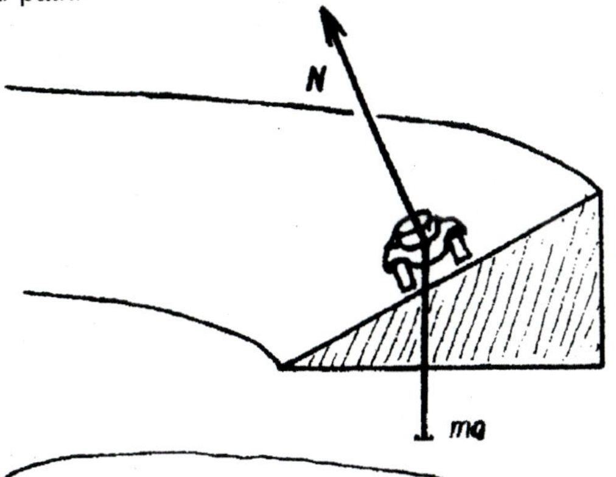

6. The resultant lies in a [horizontal] [vertical] direction, and [toward] [away from] the center of the circular path. The resultant of ***mg*** and ***N*** is a [centripetal] [centrifugal] force.

7. The resultant of ***mg*** and ***N*** is the same as [N_x] [N_y], and provides the [centripetal] [centrifugal] force.

Notice that when a component of ***N*** makes up a centripetal force, ***N*** > ***mg***.

thanx to Pablo Robinson

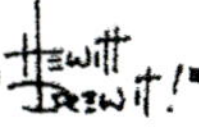

Appendix B Linear and Rotational Motion

Leaning On

When turning a corner on a bicycle, everyone knows that you've got to lean "into the curve." What is the physics of this leaning? It involves torque, friction, and centripetal force (mv^2/r).

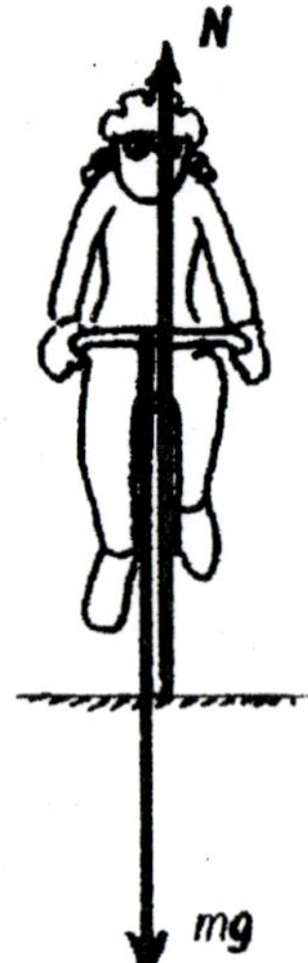

First, consider the simple case of riding a bicycle along a straight-line path. Except for the force that propels the bike forward (friction of the road in the direction of motion) and air resistance (friction of air against the direction of motion), only two significant forces act: weight ***mg*** and the normal force ***N***. (The vectors are drawn side-by-side, but actually lie along a single vertical line.)

Circle the correct answers:

1. Since there is no vertical acceleration, we can say that the magnitude of

   [$N > mg$] [$N < mg$] [$N = mg$], which means that in the vertical direction,

   [$\Sigma F_y > 0$] [$\Sigma F_y < 0$] [$\Sigma F_y = 0$].

2. Since the bike doesn't rotate or change in its rotational state, then the total

   torque is [zero] [not zero].

Now consider the same bike rounding a corner. In order to safely make the turn, the bicyclist leans in the direction of the turn. A force of friction pushes sideways on the tire toward the center of the curve.

3. The friction force, ***f***, provides the centripetal force that produces a

   curved path. Then [$f = mv^2/r$] [$f \neq mv^2/r$].

4. Consider the net torque about the center of mass (*CM*) of the bike-rider system. Gravity produces no torque about this point, but ***N*** and ***f*** do. The torque involving ***N*** tends to produce

   [clockwise] [counterclockwise] rotation, and the one involving ***f***

   tends to produce [clockwise] [counterclockwise] rotation.

   These torques cancel each other when the resultant of vectors ***N*** and ***f*** pass through the *CM*.

5. With your pencil, use the parallelogram rule and sketch in the resultant of vectors ***N*** and ***f***. Label your resultant ***R***. Note the ***R*** passes through the center of mass of the bike-rider system. That

   means that ***R*** produces [a clockwise] [a counterclockwise] [no]

   torque about the *CM*. Therefore the bike-rider system

   [topples clockwise] [topples counterclockwise] [doesn't topple].

When learning how to turn on a bike, you lean so that the sum of the torques about your CM is zero. You may not be calculating torques, but your body learns to feel them.

thanx to Pablo Robinson

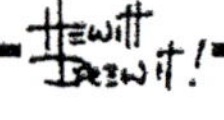

Name ______________________________ Date __________

CONCEPTUAL Physics FUNDAMENTALS PRACTICE PAGE

Appendix B Linear and Rotational Motion

Simulated Gravity and Frames of Reference

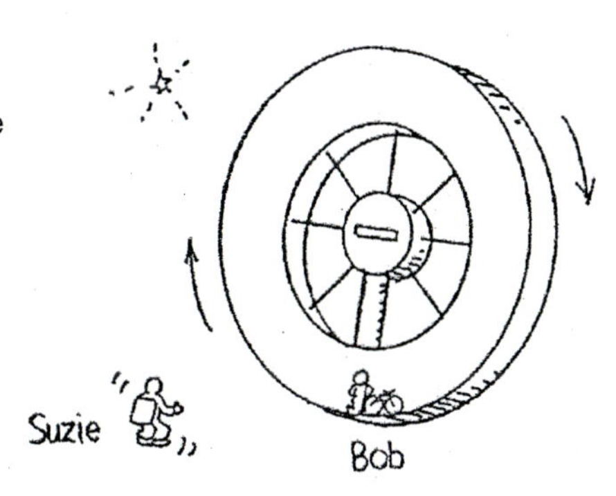

Suzie Spacewalker and Bob Biker are in outer space. Bob experiences Earth-normal gravity in a rotating habitat, where centripetal force on his feet provides a normal support force that feels like weight. Suzie hovers outside in a weightless condition, motionless relative to the stars and the center of the habitat.

1. Suzie sees Bob rotating clockwise in a circular path at a linear speed of 30 km/h. Suzie and Bob are facing each other, and from Bob's point of view, he is at rest

   and he sees Suzie moving [clockwise] [counterclockwise].

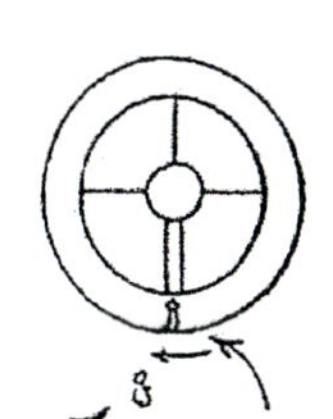
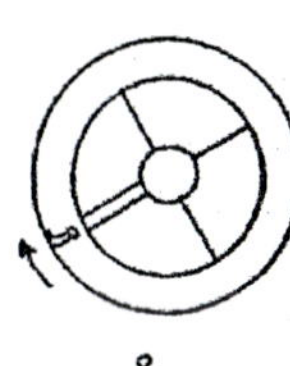
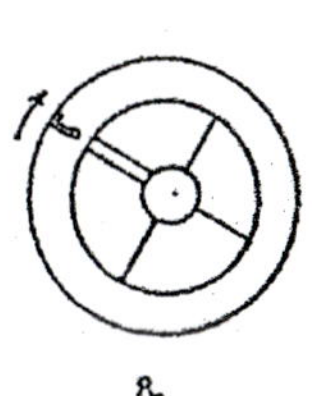
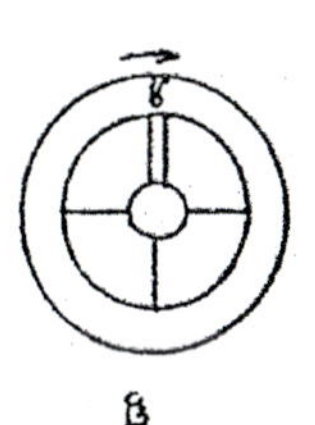
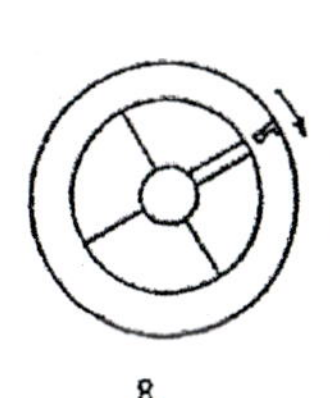
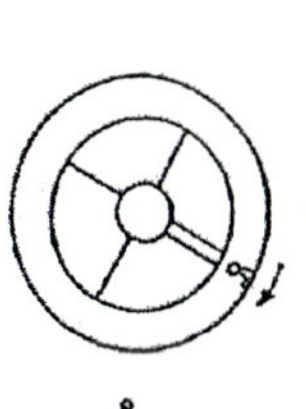
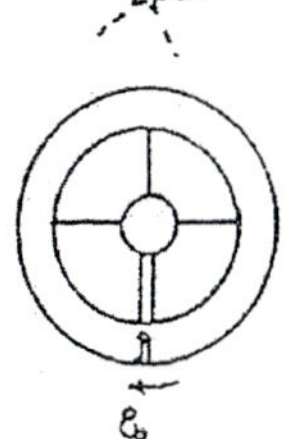

Bob at rest on the floor

Suzie hovering in space

2. The rotating habitat seems like home to Bob—until he rides his bicycle. When he rides in the

   opposite direction as the habitat rotates, Suzie sees him moving [faster] [slower].

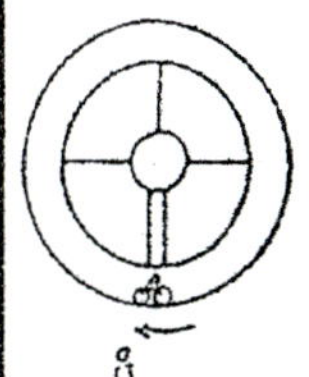
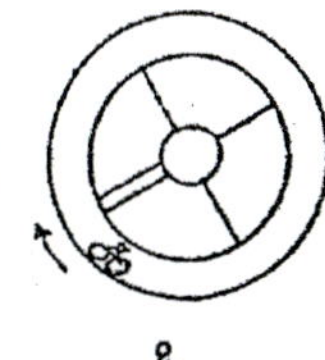
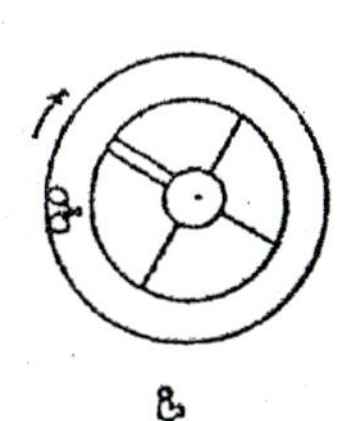
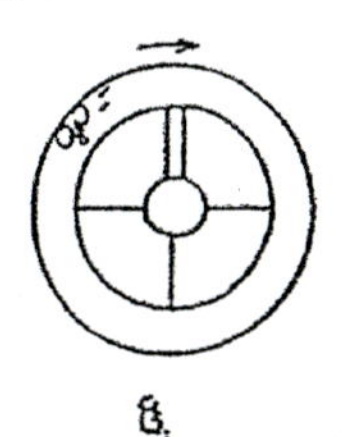
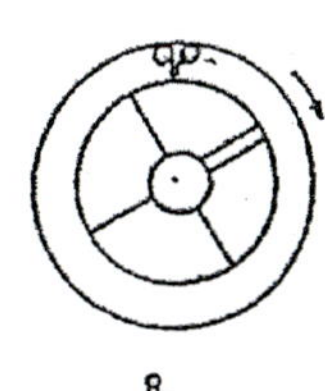
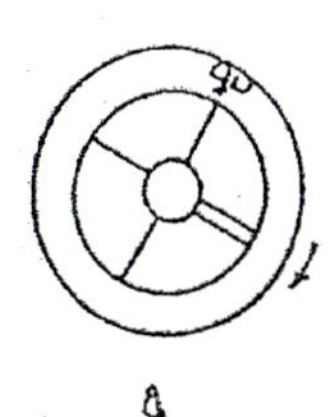
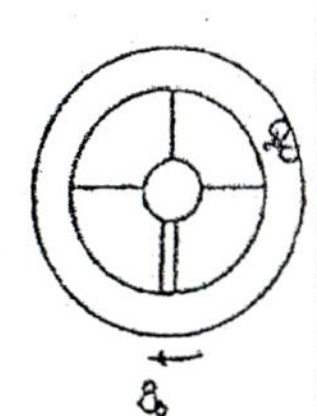

Bob rides counter-clockwise

3. As Bob's bicycle speedometer reading increases, his rotational speed

   [decreases] [remains unchanged] [increases] and the normal force that feels like weight

   [decreases] [remains unchanged] [increases]. So friction between the tires and the floor

   [decreases] [remains unchanged] [increases].

4. When Bob nevertheless gets his speed up to 30 km/h,

   as indicated on his bicycle speedometer, Suzie sees him

   [moving at 30 km/h] [motionless] [moving at 60 km/h].

thanx to Bob Becker

Hewitt Drewit!

Appendix B Linear and Rotational Motion

Simulated Gravity and Frames of Reference*—continued*

5. Bounding off the floor a bit while riding at 30 km/h, and neglecting wind effects, Bob

   [drifts toward the ceiling in midspace as the floor whizzes by him at 30 km/h]

   [falls as he would on Earth]

   [slams onto the floor with increased force]

   and finds himself

   [in the same frame of reference as Suzie]

   [as if he rode at 30 km/h on Earth's surface]

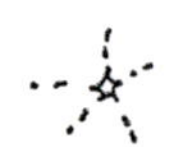

   [pressed harder against the bicyclist seat].

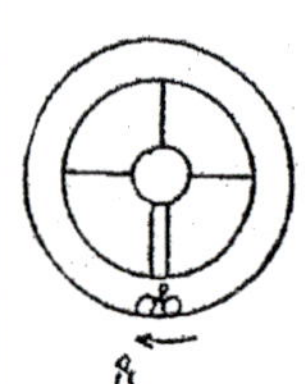
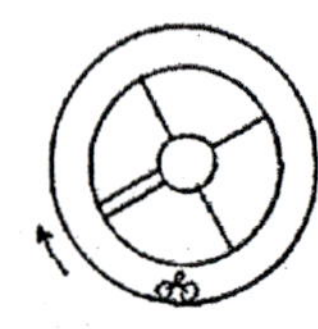
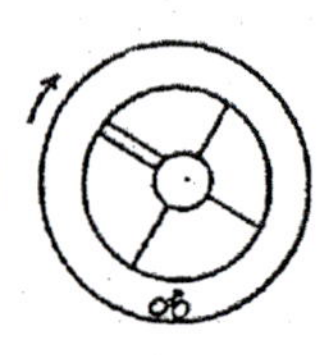
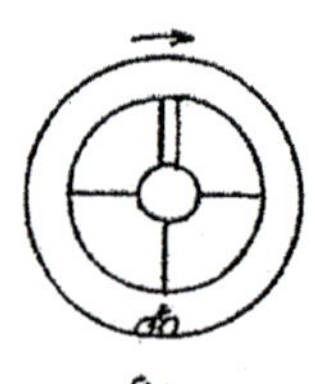
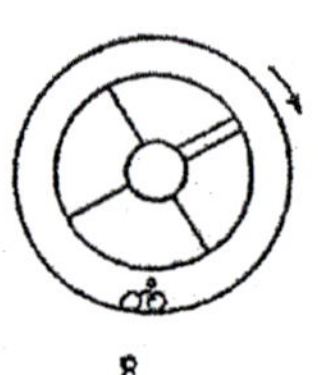
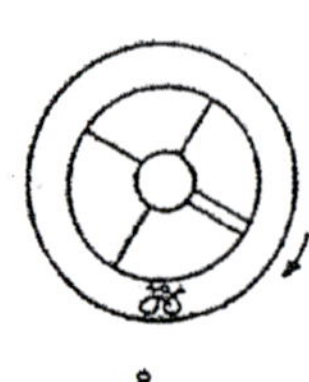
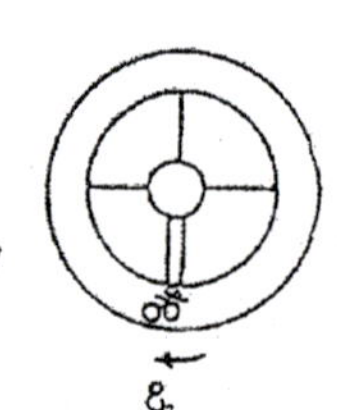

Bob rides at 30 km/h with respect to the floor

6. Bob maneuvers back to his initial condition, whirling at rest with the habitat, standing beside his bicycle. But not for long. Urged by Suzie, he rides in the opposite direction, clockwise with the rotation of the habitat.
   Now Suzie sees him moving [faster] [slower].

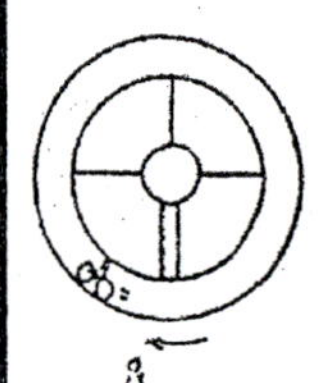
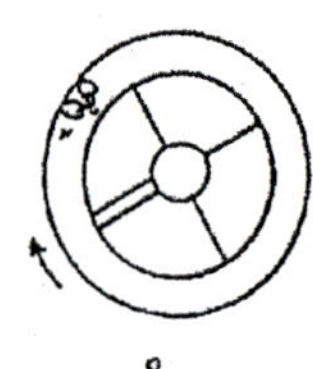
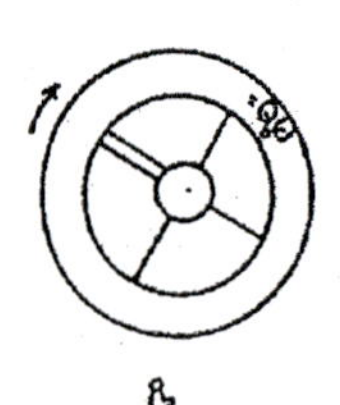
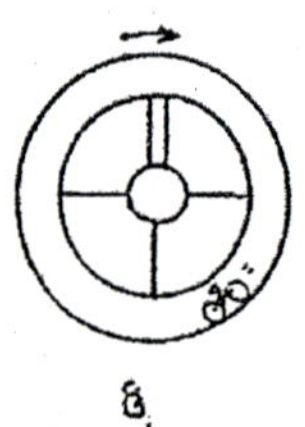
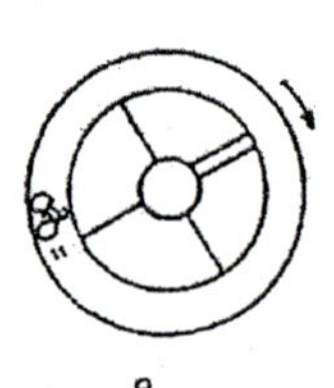
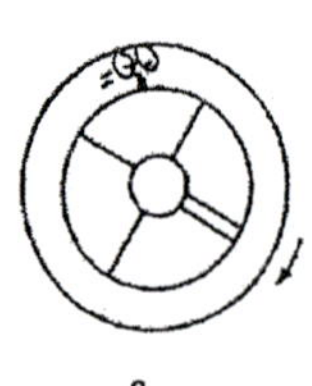
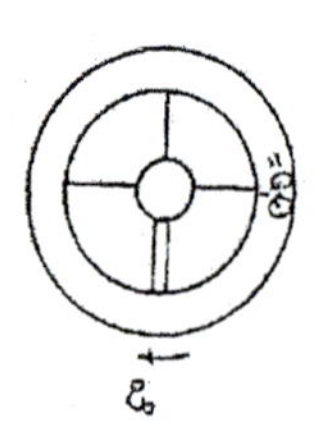

Bob rides clockwise

7. As Bob gains speed, the normal support force that feels like weight

   [decreases] [remains unchanged] [increases].

8. When Bob's speedometer reading gets up to 30 km/h, Suzie sees him moving

   [30 km/h] [not at all] [60 km/h] and Bob finds himself

   [weightless like Suzie]

   [just as if he rode at 30 km/h on Earth's surface]

   [pressed harder against the bicyclist seat].

Next, Bob goes bowling. You decide whether the game depends on which direction the ball is rolled!

Hewitt Drew it!

Name ______________________________ Date __________

CONCEPTUAL **Physics** FUNDAMENTALS PRACTICE PAGE

Appendix C Vectors

Vectors and Sailboats

(Please do not attempt this until you have studied Appendix D!)

1. The sketch shows a top view of a small railroad car pulled by a rope. The force ***F*** that the rope exerts on the car has one component along the track, and another component perpendicular to the track.

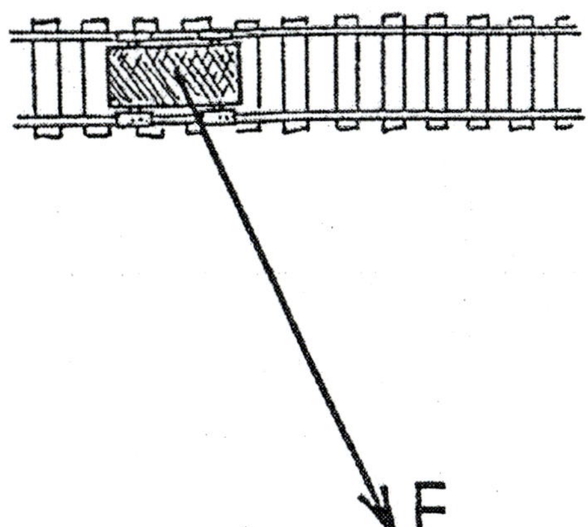

a. Draw these components on the sketch. Which component is larger?

____

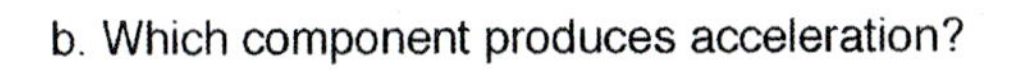

b. Which component produces acceleration?

____

c. What would be the effect of pulling on the rope if it were perpendicular to the track?

____

2. The sketches below represent simplified top views of sailboats in a cross-wind direction. The impact of the wind produces a FORCE vector on each as shown. (We do NOT consider *velocity* vectors here!)

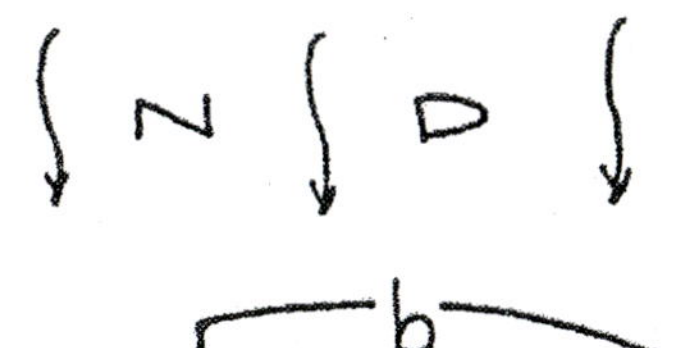

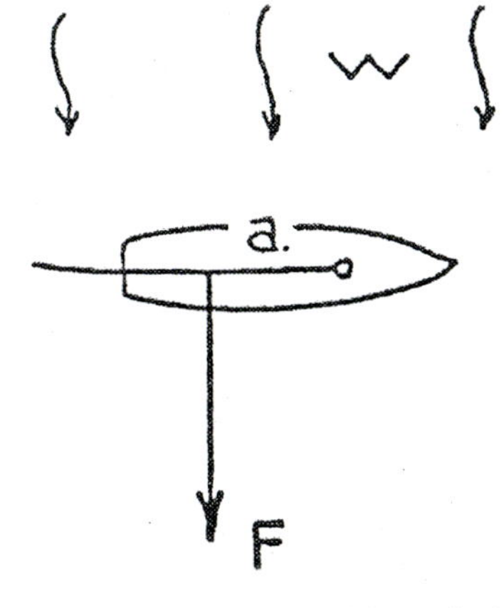

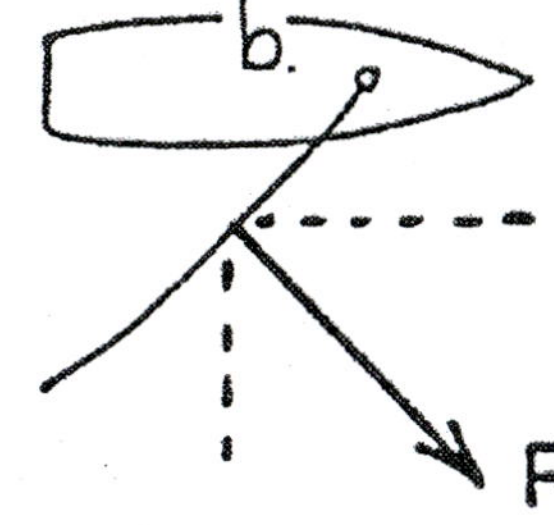

a. Why is the position of the sail above useless for propelling the boat along its forward direction? (Relate this to Question 1.c above where the train is constrained by tracks to move in one direction, and the boat is similarly constrained to move along one direction by its deep vertical fin—the *keel.*)

____

____

b. Sketch the component of force parallel to the to the direction of the boat's motion (along its keel), and the component perpendicular to its motion. Will the boat move in a forward direction? (Relate this to Question 1.b above.)

____

____

____

Hewitt Drew It!

Appendix C Vectors

Vectors and Sailboats—continued

3. The boat to the right is oriented at an angle into the wind. Draw the force vector and its forward and perpendicular components.

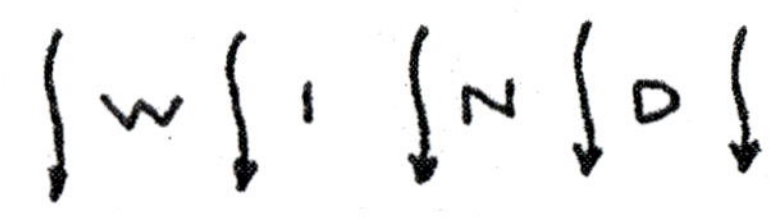

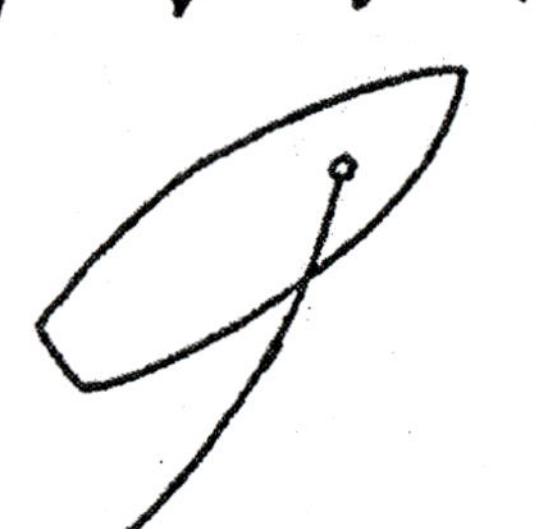

   a. Will the boat move in a forward direction and tack into the wind? Why or why not?

   ____

   ____

   ____

4. The sketch below is a top view of five identical sailboats. Where they exist, draw force vectors to represent wind impact on the sails. Then draw components parallel and perpendicular to the keels of each boat.

   a. Which boat will sail the fastest in a forward direction?

   ____

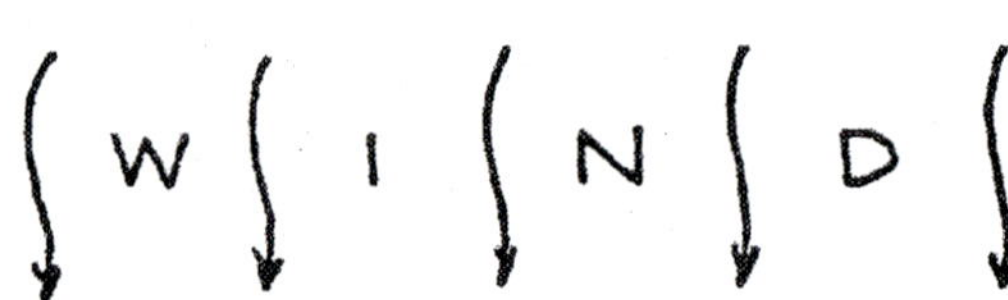

   b. Which will respond least to the wind?

   ____

   c. Which will move in a backward direction?

   ____

   d. Which will experience decreasing wind impact with increasing speed?

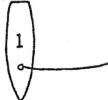

   ____

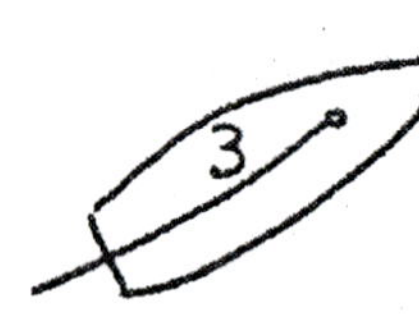

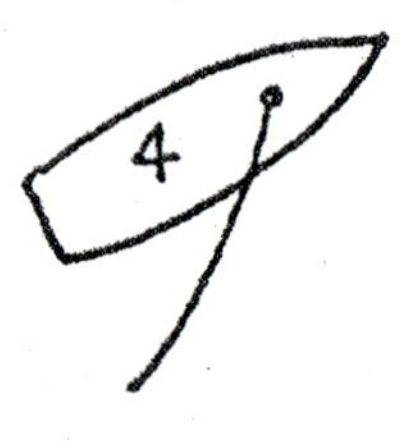

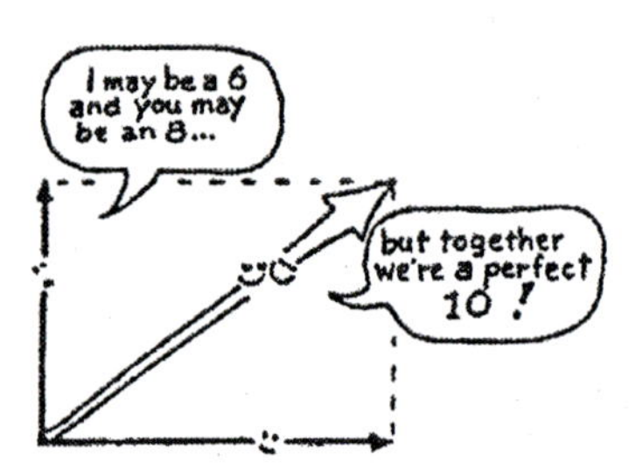

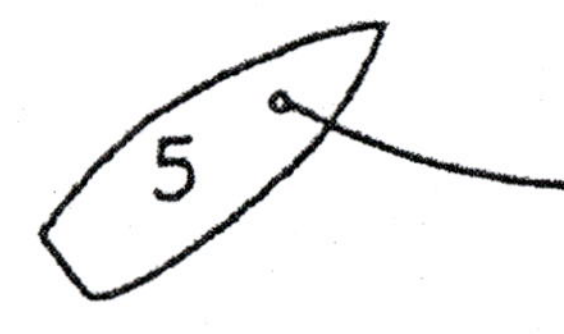

Hewitt Drew it!

ANSWERS TO THE PRACTICE PAGES

Compare your responses to the previous pages with my responses in the reduced ones that follow. You have the choice of taking a shortcut and looking at my responses first—or you can be nice to yourself and first work out your own—before looking. In working through the practice pages on your own or with friends, check your answers only *after* you've given them a good college try. Then you may experience the exhilaration that comes with doing a good thing well.

Learning can be enjoyable!

Name ______________________________ Date ____________

CONCEPTUAL **Physics** FUNDAMENTALS PRACTICE PAGE

Chapter 1 About Science
Making Hypotheses

The word science comes from Latin, meaning "to know." The word *hypothesis* comes from Greek, "under an idea." A hypothesis (an educated guess) often leads to new knowledge and may help to establish a theory.

WHICH IS AN EDUCATED GUESS... A HYPOTHESIS OR A THEORY?

WHICH RESULTS FROM A LARGE BODY OF KNOWLEDGE?

Examples:

1. It is well known that objects generally expand when heated. An iron plate gets slightly bigger, for example, when placed in an oven. But what of a hole in the middle of the plate? One friend may say the size of the hole will increase, and another may say it will decrease.

I CUT A DISK FROM THIS IRON PLATE. WHEN I HEAT THE PLATE, WILL THE HOLE GET BIGGER, OR SMALLER?

a. What is your hypothesis about hole size, and if you are wrong, is there a test for finding out?

HYP 1: HOLE GETS BIGGER. HYP 2: SMALLER. HYP 3: NO CHANGE.

TEST: HEAT IT IN AN OVEN, THEN MEASURE! (HYP 1 IS CORRECT.)

WHAT HAPPENS IF HE PLUGS THE DISK BACK INTO THE HOLE BEFORE HEATING EVERYTHING?

b. There are often several ways to test a hypothesis. For example, you can perform a physical experiment and witness the results yourself, or you can use the library or internet to find the reported results of other investigators. Which of these two methods do you favor, and why?

(IT DEPENDS ON THE SITUATION—MOST RESEARCH INVOLVES BOTH.)

2. Before the time of the printing press, books were hand-copied by scribes, many of whom were monks in monasteries. There is the story of the scribe who was frustrated to find a smudge on an important page he was copying. The smudge blotted out part of the sentence that reported the number of teeth in the head of a donkey. The scribe was very upset and didn't know what to do. He consulted with other scribes to see if any of their books stated the number of teeth in the head of a donkey. After many hours of fruitless searching through the library, it was agreed that the best thing to do was to send a messenger by donkey to the next monastery and continue the search there. What would be your advice?

ACTUALLY LOOK IN THE DONKEY'S MOUTH AND COUNT!

Making Distinctions

Many people don't seem to see the difference between a thing and the abuse of the thing. For example, a city council that bans skateboarding may not distinguish between skateboarding and reckless skateboarding. A person who advocates that a particular technology be banned may not distinguish between that technology and the abuses of that technology. There's a difference between a thing and the abuse of the thing.

On a separate sheet of paper, list other examples where use and abuse are often not distinguished. Compare your list with others in your class.

Hewitt Drew it!

CONCEPTUAL **Physics** FUNDAMENTALS PRACTICE PAGE

Chapter 1 About Science
Pinhole Formation

Look carefully on the round spots of light on the shady ground beneath trees. These are *sunballs*, which are images of the sun. They are cast by openings between leaves in the trees that act as pinholes. (Did you make a pinhole "camera" back in middle school?) Large sunballs, several centimeters in diameter or so, are cast by openings that are relatively high above the ground, while small ones are produced by closer "pinholes." The interesting point is that the ratio of the diameter of the sunball to its distance from the pinhole is the same ratio of the Sun's diameter to its distance from the pinhole. We know the Sun is approximately 150,000,000 km from the pinhole, so careful measurements of of the ratio of diameter/distance for a sunball leads you to the diameter of the Sun. That's what this page is about. Instead of measuring sunballs under the shade of trees on a sunny day, make your own easier-to-measure sunball.

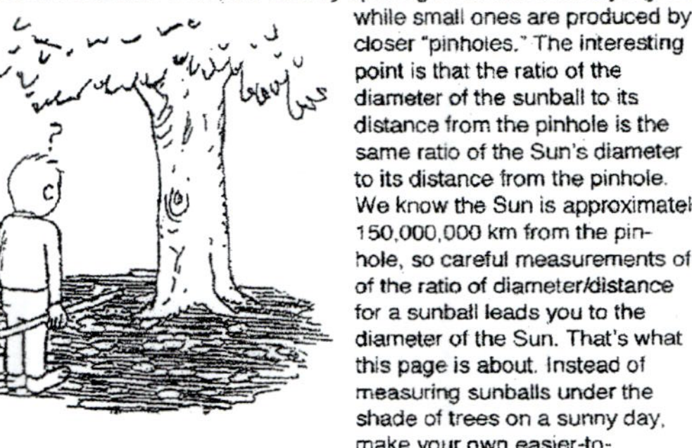

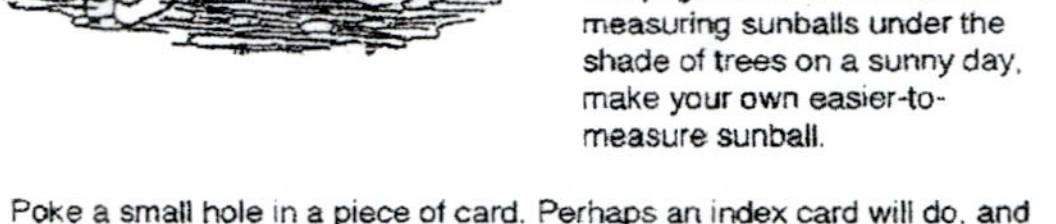

D

150,000,000 km

h

d

1. Poke a small hole in a piece of card. Perhaps an index card will do, and poke the hole with a sharp pencil or pen. Hold the card in the sunlight and note the circular image that is cast. This is an image of the Sun. Note that its size doesn't depend on the size of the hole in the card, but only on its distance. The image is a circle when cast on a surface perpendicular to the rays—otherwise it's "stretched out" as an ellipse.

2. Try holes of various shapes; say a square hole, or a triangular hole. What is the shape of the image when its distance from the card is large compared with the size of the hole? Does the shape of the pinhole make a difference?

IMAGE IS ALWAYS A CIRCLE. SHAPE OF PINHOLE IS *NOT* THE SHAPE OF THE IMAGE CAST THROUGH IT (EXCEPT WHEN VERY CLOSE).

3. Measure the diameter of a small coin. Then place the coin on a viewing area that is perpendicular to the Sun's rays. Position the card so the image of the sunball exactly covers the coin. Carefully measure the distance between the coin and the small hole in the card. Complete the following:

$$\frac{\text{Diameter of sunball}}{\text{Distance of pinhole}} = \frac{d}{h} \approx \frac{1}{110} \quad \textbf{(SO SUN'S DIAMETER} = \frac{1}{110} \times \textbf{150,000,000 km).}$$

With this ratio, estimate the diameter of the Sun. Show your work on a separate piece of paper.

WHAT SHAPE DO SUNBALLS HAVE DURING A PARTIAL ECLIPSE OF THE SUN?

4. If you did this on a day when the Sun is partially eclipsed, what shape of image would you expect to see?

UPSIDE-DOWN CRESCENT, IMAGE OF THE PARTIALLY-ECLIPSED SUN.

Hewitt Drew it!

Name ______________________ Date ________

CONCEPTUAL *Physics* FUNDAMENTALS PRACTICE PAGE

Chapter 2 Atoms

Atoms and Atomic Nuclei

ATOMS ARE CLASSIFIED BY THEIR ATOMIC NUMBER, WHICH IS THE SAME AS THE NUMBER OF **PROTONS** IN THE NUCLEUS.

TO CHANGE THE ATOMS OF ONE ELEMENT INTO THOSE OF ANOTHER, **PROTONS** MUST BE ADDED OR SUBTRACTED!

Use the periodic table in your text to help you answer the following questions.

1. When the atomic nuclei of hydrogen and lithium are squashed together (nuclear fusion) the element that is produced is

   BERYLLIUM

2. When the atomic nuclei of a pair of lithium nuclei are fused, the element produced is

   CARBON

3. When the atomic nuclei of a pair of aluminum nuclei are fused, the element produced is

   IRON

4. When the nucleus of a nitrogen atom absorbs a proton, the resulting element is

   OXYGEN

5. What element is produced when a gold nucleus gains a proton? **MERCURY**

6. What element is produced when a gold nucleus loses a proton? **PLATINUM**

7. What element is produced when a uranium nucleus ejects an elementary particle composed of two protons and two neutrons?

   THORIUM

8. If a uranium nucleus breaks into two pieces (nuclear fission) and one of the pieces is zirconium (atomic number 40), the other piece is the element

   TELLURIUM (ATOMIC NUMBER 52)

9. Which has more mass, a nitrogen molecule (N_2) or an oxygen molecule (O_2)?

   AN OXYGEN MOLECULE

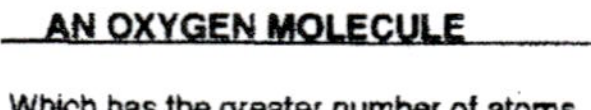

10. Which has the greater number of atoms, a gram of helium or a gram of neon?

    A GRAM OF HELIUM

I LIKE THE WAY YOUR ATOMS ARE PUT TOGETHER!

SIGH

3

Chapter 2 Atoms
Subatomic Particles

Three fundamental particles of the atom are the **PROTON**, **NEUTRON**, and **ELECTRON**. At the center of each atom lies the atomic **NUCLEUS** which consists of **PROTONS** and **NEUTRONS**. The atomic number refers to the number of **PROTONS** in the nucleus. All atoms of the same element have the same number of **PROTONS**, hence, the same atomic number.

Isotopes are atoms that have the same number of **PROTONS** but a different number of **NEUTRONS**. An isotope is identified by its atomic mass number, which is the total number of **PROTONS** and **NEUTRONS** in the nucleus. A carbon isotope that has 6 **PROTONS** and **NEUTRONS** is identified as carbon-12, where 12 is the atomic mass number. A carbon isotope having 6 **PROTONS** and 8 **NEUTRONS**, on the other hand is carbon-14.

1. *Complete the following table:*

ISOTOPE	ELECTRONS	NUMBER OF PROTONS	NEUTRONS
Hydrogen-1	1	1	0
Chlorine-36	17	17	19
Nitrogen-14	7	7	7
Potassium-40	19	19	21
Arsenic-75	33	33	42
Gold-197	79	79	118

2. Which results in a more valuable product—*adding* or *subtracting* protons from gold nuclei?
SUBTRACT FROM PLATINUM (MORE VALUABLE)

3. Which has more mass, a lead atom or a uranium atom?
NEON

4. Which has a greater number of atoms, a gram of lead or a gram of uranium?
HELIUM!

Of every 200 atoms in our bodies, 126 are hydrogen, 51 are oxygen, and just 19 are carbon. In addition to carbon we need iron to manufacture hemoglobin, cobalt for the creation of vitamin B-12, potassium and a little sodium for our nerves, and molybdenum, manganese, and vanadium to keep our enzymes purring. Ah, we'd be nothing without atoms!

Hewitt Drew it!

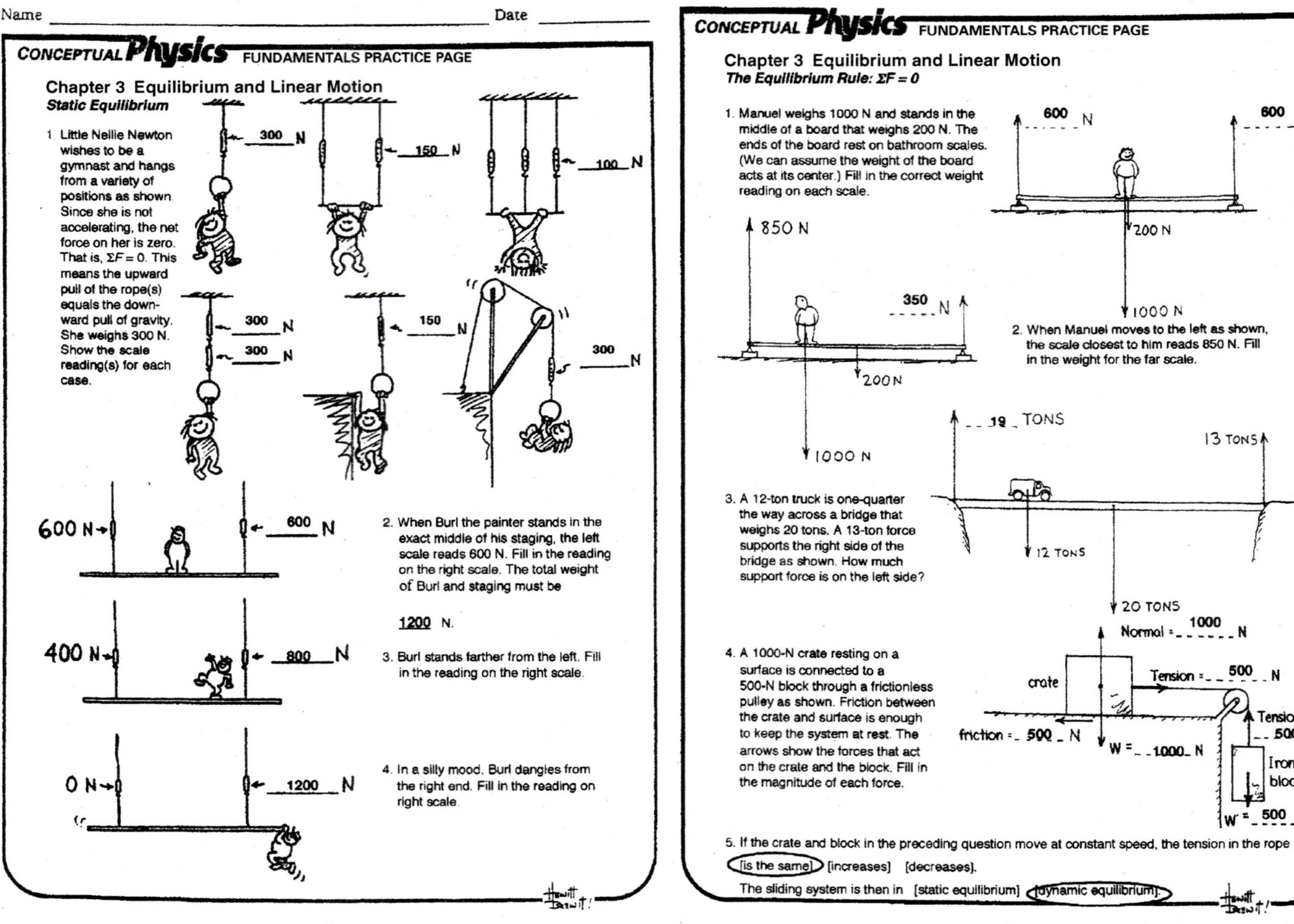

Name ____________________ Date ________

CONCEPTUAL Physics FUNDAMENTALS PRACTICE PAGE

Chapter 3 Equilibrium and Linear Motion

Static Equilibrium

1. Little Nellie Newton wishes to be a gymnast and hangs from a variety of positions as shown. Since she is not accelerating, the net force on her is zero. That is, $\Sigma F = 0$. This means the upward pull of the rope(s) equals the downward pull of gravity. She weighs 300 N. Show the scale reading(s) for each case.

2. When Burl the painter stands in the exact middle of his staging, the left scale reads 600 N. Fill in the reading on the right scale. The total weight of Burl and staging must be

1200 N.

3. Burl stands farther from the left. Fill in the reading on the right scale.

4. In a silly mood, Burl dangles from the right end. Fill in the reading on right scale.

5

CONCEPTUAL Physics FUNDAMENTALS PRACTICE PAGE

Chapter 3 Equilibrium and Linear Motion

The Equilibrium Rule: $\Sigma F = 0$

1. Manuel weighs 1000 N and stands in the middle of a board that weighs 200 N. The ends of the board rest on bathroom scales. (We can assume the weight of the board acts at its center.) Fill in the correct weight reading on each scale.

2. When Manuel moves to the left as shown, the scale closest to him reads 850 N. Fill in the weight for the far scale.

3. A 12-ton truck is one-quarter the way across a bridge that weighs 20 tons. A 13-ton force supports the right side of the bridge as shown. How much support force is on the left side?

4. A 1000-N crate resting on a surface is connected to a 500-N block through a frictionless pulley as shown. Friction between the crate and surface is enough to keep the system at rest. The arrows show the forces that act on the crate and the block. Fill in the magnitude of each force.

5. If the crate and block in the preceding question move at constant speed, the tension in the rope [is the same] [increases] [decreases].

The sliding system is then in [static equilibrium] [dynamic equilibrium].

6

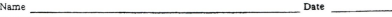

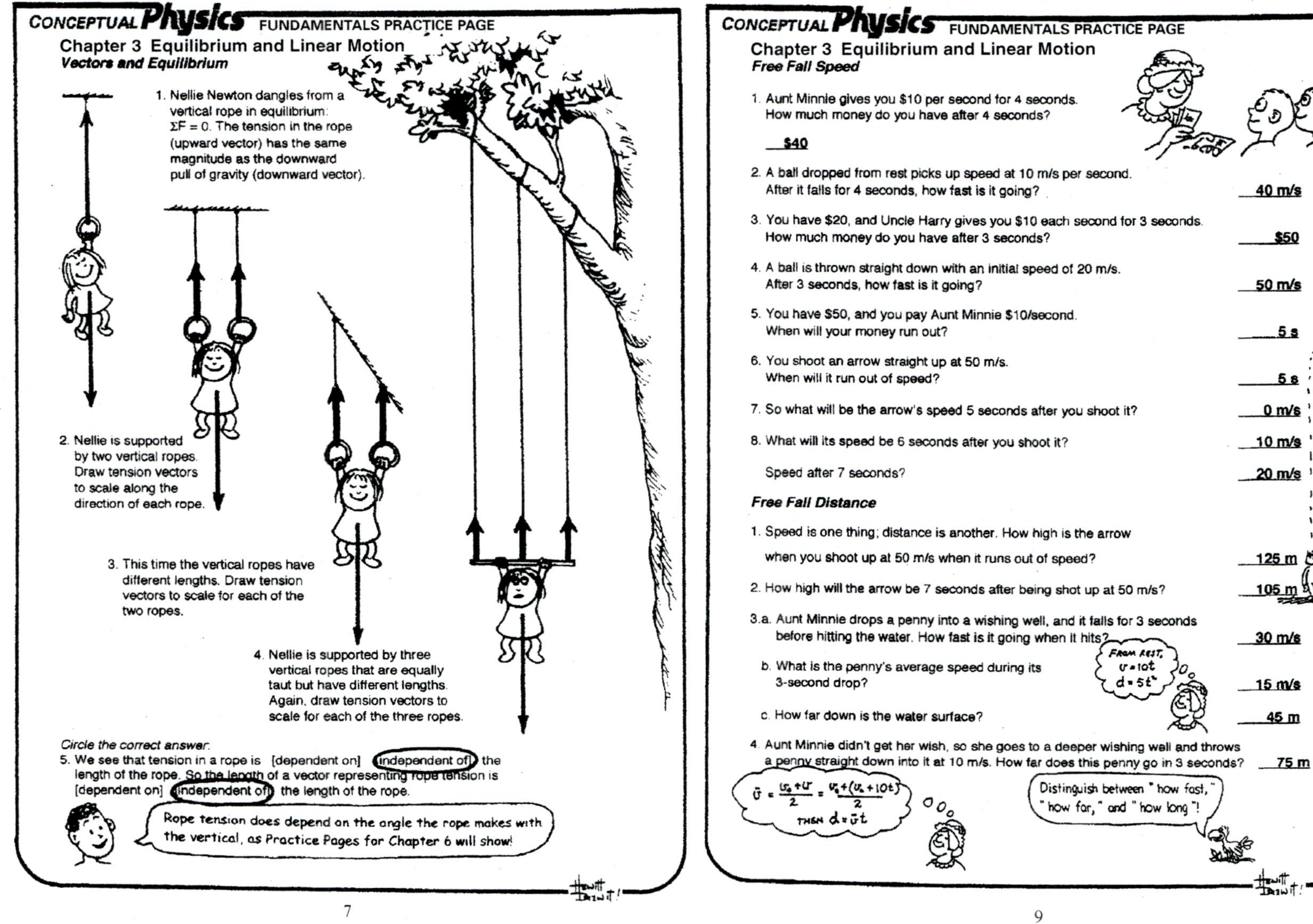

Name ______________________ Date __________

CONCEPTUAL Physics FUNDAMENTALS PRACTICE PAGE

Chapter 3 Equilibrium and Linear Motion

Vectors and Equilibrium

1. Nellie Newton dangles from a vertical rope in equilibrium: $\Sigma F = 0$. The tension in the rope (upward vector) has the same magnitude as the downward pull of gravity (downward vector).

2. Nellie is supported by two vertical ropes. Draw tension vectors to scale along the direction of each rope.

3. This time the vertical ropes have different lengths. Draw tension vectors to scale for each of the two ropes.

4. Nellie is supported by three vertical ropes that are equally taut but have different lengths. Again, draw tension vectors to scale for each of the three ropes.

Circle the correct answer.

5. We see that tension in a rope is [dependent on] [independent of] the length of the rope. So the length of a vector representing rope tension is [dependent on] [independent of] the length of the rope.

Rope tension does depend on the angle the rope makes with the vertical, as Practice Pages for Chapter 6 will show!
Hewitt Drew it!

7

Name ______________________ Date __________

CONCEPTUAL Physics FUNDAMENTALS PRACTICE PAGE

Chapter 3 Equilibrium and Linear Motion

Free Fall Speed

1. Aunt Minnie gives you $10 per second for 4 seconds. How much money do you have after 4 seconds? **$40**

2. A ball dropped from rest picks up speed at 10 m/s per second. After it falls for 4 seconds, how fast is it going? **40 m/s**

3. You have $20, and Uncle Harry gives you $10 each second for 3 seconds. How much money do you have after 3 seconds? **$50**

4. A ball is thrown straight down with an initial speed of 20 m/s. After 3 seconds, how fast is it going? **50 m/s**

5. You have $50, and you pay Aunt Minnie $10/second. When will your money run out? **5 s**

6. You shoot an arrow straight up at 50 m/s. When will it run out of speed? **5 s**

7. So what will be the arrow's speed 5 seconds after you shoot it? **0 m/s**

8. What will its speed be 6 seconds after you shoot it? **10 m/s**

   Speed after 7 seconds? **20 m/s**

Free Fall Distance

1. Speed is one thing; distance is another. How high is the arrow when you shoot up at 50 m/s when it runs out of speed? **125 m**

2. How high will the arrow be 7 seconds after being shot up at 50 m/s? **105 m**

3.a. Aunt Minnie drops a penny into a wishing well, and it falls for 3 seconds before hitting the water. How fast is it going when it hits? **30 m/s**

b. What is the penny's average speed during its 3-second drop? **15 m/s**

c. How far down is the water surface? **45 m**

4. Aunt Minnie didn't get her wish, so she goes to a deeper wishing well and throws a penny straight down into it at 10 m/s. How far does this penny go in 3 seconds? **75 m**

FROM REST, $v = 10t$, $d = 5t^2$
$\bar{v} = \frac{v_0 + v}{2} = \frac{v_0 + (v_0 + 10t)}{2}$ THEN $d = \bar{v}t$
Distinguish between "how fast," "how far," and "how long"!
Hewitt Drew it!

9

CONCEPTUAL **Physics** FUNDAMENTALS PRACTICE PAGE

Chapter 3 Equilibrium and Linear Motion

Acceleration of Free Fall

A rock dropped from the top of a cliff picks up speed as it falls. Pretend that a speedometer and odometer are attached to the rock to indicate readings of speed and distance at 1-second intervals. Both speed and distance are zero at time = zero (see sketch). Note that after falling 1 second, the speed reading is 10 m/s and the distance fallen is 5 m. The readings of succeeding seconds of fall are not shown and are left for you to complete. So draw the position of the speedometer pointer and write in the correct odometer reading for each time. Use $g = 10$ m/s^2 and neglect air resistance.

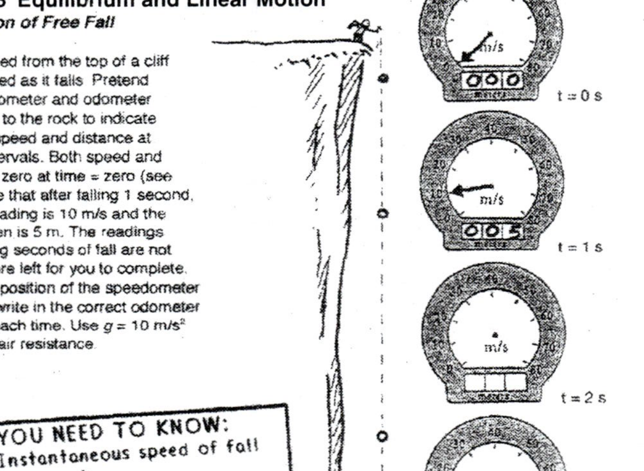
t = 0 s
t = 1 s
t = 2 s
t = 3 s
t = 4 s
t = 5 s
t = 6 s
YOU NEED TO KNOW:
Instantaneous speed of fall from rest:
$v = gt$
Distance fallen from rest:
$d = v_{average}t$
or
$d = \frac{1}{2}gt^2$

1. The speedometer reading increases the same amount, <u>10</u> m/s, each second.

   This increase in speed per second is called

   <u>**ACCELERATION**</u>.

2. The distance fallen increases as the square of the <u>**TIME**</u>.

3. If it takes 7 seconds to reach the ground, then its speed at impact is <u>70</u> m/s, the total distance fallen is <u>245</u> m, and its acceleration of fall just before impact is <u>10</u> m/s^2.

Name ______________________ Date ________

CONCEPTUAL **Physics** FUNDAMENTALS PRACTICE PAGE

Chapter 3 Equilibrium and Linear Motion

Hang Time

Some athletes and dancers have great jumping ability. When leaping, they seem to momentarily "hang in the air" and defy gravity. The time that a jumper is airborne with feet off the ground is called hang time. Ask your friends to estimate the hang time of the great jumpers. They may say two or three seconds. But surprisingly, the hang time of the greatest jumpers is most always less than 1 second! A longer time is one of many illusions we have about nature.

To better understand this, find the answers to the following questions:

1. If you step off a table and it takes one-half second to reach the floor, what will be the speed when you meet the floor?

   $v = gt = 10 \text{ m/s}^2 \times \frac{1}{2} = 5 \text{ m/s}$

2. What will be your average speed of fall?

   $v = \frac{0 + 5 \text{ m/s}}{2} = 2.5 \text{ m/s}$

3. What will be the distance of fall?

   $d = vt = 2.5 \text{ m/s} \times \frac{1}{2}\text{s} = 1.25 \text{ m}$

4. So how high is the surface of the table above the floor? <u>**1.25 m**</u>

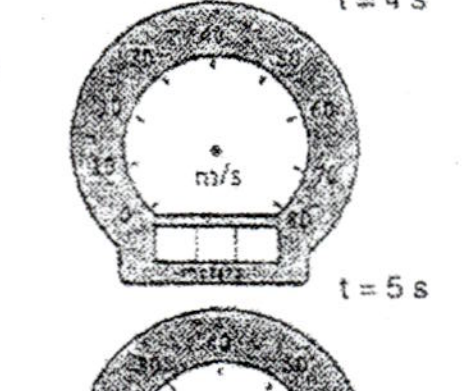
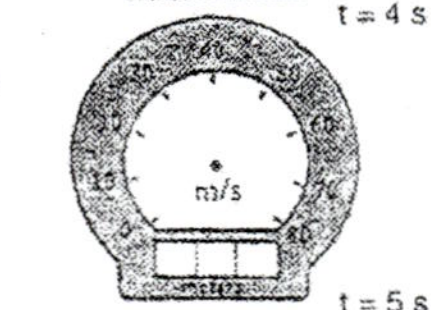
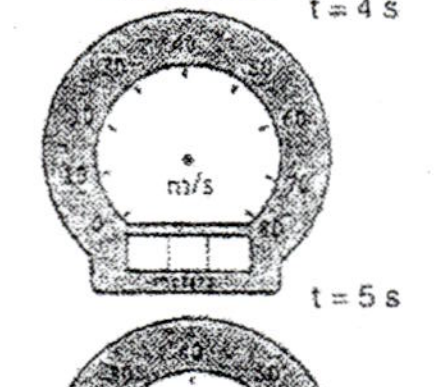
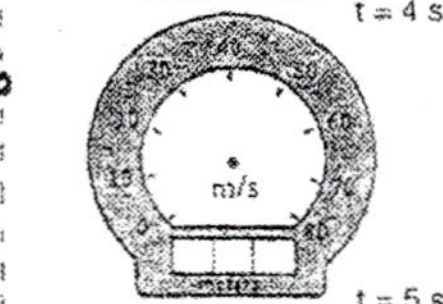
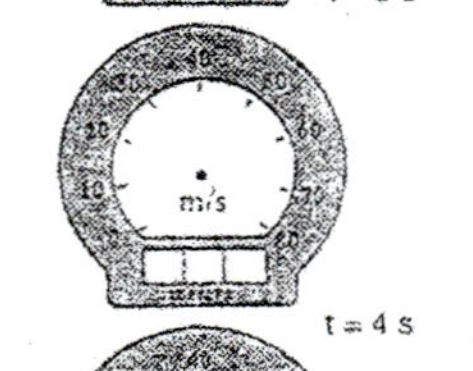
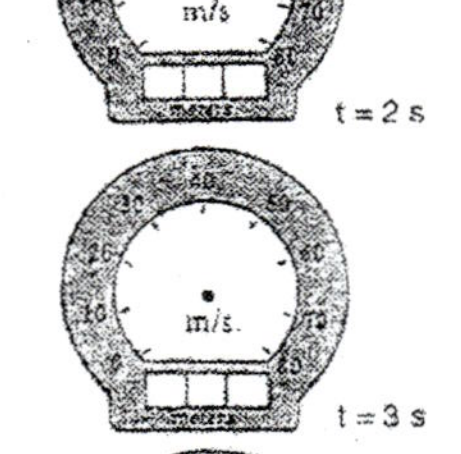
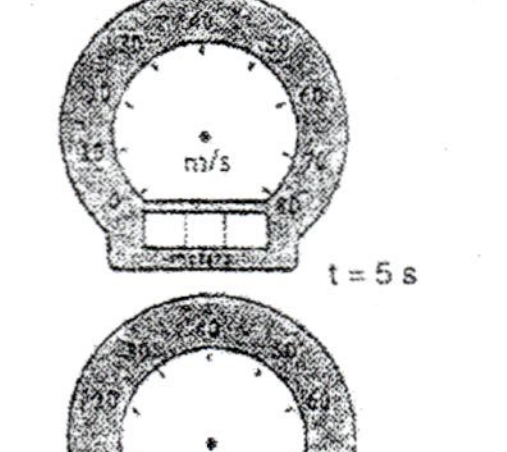
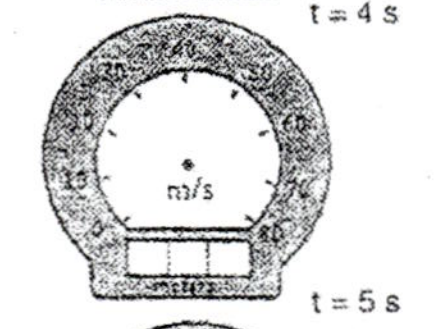
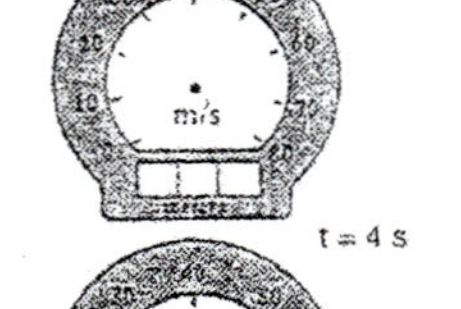
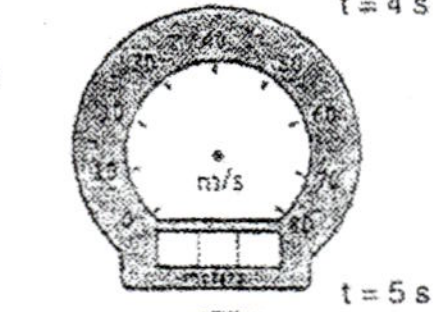
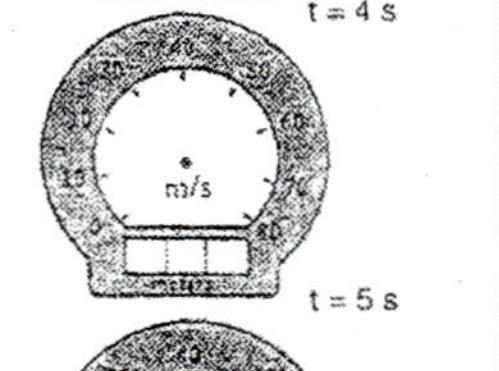
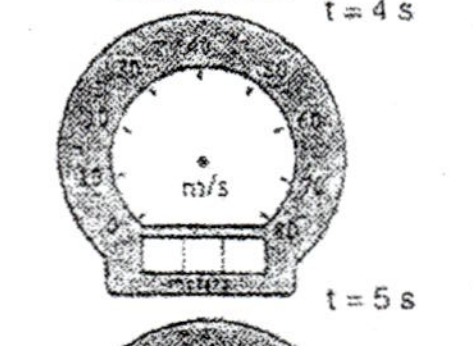
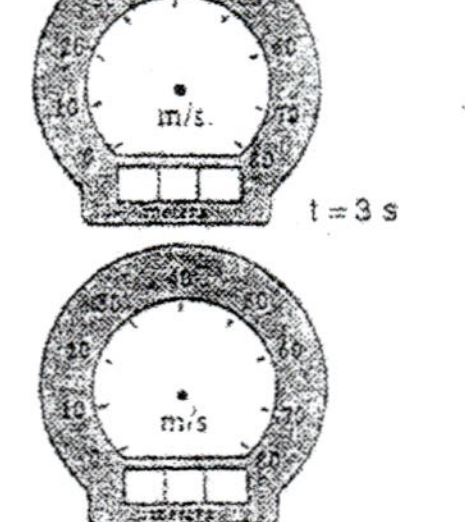
Speed of free fall = acceleration × time
= 10 m/s² × number of seconds
= 10t m.
Average speed = (initial speed + final speed)/2
Distance = average speed × time.
Almost anybody can safely step off a 1.25-m (4-feet) high table. Can anybody in your school jump from the floor up onto the same table?
No way!
There's a big difference in how high you can reach and how high you raise your "center of gravity" when you jump. Even basketball star Michael Jordan in his prime couldn't quite raise his body 1.25 meters high, although he could easily reach higher than the more-than-3-meter high basket.

Jumping ability is best measured by a standing vertical jump. Stand facing a wall with feet flat on the floor and arms extended upward. Make a mark on the wall at the top of your reach. Then make your jump and at the peak make another mark. The distance between these two marks measures your vertical leap. If it's more than 0.6 meters (2 feet), you're exceptional.

5. What is your vertical jumping distance? <u>**(VARIES)**</u>

6. Calculate your personal hang time using the formula $d = 1/2\ gt^2$. (Remember that hang time is the time that you move upward + the time you return downward.)

Here we're talking about vertical motion. How about running jumps? We'll see in Chapter 10 that the height of a jump depends only on the jumper's vertical speed at launch. While airborne, the jumper's horizontal speed remains constant while the vertical speed undergoes acceleration due to gravity. While airborne, no amount of leg or arm pumping or other bodily motions can change your hang time.

CONCEPTUAL Physics FUNDAMENTALS PRACTICE PAGE

Chapter 3 Equilibrium and Linear Motion
Non-Accelerated Motion

1. The sketch shows a ball rolling at constant velocity along a level floor. The ball rolls from the first position shown to the second in 1 second. The two positions are 1 meter apart. Sketch the ball at successive 1-second intervals all the way to the wall (neglect resistance).

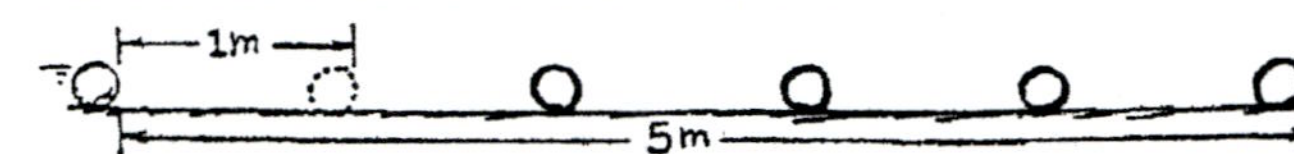

a. Did you draw successive ball positions evenly spaced, farther apart, or closer together? Why?
EVENLY SPACED-EQUAL DISTANCE IN EQUAL TIME —> CONSTANT v

b. The ball reaches the wall with a speed of __1__ m/s and takes a time of __5__ seconds.

2. Table I shows data of sprinting speeds of some animals. Make whatever computations necessary to complete the table.

TABLE I

ANIMAL	DISTANCE	TIME	SPEED
CHEETAH	75 m	3 s	25 m/s
GREYHOUND	160 m	10 s	16 m/s
GAZELLE	1 km	0.01 h	100 km/h
TURTLE	30 cm	30 s	1 cm/s

Accelerated Motion

3. An object starting from rest gains a speed $v = at$ when it undergoes uniform acceleration. The distance it covers is $d = 1/2\ at^2$. Uniform acceleration occurs for a ball rolling down an inclined plane. The plane below is tilted so a ball picks up a speed of 2 m/s each second; then its acceleration $a = 2\ m/s^2$. The positions of the ball are shown at 1-second intervals. Complete the six blank spaces for distance covered and the four blank spaces for speeds.

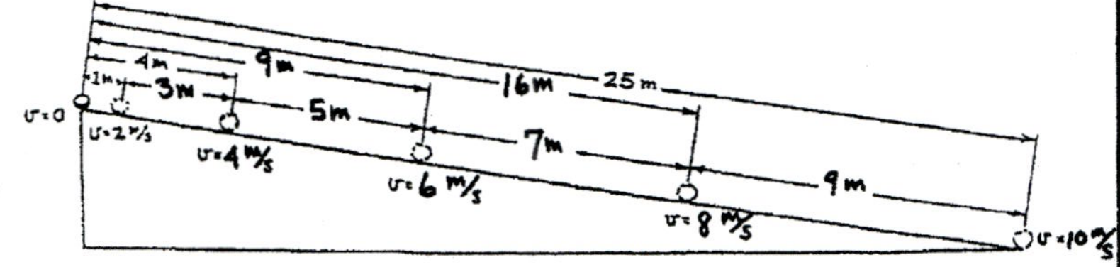

a. Do you see that the total distance from the starting point increases as the square of the time? This was discovered by Galileo. If the incline were to continue, predict the ball's distance from the starting point for the next 3 seconds.

YES: DISTANCE INCREASES AS SQUARE OF TIME; 36 m, 49 m, 64 m.

b. Note the increase of distance between ball positions with time. Do you see an odd-integer pattern (also discovered by Galileo) for this increase? If the incline were to continue, predict the successive distances between ball positions for the next 3 seconds.

YES: 11 m, 13 m, 15 m.

Hewitt Drew it!

Name ____________________ Date ________

CONCEPTUAL Physics FUNDAMENTALS PRACTICE PAGE

Chapter 4 Newton's Laws of Motion
Mass and Weight

Learning physics is learning the connections among concepts in nature, and also learning to distinguish between closely-related concepts. Velocity and acceleration, previously treated, are often confused. Similarly in this chapter, we find that mass and weight are often confused. They aren't the same! Please review the distinction between mass and weight in your textbook. To reinforce your understanding of this distinction, circle the correct answers below.

Comparing the concepts of mass and weight, one is basic—fundamental—depending only on the internal makeup of an object and the number and kind of atoms that compose it. The concept that is fundamental is [mass] [weight].

The concept that additionally depends on location in a gravitational field is [mass] [weight].

[Mass] [Weight] is a measure of the amount of matter in an object and only depends on the number and kind of atoms that compose it.

It can correctly be said that [mass] [weight] is a measure of "laziness" of an object.

[Mass] [Weight] is related to the gravitational force acting on the object.

[Mass] [Weight] depends on an object's location, whereas [mass] [weight] does not.

In other words, a stone would have the same [mass] [weight] whether it is on the surface of Earth or on the surface of the Moon. However, its [mass] [weight] depends on its location.

On the Moon's surface, where gravity is only about 1/6th Earth gravity [mass] [weight] [both the mass and the weight] of the stone would be the same as on Earth.

While mass and weight are not the same, they are [directly proportional] [inversely proportional] to each other. In the same location, twice the mass has [twice] [half] the weight.

The Standard International (SI) unit of mass is the [kilogram] [newton], and the SI unit of force is the [kilogram] [newton].

In the United States, it is common to measure the mass of something by measuring its gravitational pull to Earth, its weight. The common unit of weight in the U.S. is the [pound] [kilogram] [newton].

Pull of gravity

Support Force

When I step on a weighing scale, two forces act on it; a downward pull of gravity, and an upward support force. These equal and opposite forces effectively compress a spring inside the scale that is calibrated to show weight. When in equilibrium, my weight = *mg*.

thanx to Daniela Taylor

Hewitt Drew it!

CONCEPTUAL Physics FUNDAMENTALS PRACTICE PAGE

Chapter 4 Newton's Laws of Motion
Converting Mass to Weight

Objects with mass also have weight (although they can be weightless under special conditions). If you know the mass of something in **kilograms** and want its weight in **newtons**, at Earth's surface, you can take advantage of the formula that relates weight and mass.

Weight = mass × acceleration due to gravity
$W = mg$

This is in accord with Newton's 2nd law, written as $F = ma$. When the force of gravity is the only force, the acceleration of any object of mass m will be g, the acceleration of free fall. Importantly, g acts as a proportionality constant, 9.8 N/kg, which is equivalent to 9.8 m/s².

Sample Question:
How much does a 1-kg bag of nails weigh on Earth?

$W = mg = (1\ \text{kg})(9.8\ \text{m/s}^2) = 9.8\ \text{m/s}^2 = 9.8\ \text{N}$.
or simply, $W = mg = (1\ \text{kg})(9.8\ \text{N/kg}) = 9.8\ \text{N}$.

From $F = ma$, we see that the unit of force equals the units [kg x m/s²]. Can you see the units [m/s²] = [N/kg]?

Answer the following questions:
Felicia the ballet dancer has a mass of 45.0 kg.

1. What is Felicia's weight in newtons at Earth's surface? **441 N**
2. Given that 1 kilogram of mass corresponds to 2.2 pounds at Earth's surface, what is Felicia's weight in pounds on Earth? **99 LB**
3. What would be Felicia's mass on the surface of Jupiter? **45.0 kg**
4. What would be Felicia's weight on Jupiter's surface, where the acceleration due to gravity is 25.0 m/s²? **1125 N**

Different masses are hung on a spring scale calibrated in newtons.
The force exerted by gravity on 1 kg = 9.8 N.

5. The force exerted by gravity on 5 kg = __49__ N.
6. The force exerted by gravity on __10__ kg = 98 N.

Make up your own mass and show the corresponding weight:
The force exerted by gravity on __*__ kg = __*__ N.
***ANY VALUE FOR kg AS LONG AS THE SAME VALUE IS MULTIPLIED BY 9.8 FOR N.**

9.8 N
1 kg

By whatever means (spring scales, measuring balances, etc.), find the mass of your physics book. Then complete the table.

OBJECT	MASS	WEIGHT
MELON	1 kg	9.8 N
APPLE	0.1 kg	1 N
BOOK		
A FRIEND	60 kg	588 N

Name ______________________ Date ________

CONCEPTUAL Physics FUNDAMENTALS PRACTICE PAGE

Chapter 4 Newton's Laws of Motion
A Day at the Races with a = F/m

In each situation below, Cart A has a mass of **1 kg**. *Circle the correct answer* (A, B, or Same for both).

1. Cart A is pulled with a force of **1 N**.
Cart B also has a mass of **1 kg** and is pulled with a force of **2 N**.
Which undergoes the greater acceleration?

[A] **(B) circled** [Same for both]

2. Cart A is pulled with a force of **1 N**.
Cart B has a mass of **2 kg** and is also pulled with a force of **1 N**.
Which undergoes the greater acceleration?

(A) circled [B] [Same for both]

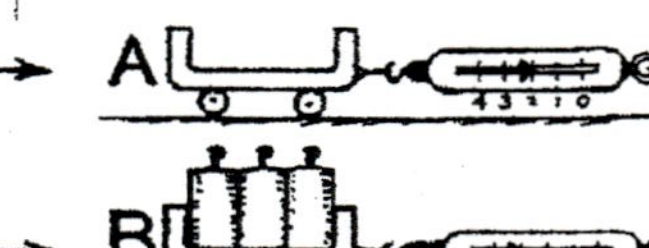

3. Cart A is pulled with a force of **1 N**.
Cart B has a mass of **2 kg** and is pulled with a force of **2 N**.
Which undergoes the greater acceleration?

[A] [B] **(Same for both) circled**

4. Cart A is pulled with a force of **1 N**.
Cart B has a mass of **3 kg** and is pulled with a force of **3 N**.
Which undergoes the greater acceleration?

[A] [B] **(Same for both) circled**

5. This time Cart A is pulled with a force of **4 N**.
Cart B has a mass of **4 kg** and is pulled with a force of **4 N**.
Which undergoes the greater acceleration?

(A) circled [B] [Same for both]

6. Cart A is pulled with a force of **2 N**.
Cart B has a mass of **4 kg** and is pulled with a force of **3 N**.
Which undergoes the greater acceleration?

(A) circled [B] [Same for both]

thanx to Dean Baird

CONCEPTUAL *Physics* FUNDAMENTALS PRACTICE PAGE

Chapter 4 Newton's Laws of Motion
Dropping Masses and Accelerating Cart

1. Consider a 1-kg cart being pulled by a 10-N applied force. According to Newton's 2nd law, acceleration of the cart is

$$a = \frac{F}{m} = \frac{10\text{ N}}{1\text{ kg}} = 10\text{ m/s}^2.$$

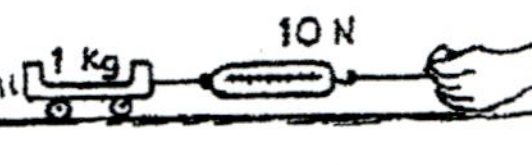

This is the same as the acceleration of free fall, g—because a force equal to the cart's weight accelerates it.

2. Consider the acceleration of the cart when the applied force is due to a 10-N iron weight attached to a string draped over a pulley. Will the cart accelerate as before, at 10 m/s²? The answer is no, because the mass being accelerated is the mass of the cart *plus* the mass of the piece of iron that pulls it. Both masses accelerate. The mass of the 10-N iron weight is 1 kg—so the total mass being accelerated (cart + iron) is 2 kg. Then,

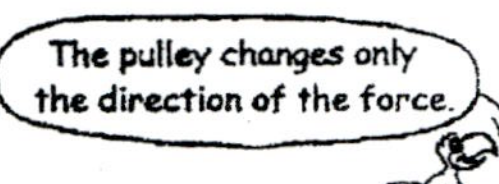

The pulley changes only the direction of the force.

$$a = \frac{F}{m} = \frac{10\text{ N}}{2\text{ kg}} = 5\text{ m/s}^2.$$

Don't forget; the total mass of a system includes the mass of the hanging iron.

Note this is half the acceleration due to gravity alone, g. So the acceleration of 2 kg produced by the weight of 1 kg is $g/2$.

a. Find the acceleration of the 1-kg cart when two identical 10-N weights are attached to the string.

$$a = \frac{F}{m} = \frac{\text{applied force}}{\text{total mass}} = \frac{20\text{ N}}{3\text{ kg}} = 6.7\text{ m/s}^2.$$

CONCEPTUAL *Physics* FUNDAMENTALS PRACTICE PAGE

Chapter 4 Newton's Laws of Motion
Dropping Masses and Accelerating Cart*—continued*

b. Find the acceleration of the 1-kg cart when the three identical 10-N weights are attach to the string.

$$a = \frac{F}{m} = \frac{\text{applied force}}{\text{total mass}} = \frac{30\text{ N}}{4\text{ kg}} = 7.5\text{ m/s}^2.$$

c. Find the acceleration of the 1-kg cart when four identical 10-N weights (not shown) are attached to the string.

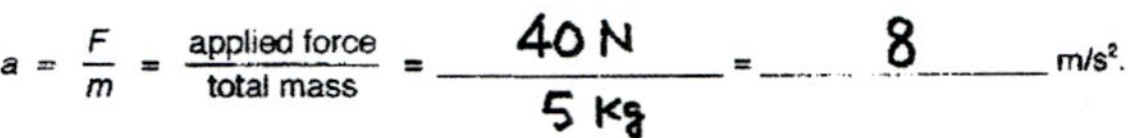

$$a = \frac{F}{m} = \frac{\text{applied force}}{\text{total mass}} = \frac{40\text{ N}}{5\text{ kg}} = 8\text{ m/s}^2.$$

d. This time 1 kg of iron is added to the cart, and only one iron piece dangles from the pulley. Find the acceleration of the cart.

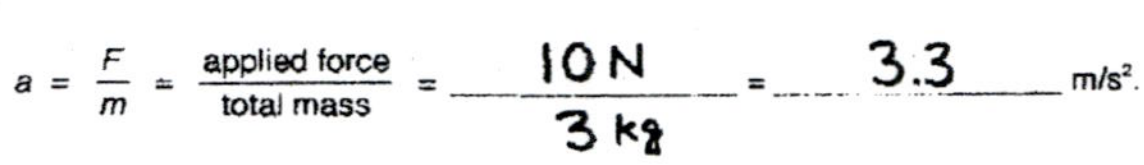

$$a = \frac{F}{m} = \frac{\text{applied force}}{\text{total mass}} = \frac{10\text{ N}}{3\text{ kg}} = 3.3\text{ m/s}^2.$$

The force due to gravity on a mass m is mg. So gravitational force on 1 kg is (1 kg)(10 m/s²) = 10 N.

e. Find the acceleration of the cart when it carries two pieces of iron and only one iron piece dangles from the pulley.

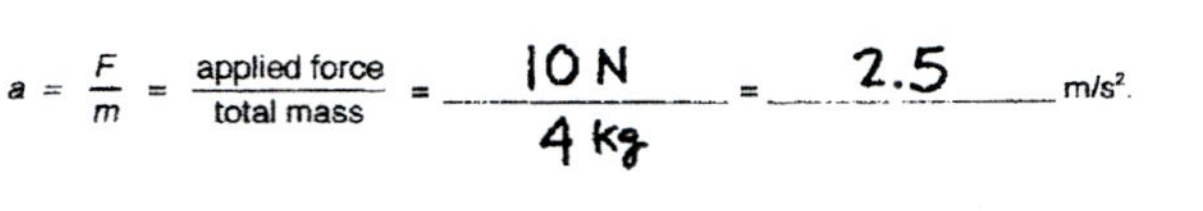

$$a = \frac{F}{m} = \frac{\text{applied force}}{\text{total mass}} = \frac{10\text{ N}}{4\text{ kg}} = 2.5\text{ m/s}^2.$$

CONCEPTUAL **Physics** FUNDAMENTALS PRACTICE PAGE

Chapter 4 Newton's Laws of Motion

***Dropping Masses and Accelerating Cart**—continued*

f. Find the acceleration of the cart when it carries 3 pieces of iron and only one iron piece dangles from the pulley.

$a = \frac{F}{m} = \frac{\text{applied force}}{\text{total mass}} = \frac{10\text{ N}}{5\text{ kg}} = $ 2 m/s².

g. Find the acceleration of the cart when it carries 3 pieces of iron and 4 pieces of iron dangle from the pulley.

$a = \frac{F}{m} = \frac{\text{applied force}}{\text{total mass}} = \frac{40\text{ N}}{8\text{ kg}} = $ 5 m/s².

Mass of cart is 1 kg. Mass of 10-N iron is also 1 kg.

h. Draw your own combination of masses and find the acceleration.

$a = \frac{F}{m} = \frac{\text{applied force}}{\text{total mass}} = $ ________ = ________ m/s².

Name ____________________ Date ________

CONCEPTUAL **Physics** FUNDAMENTALS PRACTICE PAGE

Chapter 4 Newton's Laws of Motion

Force and Acceleration

1. Skelly the skater, total mass 25 kg, is propelled by rocket power.

a. Complete Table I (neglect resistance).

TABLE I

FORCE	ACCELERATION
100 N	4 m/s
200 N	8 m/s
250 N	10 m/s²

$a = \frac{F}{25 \text{ mg}}$

b. Complete Table II for a constant 50-N resistance.

TABLE II

FORCE	ACCELERATION
50 N	0 m/s²
100 N	2 m/s
200 N	6 m/s

$a = \frac{F - 50\text{ N}}{25\text{ kg}}$

2. Block A on a horizontal friction-free table is accelerated by a force from a string attached to Block B of the same mass. Block B falls vertically and drags Block A horizontally. (Neglect the string's mass).

Circle the correct answers:

a. The *mass* of the system (A + B) is [*m*] [2 *m*].

b. The *force* that accelerates (A + B) is the weight of [A] [B] [A + B].

c. The weight of B is [*mg*/2] [*mg*] [2 *mg*].

d. Acceleration of (A + B) is [less than *g*] [*g*] [more than *g*].

e. Use $a = \frac{F}{m}$ to show the acceleration of (A + B) as a fraction of *g*. $a = \frac{mg}{2m} = \frac{g}{2}$

If B were allowed to fall by itself, not dragging A, then wouldn't its acceleration be *g*?

Yes, because the force that accelerates it would only be acting on its own mass – not twice the mass!

To better understand this, consider 3 and 4 on the other side!

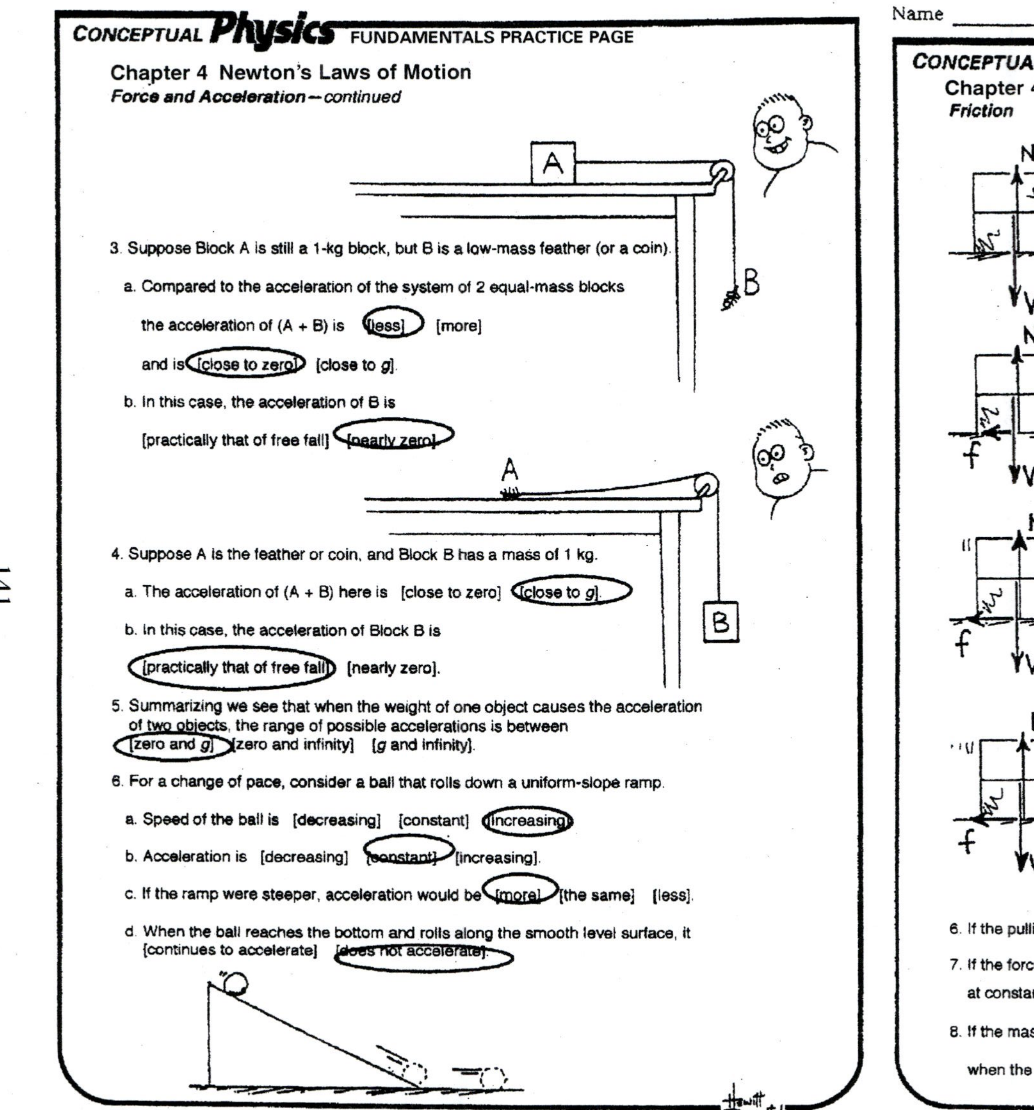

CONCEPTUAL Physics FUNDAMENTALS PRACTICE PAGE

Chapter 4 Newton's Laws of Motion

***Force and Acceleration**—continued*

3. Suppose Block A is still a 1-kg block, but B is a low-mass feather (or a coin).

   a. Compared to the acceleration of the system of 2 equal-mass blocks the acceleration of (A + B) is **[less]** [more] and is **[close to zero]** [close to *g*].

   b. In this case, the acceleration of B is [practically that of free fall] **[nearly zero]**.

4. Suppose A is the feather or coin, and Block B has a mass of 1 kg.

   a. The acceleration of (A + B) here is [close to zero] **[close to *g*]**.

   b. In this case, the acceleration of Block B is **[practically that of free fall]** [nearly zero].

5. Summarizing we see that when the weight of one object causes the acceleration of two objects, the range of possible accelerations is between **[zero and *g*]** [zero and infinity] [*g* and infinity].

6. For a change of pace, consider a ball that rolls down a uniform-slope ramp.

   a. Speed of the ball is [decreasing] [constant] **[increasing]**.

   b. Acceleration is [decreasing] **[constant]** [increasing].

   c. If the ramp were steeper, acceleration would be **[more]** [the same] [less].

   d. When the ball reaches the bottom and rolls along the smooth level surface, it [continues to accelerate] **[does not accelerate]**.

Name ______________________ Date __________

CONCEPTUAL Physics FUNDAMENTALS PRACTICE PAGE

Chapter 4 Newton's Laws of Motion

Friction

1. A crate filled with delicious junk food rests on a horizontal floor. Only gravity and the support force of the floor act on it, as shown by the vectors for weight **W** and normal force **N**.

   a. The net force on the crate is **[zero]** [greater than zero].

   b. Evidence for this is __NO ACCELERATION__.

2. A slight pull **P** is exerted on the crate, not enough to move it. A force of friction ***f*** now acts,

   a. which is [less than] **[equal to]** [greater than] **P**.

   b. Net force on the crate is **[zero]** [greater than zero].

3. Pull **P** is increased until the crate begins to move. It is pulled so that it moves with constant velocity across the floor.

   a. Friction ***f*** is [less than] **[equal to]** [greater than] **P**.

   b. Constant velocity means acceleration is **[zero]** [more than zero].

   c. Net force on the crate is [less than] **[equal to]** [more than] zero.

4. Pull **P** is further increased and is now greater than friction ***f***.

   a. Net force on the crate is [less than] [equal to] **[greater than]** zero.

   b. The net force acts toward the right, so acceleration acts toward the [left] **[right]**.

5. If the pulling force **P** is 150 N and the crate doesn't move, what is the magnitude of ***f***? __150 N__

6. If the pulling force **P** is 200 N and the crate doesn't move, what is the magnitude of ***f***? __200 N__

7. If the force of sliding friction is 250 N, what force is necessary to keep the crate sliding at constant velocity? __250 N__

8. If the mass of the crate is 50 kg and sliding friction is 250 N, what is the acceleration of the crate when the pulling force is 250 N? __0 m/s²__ 300 N? __1 m/s²__ 500 N? __5 m/s²__

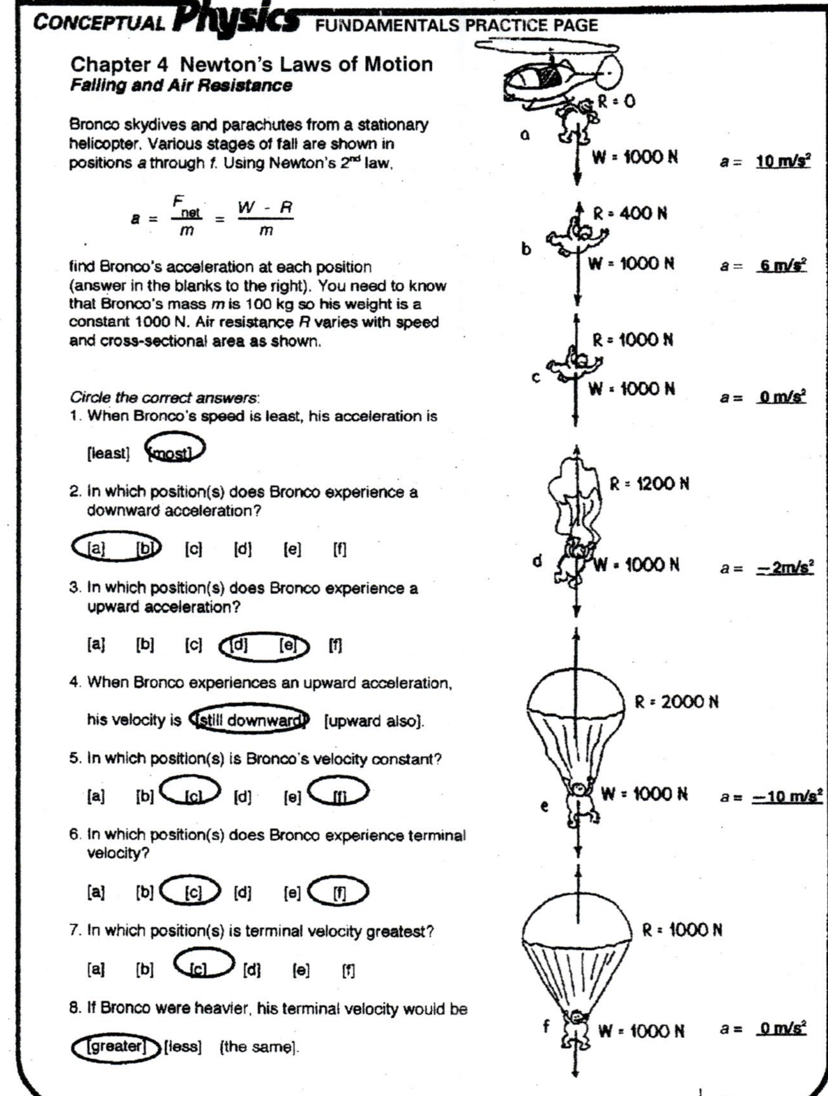

CONCEPTUAL **Physics** FUNDAMENTALS PRACTICE PAGE

Chapter 4 Newton's Laws of Motion
Falling and Air Resistance

Bronco skydives and parachutes from a stationary helicopter. Various stages of fall are shown in positions *a* through *f*. Using Newton's 2nd law,

$$a = \frac{F_{net}}{m} = \frac{W - R}{m}$$

find Bronco's acceleration at each position (answer in the blanks to the right). You need to know that Bronco's mass *m* is 100 kg so his weight is a constant 1000 N. Air resistance *R* varies with speed and cross-sectional area as shown.

Circle the correct answers:

1. When Bronco's speed is least, his acceleration is

   [least] **(circled: [most])**

2. In which position(s) does Bronco experience a downward acceleration?

   (circled: [a] [b]) [c] [d] [e] [f]

3. In which position(s) does Bronco experience a upward acceleration?

   [a] [b] [c] **(circled: [d] [e])** [f]

4. When Bronco experiences an upward acceleration, his velocity is **(circled: [still downward])** [upward also].

5. In which position(s) is Bronco's velocity constant?

   [a] [b] **(circled: [c])** [d] [e] **(circled: [f])**

6. In which position(s) does Bronco experience terminal velocity?

   [a] [b] **(circled: [c])** [d] [e] **(circled: [f])**

7. In which position(s) is terminal velocity greatest?

   [a] [b] **(circled: [c])** [d] [e] [f]

8. If Bronco were heavier, his terminal velocity would be

   (circled: [greater]) [less] [the same].

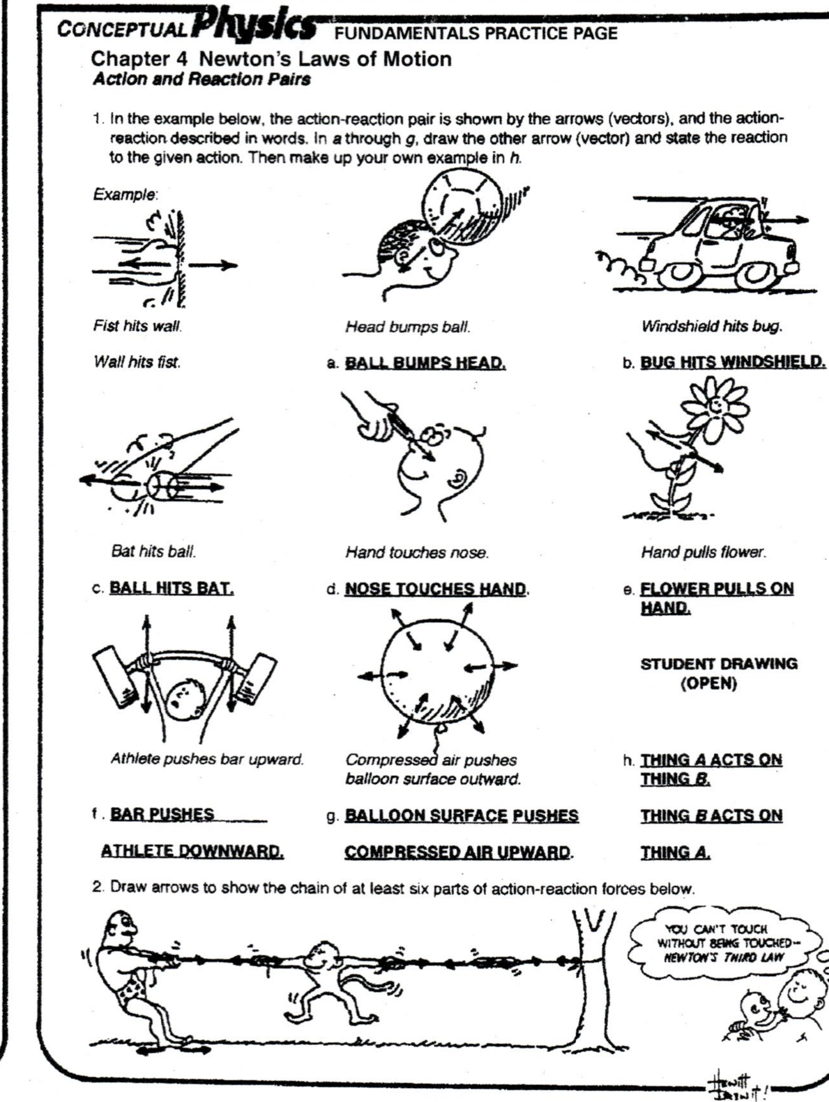

Name ______________________ Date __________

CONCEPTUAL **Physics** FUNDAMENTALS PRACTICE PAGE

Chapter 4 Newton's Laws of Motion
Action and Reaction Pairs

1. In the example below, the action-reaction pair is shown by the arrows (vectors), and the action-reaction described in words. In *a* through *g*, draw the other arrow (vector) and state the reaction to the given action. Then make up your own example in *h*.

Example:

Fist hits wall.	*Head bumps ball.*	*Windshield hits bug.*
Wall hits fist.	a. **BALL BUMPS HEAD.**	b. **BUG HITS WINDSHIELD.**
Bat hits ball.	*Hand touches nose.*	*Hand pulls flower.*
c. **BALL HITS BAT.**	d. **NOSE TOUCHES HAND.**	e. **FLOWER PULLS ON HAND.**
		STUDENT DRAWING (OPEN)
Athlete pushes bar upward.	*Compressed air pushes balloon surface outward.*	h. **THING A ACTS ON THING B.**
f. **BAR PUSHES ATHLETE DOWNWARD.**	g. **BALLOON SURFACE PUSHES COMPRESSED AIR UPWARD.**	**THING B ACTS ON THING A.**

2. Draw arrows to show the chain of at least six parts of action-reaction forces below.

CONCEPTUAL Physics FUNDAMENTALS PRACTICE PAGE

Chapter 4 Newton's Laws of Motion
Interactions

1. Nellie Newton holds an apple weighing 1 newton at rest on the palm of her hand. The force vectors shown are the forces that act on the apple.

   a. To say the weight of the apple is 1 N is to say that a downward gravitational force of 1 N is exerted on the apple by [Earth] [her hand].

   b. Nellie's hand supports the apple with normal force **N**, which acts in a direction opposite to **W**. We can say **N** [equals **W**] [has the same magnitude as **W**].

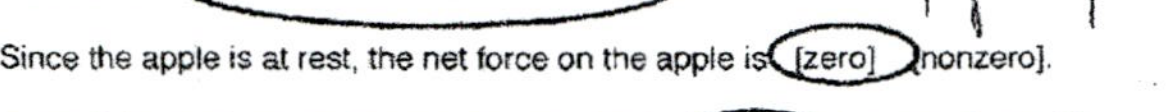

   c. Since the apple is at rest, the net force on the apple is [zero] [nonzero].

   d. Since **N** is equal and opposite to **W**, we [can] [cannot] say that **N** and **W** comprise an action-reaction pair. The reason is because action and reaction always [act on the same object] [act on different objects], and here we see **N** and **W** [both acting on the apple] [acting on different objects].

   e. In accord with the rule, "If ACTION is A acting on B, then REACTION is B acting on A," if we say *action* is Earth pulling down on the apple, then *reaction* is [the apple pulling up on Earth] [**N**, Nellie's hand pushing up on the apple].

   f. To repeat for emphasis, we see that **N** and **W** are equal and opposite to each other [and comprise an action-reaction pair] [but do not comprise an action-reaction pair].

To identify a pair of action-reaction forces in any situation, first identify the pair of interacting objects involved. Something is interacting with something else. In this case the whole Earth is interacting (gravitationally) with the apple. So Earth pulls downward on the apple (call it action), while the apple pulls upward on Earth (reaction).

Simply put, Earth pulls on apple (action); apple pulls on Earth (reaction).

Better put, apple and Earth *pull* on *each other* with equal and opposite forces that comprise a *single* interaction.

   g. Another pair of forces is **N** as shown, and the downward force of the apple against Nellie's hand, not shown. This force pair [is] [isn't] an action-reaction pair.

   h. Suppose Nellie now pushes upward on the apple with a force of 2 N. The apple [is still in equilibrium] [accelerates upward], and compared to **W**, the magnitude of **N** is [the same] [twice] [not the same, and not twice].

   i. Once the apple leaves Nellie's hand, **N** is [zero] [still twice the magnitude of **W**], and the net force on the apple is [zero] [only **W**] [still **W** - **N**, a negative force].

Hewitt Drew it!

Name ______________________ Date ________

CONCEPTUAL Physics FUNDAMENTALS PRACTICE PAGE

Chapter 4 Newton's Laws of Motion
Vectors and the Parallelogram Rule

1. When two vectors **A** and **B** are at an angle to each other, they add to produce the resultant **C** by the *parallelogram rule*. Note that **C** is the diagonal of a parallelogram where **A** and **B** are adjacent sides. Resultant **C** is shown in the first two diagrams, *a* and *b*. Construct resultant **C** in diagrams *c* and *d*. Note that in diagram *d* you form a rectangle (a special case of a parallelogram).

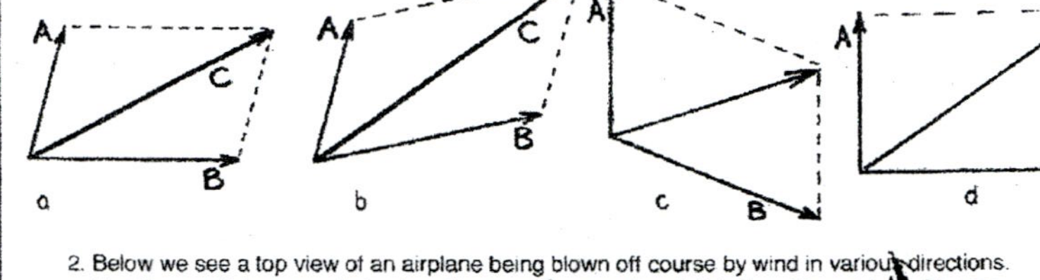

2. Below we see a top view of an airplane being blown off course by wind in various directions. Use the parallelogram rule to show the resulting speed and direction of travel for each case. In which case does the airplane travel fastest across the ground? __d__ Slowest? __c__

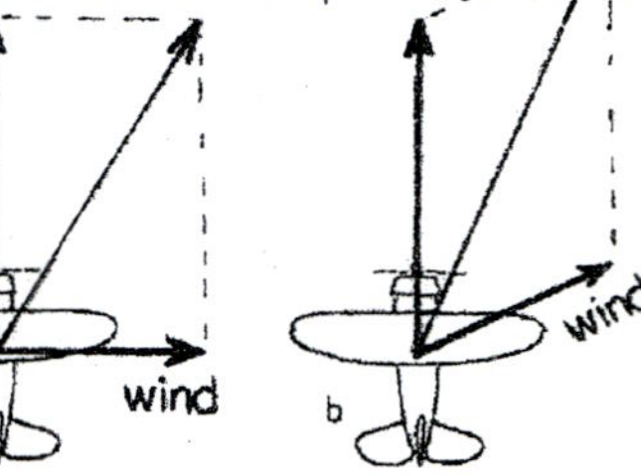

3. To the right we see the top views of 3 motorboats crossing a river. All have the same speed relative to the water, and all experience the same water flow.

   Construct resultant vectors showing the speed and direction of the boats.

   a. Which boat takes the shortest path to the opposite shore? __a__

   b. Which boat reaches the opposite shore first? __b__

   c. Which boat provides the fastest ride? __c__

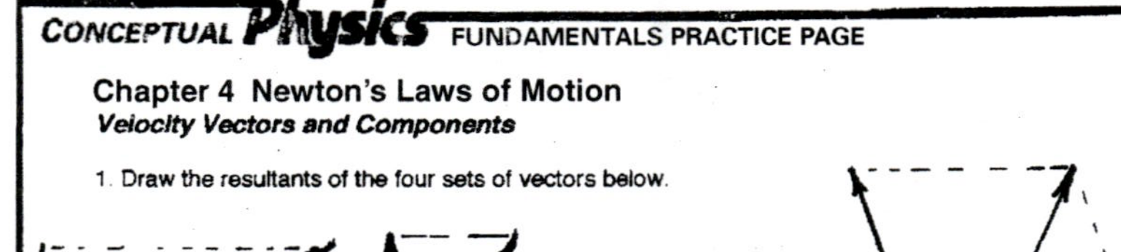

CONCEPTUAL Physics FUNDAMENTALS PRACTICE PAGE

Chapter 4 Newton's Laws of Motion

Velocity Vectors and Components

1. Draw the resultants of the four sets of vectors below.

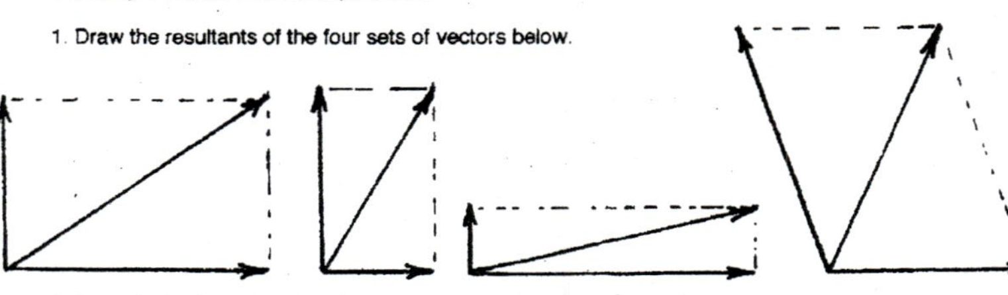

2. Draw the horizontal and vertical components of the four vectors below.

I was only a scalar until you came along and gave me direction! ≥sigh≤

3. She tosses the ball along the dashed path. The velocity vector, complete with its horizontal and vertical components, is shown at position A. Carefully sketch the appropriate components for positions B and C.

   a. Since there is no acceleration in the horizontal direction, how does the horizontal component of velocity compare for positions A, B, and C? **SAME**

   b. What is the value of the vertical component of velocity at position B? **0 m/s**

   c. How does the vertical component of velocity at position C compare with that of position A? **EQUAL AND OPPOSITE**

Velocity of stone

Vertical component of stone's velocity

Horizontal component of stone's velocity

A B C

Name ________________________ Date ________

CONCEPTUAL Physics FUNDAMENTALS PRACTICE PAGE

Chapter 4 Newton's Laws of Motion

Force and Velocity Vectors

1. Draw sample vectors to represent the force of gravity on the ball in the positions shown below after it leaves the thrower's hand. (Neglect air resistance.)

2. Draw sample bold vectors to represent the velocity of the ball in the positions shown below. With lighter vectors, show the horizontal and vertical components of velocity for each position.

3.a. Which velocity component in the previous question remains constant? Why?

HORIZONTAL COMPONENT CONSTANT—NO HORIZONTAL FORCE ON BALL

b. Which velocity component changes along the path? Why?

VERTICAL COMPONENT CHANGES DUE TO GRAVITY IN VERTICAL DIRECTION

4. It is important to distinguish between force and velocity vectors. Force vectors combine with other force vectors, and velocity vectors combine with other velocity vectors. Do velocity vectors combine with force vectors?

NO

5. All forces on the bowling ball, weight (down) and support of alley (up), are shown by vectors at its center before it strikes the pin *a*. Draw vectors of all the forces that act on the ball *b* when it strikes the pin, and *c* after it strikes the pin.

a b c

thanx to Howie Brand

CONCEPTUAL Physics FUNDAMENTALS PRACTICE PAGE

Chapter 4 Newton's Laws of Motion

Force Vectors and the Parallelogram Rule

1. The heavy ball is supported in each case by two strands of rope. The tension in each strand is shown by the vectors. Use the parallelogram rule to find the resultant of each vector pair.

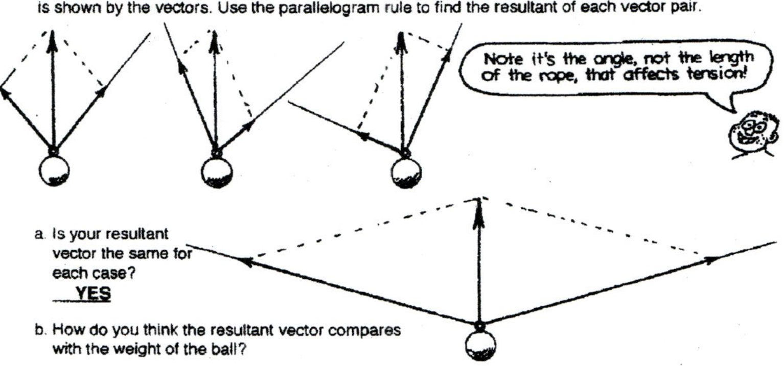

Note it's the angle, not the length of the rope, that affects tension!

a. Is your resultant vector the same for each case?
YES

b. How do you think the resultant vector compares with the weight of the ball?
SAME (BUT OPPOSITE DIRECTION)

2. Now let's do the opposite of what we've done above. More often, we know the weight of the suspended object, but we don't know the rope tensions. In each case below, the weight of the ball is shown by the vector **W**. Each dashed vector represents the resultant of the pair of rope tensions. Note that each is equal and opposite to vectors **W** (they must be; otherwise the ball wouldn't be at rest).

a. Construct parallelograms where the ropes define adjacent sides and the dashed vectors are the diagonals.
b. How do the relative lengths of the sides of each parallelogram compare to rope tension?
c. Draw rope-tension vectors, clearly showing their relative magnitudes.

No wonder that hanging from a horizontal tightly-stretched clothesline breaks it!

W W W W

A B C

3. A lantern is suspended as shown. Draw vectors to show the relative tensions in ropes A, B, and C. Do you see a relationship between your vectors **A** + **B** and vector **C**? Between vectors **A** + **C** and vector **B**?
YES: A + B = −C   A + B = −B

Hewitt Drew it!

Name ______________________ Date __________

CONCEPTUAL Physics FUNDAMENTALS PRACTICE PAGE

Chapter 4 Newton's Laws of Motion

Force-Vector Diagrams

In each case, a rock is acted on by one or more forces. Using a pencil and a ruler, draw an accurate vector diagram showing all forces acting on the rock, and no other forces. The first two cases are done as examples. The parallelogram rule in case 2 shows that the vector sum of **A** + **B** is equal and opposite to **W** (that is, **A** + **B** = -**W**). Do the same for cases 3 and 4. Draw and label vectors for the weight and normal support forces in cases 5 to 10, and for the appropriate forces in cases 11 and 12.

1. Static — T, W
2. Static — A, B, W (Yup! A+B=-W)
3. Static — A, B, W
4. Static — A, B, W
5. Static — N, W
6. Sliding at constant speed without friction — N, W
7. Decelerating due to friction — N, W
8. Static (Friction prevents sliding) — f, N, W
9. Rock slides (No friction) — N, W
10. Static — N_1, N_2, W
11. Rock in free fall — W
12. Falling at terminal velocity — R, W

thanx to Jim Court

Hewitt Drew it!

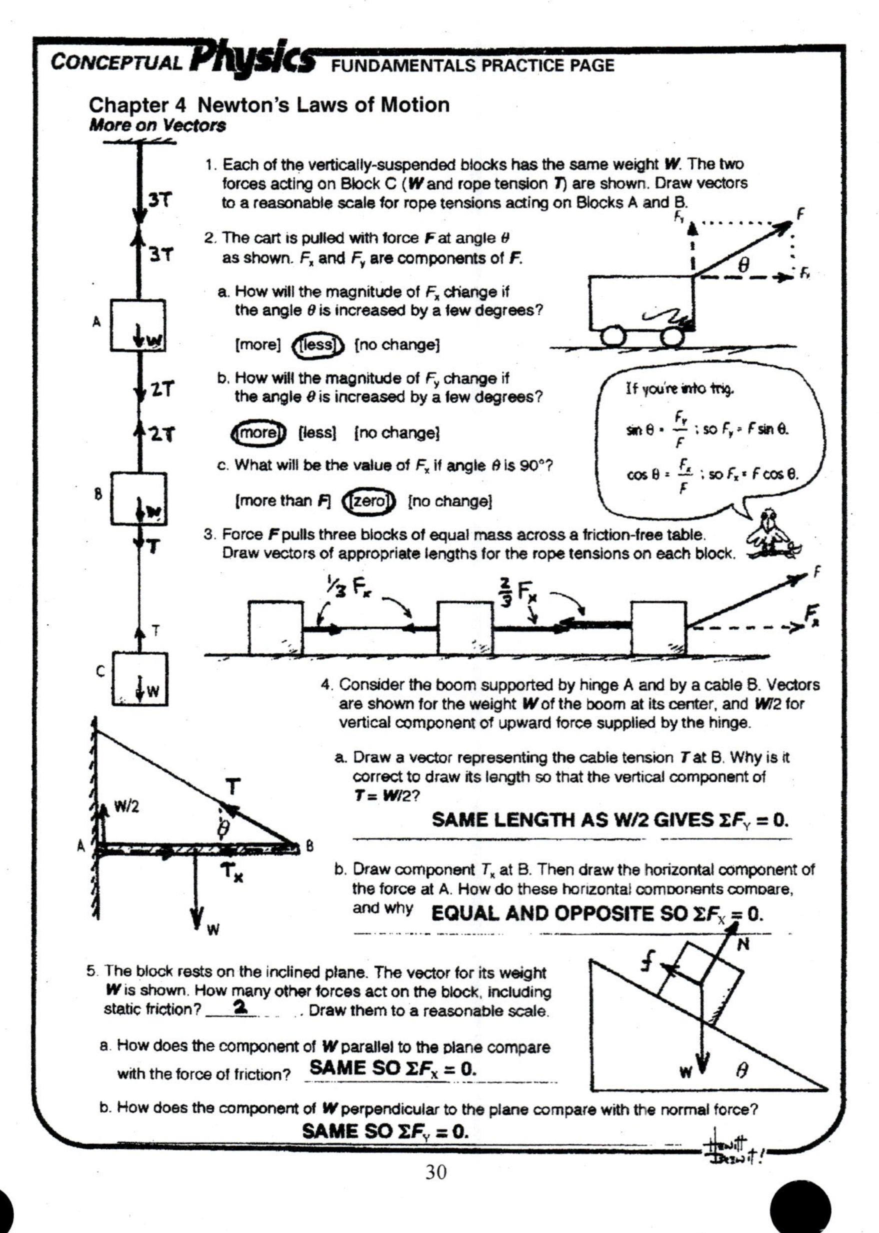

CONCEPTUAL Physics FUNDAMENTALS PRACTICE PAGE

Chapter 4 Newton's Laws of Motion
More on Vectors

1. Each of the vertically-suspended blocks has the same weight ***W***. The two forces acting on Block C (***W*** and rope tension ***T***) are shown. Draw vectors to a reasonable scale for rope tensions acting on Blocks A and B.

2. The cart is pulled with force ***F*** at angle θ as shown. F_x and F_y are components of ***F***.

   a. How will the magnitude of F_x change if the angle θ is increased by a few degrees?

   [more] **(less)** [no change]

   b. How will the magnitude of F_y change if the angle θ is increased by a few degrees?

   (more) [less] [no change]

   c. What will be the value of F_x if angle θ is 90°?

   [more than ***F***] **(zero)** [no change]

If you're into trig.

$\sin\theta = \frac{F_y}{F}$; so $F_y = F\sin\theta$.

$\cos\theta = \frac{F_x}{F}$; so $F_x = F\cos\theta$.

3. Force ***F*** pulls three blocks of equal mass across a friction-free table. Draw vectors of appropriate lengths for the rope tensions on each block.

4. Consider the boom supported by hinge A and by a cable B. Vectors are shown for the weight ***W*** of the boom at its center, and ***W***/2 for vertical component of upward force supplied by the hinge.

   a. Draw a vector representing the cable tension ***T*** at B. Why is it correct to draw its length so that the vertical component of ***T*** = ***W***/2?

   SAME LENGTH AS W/2 GIVES $\Sigma F_Y = 0$.

   b. Draw component T_x at B. Then draw the horizontal component of the force at A. How do these horizontal components compare, and why **EQUAL AND OPPOSITE SO $\Sigma F_X = 0$.**

5. The block rests on the inclined plane. The vector for its weight ***W*** is shown. How many other forces act on the block, including static friction? **2** . Draw them to a reasonable scale.

   a. How does the component of ***W*** parallel to the plane compare with the force of friction? **SAME SO $\Sigma F_X = 0$.**

   b. How does the component of ***W*** perpendicular to the plane compare with the normal force? **SAME SO $\Sigma F_Y = 0$.**

Hewitt Drew it!

Name ______________________ Date __________

CONCEPTUAL *Physics* FUNDAMENTALS PRACTICE PAGE

Chapter 5 Momentum and Energy
Changing Momentum

1. A moving car has momentum. If it moves twice as fast, its momentum is

   TWICE as much.

2. Two cars, one twice as heavy as the other, move down a hill at the same speed. Compared with the lighter car, the momentum of the heavier car is TWICE as much.

3. The recoil momentum of a cannon that kicks is

   [more than] [less than] [the same as]

   the momentum of the cannonball it fires. (Here we neglect friction and the momentum of the gases.)

   ALTHOUGH SPEED AND ACCELERATION OF BULLET GREATER

4. Suppose you are traveling in a bus at highway speed on a nice summer day and the momentum of an unlucky bug is suddenly changed as it splatters onto the front window.

   a. Compared to the force that acts on the bug, how much force acts on the bus?

   [more] [less] [the same]

   b. The time of impact is the same for both the bug and the bus. Compared to the impulse on the bug, this means the impulse on the bus is

   [more] [less] [the same]

   c. Although the momentum of the bus is very large compared to the momentum of the bug, the *change* in momentum of the bus, compared to the *change* of momentum of the bug is

   [more] [less] [the same]

   d. Which undergoes the greater acceleration?

   [bus] [both the same] [bug]

   e. Which therefore, suffers the greater damage?

   [bus] [both the same] [bug of course!]

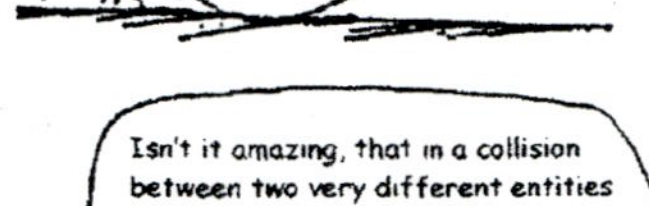

Isn't it amazing, that in a collision between two very different entities — a bug and a bus, that three opposite quantities remain equal: impact forces, impulses, and changes in momentum!

CONCEPTUAL Physics FUNDAMENTALS PRACTICE PAGE

Chapter 5 Momentum and Energy

Changing Momentum—continued

5. Granny whizzes around the rink and is suddenly confronted with Ambrose at rest directly in her path. Rather than knock him over, she picks him up and continues in motion without "braking."

Consider both Granny and Ambrose as two parts of one system. Since no outside forces act on the system, the momentum of the system before collision equals the momentum of the system after collision.

a. Complete the before-collision data in the table below.

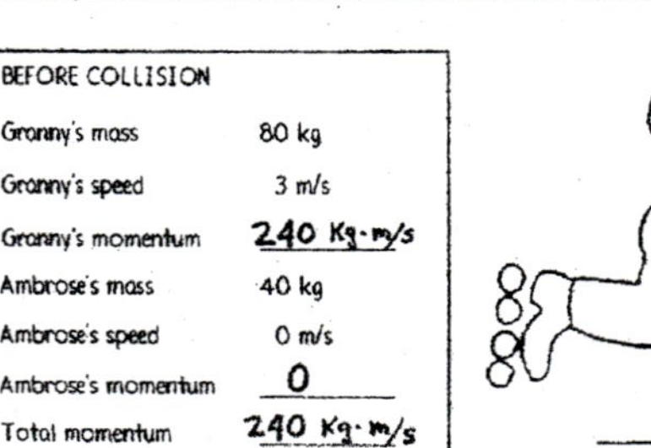

BEFORE COLLISION	
Granny's mass	80 kg
Granny's speed	3 m/s
Granny's momentum	240 Kg·m/s
Ambrose's mass	40 kg
Ambrose's speed	0 m/s
Ambrose's momentum	0
Total momentum	240 Kg·m/s

b. After collision, Granny's speed [increases] [decreases]

c. After collision, Ambrose's speed [increases] [decreases].

d. After collision, the total mass of Granny + Ambrose is 120 kg.

e. After collision, the total momentum of Granny + Ambrose is

240 kg•m/s

f. Use the conservation of momentum law to find the speed of Granny and Ambrose together after collision. (Show your work in the space below.)

$$Mv + mv' = (M + m) V$$
$$(80 \text{ kg})(3 \text{ m/s}) + 0 = (80 \text{ kg} + 40 \text{ kg})V$$
$$240 \text{ kg•m/s} = (120 \text{ kg}) V$$
$$V = 2 \text{ m/s}$$

New speed 2 m/s

CONCEPTUAL Physics FUNDAMENTALS PRACTICE PAGE

Chapter 5 Momentum and Energy

Systems

1. When the compressed spring is released, Blocks A and B will slide apart. There are 3 systems to consider, indicated by the closed dashed lines below—A, B, and A + B. Ignore the vertical forces of gravity and the support force of the table.

a. Does an external force act on System A? [Y] [N]

Will the momentum of System A change? [Y] [N]

b. Does an external force act on System B? [Y] [N]

Will the momentum of System B change? [Y] [N]

c. Does an external force act on System A + B? [Y] [N]

Will the momentum of System A + B change? [Y] [N]

System A System B System A + B

2. Billiard ball A collides with billiard ball B at rest. Isolate each system with a closed dashed line. Draw only the external force vectors that act on each system.

Note that external forces on System A and System B are internal to System A+B, so they cancel!

System A System B System A + B

a. Upon collision, the momentum of System A [increases] [decreases] [remains unchanged].

b. Upon collision, the momentum of System B [increases] [decreases] [remains unchanged].

c. Upon collision, the momentum of System A + B [increases] [decreases] [remains unchanged].

3.a. A girl jumps upward. In the left sketch, draw a closed dashed line to indicate the system of the girl.

Is there an external force acting on her? [Y] [N]

Does her momentum change? [Y] [N]

Is the girl's momentum conserved? [Y] [N]

b. In the right sketch, draw a closed dashed line to indicate the system (girl + Earth). Is there an external force acting on the system due to the interaction between the girl and Earth? [Y] [N]

4. A block strikes a blob of jelly. Isolate 3 systems with a closed dashed line and show the external force on each. In which system is momentum conserved?

ONE ON RIGHT

5. A truck crashes into a wall. Isolate 3 systems with a closed dashed line and show the external force on each. In which system is momentum conserved?

ONE ON RIGHT

thanx to Cedric Linder

Name ______________________ Date __________

CONCEPTUAL Physics FUNDAMENTALS PRACTICE PAGE

Chapter 5 Momentum and Energy

Work and Energy

1. How much work (energy) is needed to lift an object that weighs 200 N to a height of 4 meters?

   __800 J__

2. How much power is needed to lift the 200-N object to a height of 4 m in 4 seconds? __200 W__

3. What is the power output of an engine that does 60,000 J of work in 10 seconds? __6 kW__

4. The block of ice weighs 500 newtons. (Neglect friction.)

   a. How much force parallel to the incline is needed to push it to the top? __250 N__

   b. How much work is required to push it to the top of the incline? __1500 J__

   c. What is the potential energy of the block relative to ground level? __1500 J__

   d. What would be the potential energy if the block were simply lifted vertically 3 m? __1500 J__

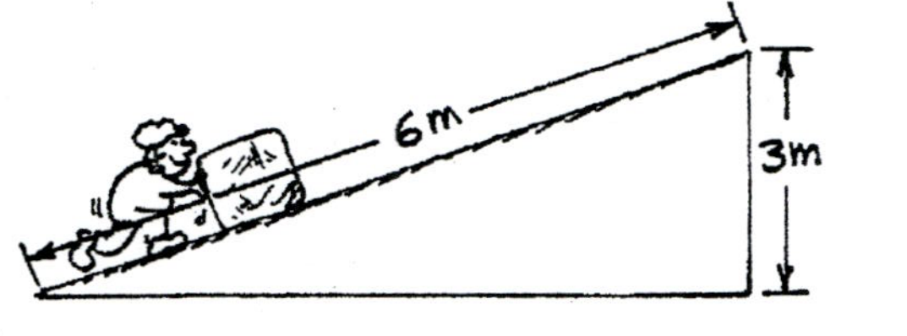

5. All the ramps below are 5 meters high. We know that the KE of the block at the bottom of each ramp will be equal to the loss of PE (conservation of energy).
   Find the speed of the block at ground level in each case. (Hint: Do you recall from earlier chapters how much time it takes something to fall a vertical distance of 5 m from a position of rest assuming $g = 10\ m/s^2$ and how much speed a falling object acquires in this time?) This gives you the answer to Case 1.
   Discuss with your classmates how energy conservation provides the answers to Cases 2 and 3.

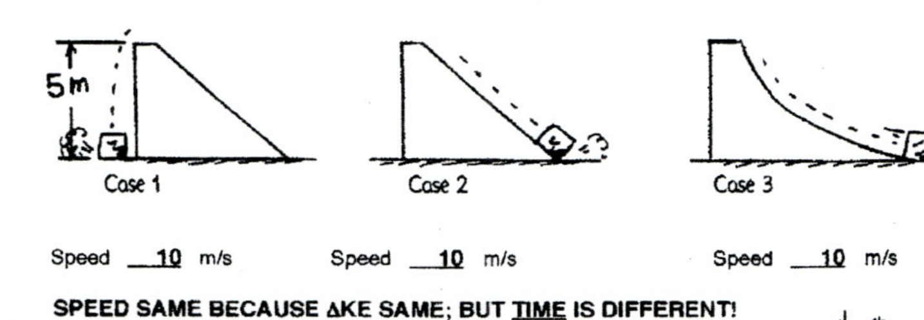

Speed __10__ m/s   Speed __10__ m/s   Speed __10__ m/s

SPEED SAME BECAUSE ΔKE SAME; BUT TIME IS DIFFERENT!

CONCEPTUAL Physics FUNDAMENTALS PRACTICE PAGE

Chapter 5 Momentum and Energy

Work and Energy—continued

6. Which block reaches the bottom of the incline first? Assume no friction. (Be careful!) Explain your answer.

   BLOCK A BECAUSE GREATER ACCELERATION AND LESS RAMP DISTANCE, SO A HAS SHORTER SLIDING TIME—BUT SAME SPEED

7. Both the KE and PE of a block freely sliding down a ramp are shown below only at the bottom position in the sketch. Fill in the missing values for the other positions.

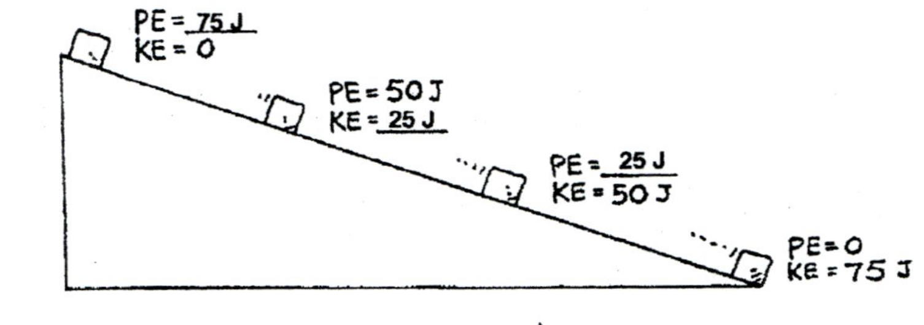

8. A big metal bead slides due to gravity along an upright friction-free wire. It starts from rest at the top of the wire as shown in the sketch.

   How fast is it traveling as it passes

   Point B? __10 m/s__

   Point D? __10 m/s__

   Point E? __10 m/s__

   Maximum speed at Point __C__

THINK ENERGY CONSERVATION

9. Rows of wind-powered generators are used in various windy locations to generate electric power. Does the power generated affect the speed of the wind? Would locations behind the "windmills" be windier if they weren't there. Discuss this in terms of energy conservation with your classmates.

   YES, BY CONSERVATION OF ENERGY, ENERGY GAINED BY WINDMILLS IS TAKEN FROM WIND KE, SO WIND MUST SLOW DOWN. LOCATIONS BEHIND WOULD BE WINDIER WITHOUT THE WINDMILLS!

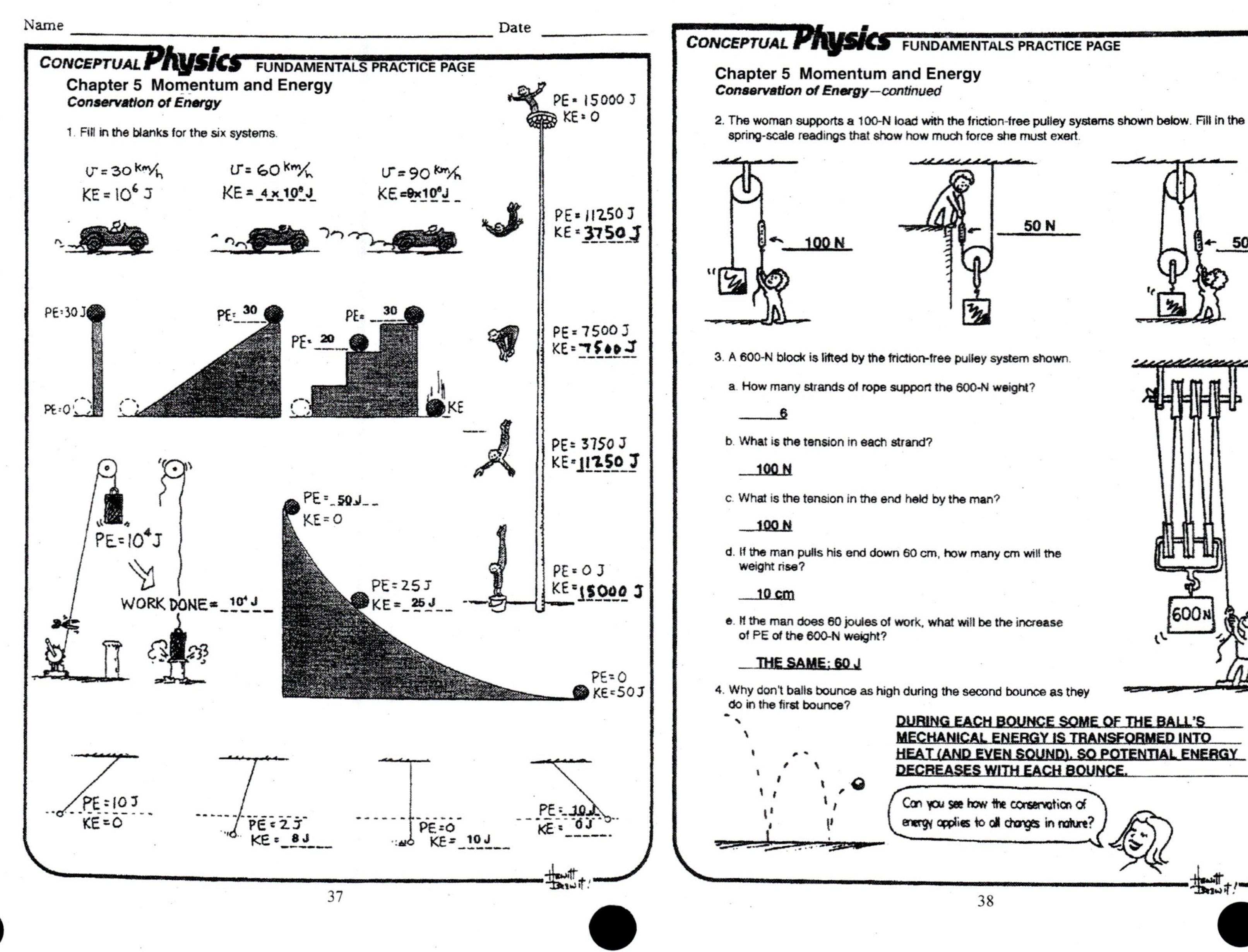

Name ____________________ Date ________

CONCEPTUAL **Physics** FUNDAMENTALS PRACTICE PAGE

Chapter 5 Momentum and Energy

Conservation of Energy

1. Fill in the blanks for the six systems.

CONCEPTUAL **Physics** FUNDAMENTALS PRACTICE PAGE

Chapter 5 Momentum and Energy

***Conservation of Energy**—continued*

2. The woman supports a 100-N load with the friction-free pulley systems shown below. Fill in the spring-scale readings that show how much force she must exert.

3. A 600-N block is lifted by the friction-free pulley system shown.

a. How many strands of rope support the 600-N weight?

6

b. What is the tension in each strand?

100 N

c. What is the tension in the end held by the man?

100 N

d. If the man pulls his end down 60 cm, how many cm will the weight rise?

10 cm

e. If the man does 60 joules of work, what will be the increase of PE of the 600-N weight?

THE SAME; 60 J

4. Why don't balls bounce as high during the second bounce as they do in the first bounce?

DURING EACH BOUNCE SOME OF THE BALL'S MECHANICAL ENERGY IS TRANSFORMED INTO HEAT (AND EVEN SOUND), SO POTENTIAL ENERGY DECREASES WITH EACH BOUNCE.

Can you see how the conservation of energy applies to all changes in nature?

Name ______________________ Date ________

CONCEPTUAL **Physics** FUNDAMENTALS PRACTICE PAGE

Chapter 5 Momentum and Energy

Momentum and Energy

Bronco Brown wants to put $Ft = \Delta mv$ to the test and try bungee jumping. Bronco leaps from a high cliff and experiences 3 s of free fall. Then the bungee cord begins to stretch, reducing his speed to zero in 2 s. Fortunately, the cord stretches to its maximum length just short of the ground below.

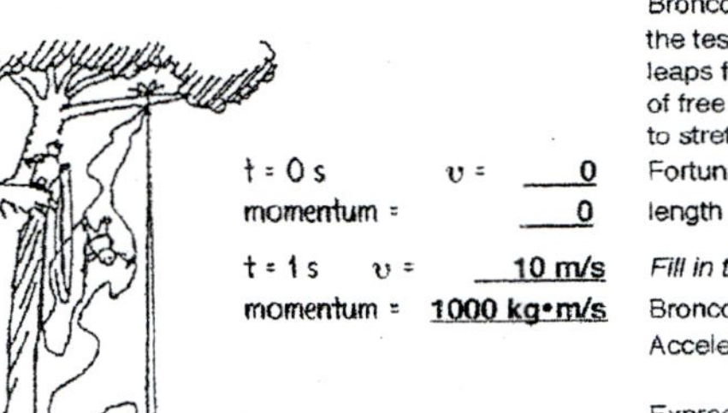

t = 0 s v = **0**
momentum = **0**

t = 1 s v = **10 m/s**
momentum = **1000 kg•m/s**

t = 2 s v = **20 m/s**
momentum = **2000 kg•m/s**

t = 3 s v = **30 m/s**
momentum = **3000 kg•m/s**

t = 5 s v = **0**
momentum = **0**

Fill in the blanks:

Bronco's mass is 100 kg.
Acceleration of free fall is 10 m/s^2.

Express values in SI units (*distance* in m, *velocity* in m/s, *momentum* in kg•m/s, *impulse* in N•s, and *deceleration* in m/s^2).

The 3-s free-fall distance of Bronco just before the bungee cord begins to stretch
= **45 m**

Δmv during the 3 to 5-s interval of free fall
= **3000 kg•m/s**

Δmv during the 3 to 5-s of slowing down
= **3000 kg•m/s**

Impulse during the 3 to 5-s of slowing down
= **3000 N•s**

Average force exerted by the cord during the 3 to 5-s interval of slowing down
= **1500 N**

How about *work* and *energy*? How much KE does Bronco have 3 s after he first jumps?

45,000 J

How much does gravitational PE decrease during this 3 s?

45,000 J

What two kinds of PE are changing during the 3 to 5-s slowing-down interval?

GRAVITATIONAL AND ELASTIC

CONCEPTUAL **Physics** FUNDAMENTALS PRACTICE PAGE

Chapter 5 Momentum and Energy

Energy and Momentum

A Mini Cooper and a Lincoln Town Car are initially at rest on a horizontal parking lot at the edge of a steep cliff. For simplicity, we assume that the Town Car has twice as much mass as the Mini Cooper. Equal constant forces are applied to each car and they accelerate across equal distances (we ignore the effects of friction). When they reach the far end of the lot, the force is suddenly removed, whereupon they sail through the air and crash to the ground below. (The cars are wrecks to begin with, and this is a scientific experiment!)

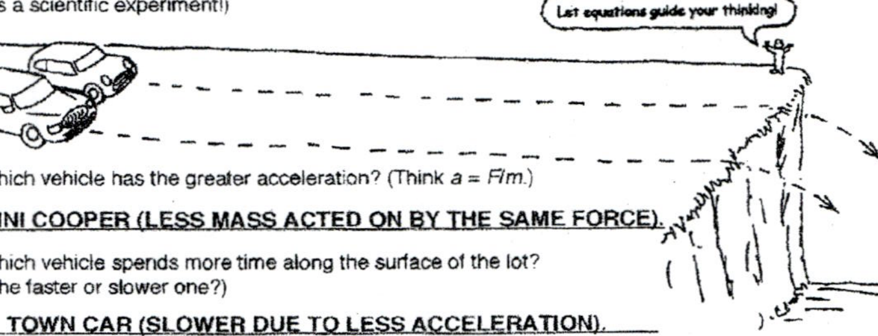

Let equations guide your thinking!
Impulse = Δ momentum
Ft = Δmv
Work = Fd = ΔKE = Δ½mv²
Making the distinction between momentum and kinetic energy is high-level physics.

1. Which vehicle has the greater acceleration? (Think $a = F/m$.)

   MINI COOPER (LESS MASS ACTED ON BY THE SAME FORCE).

2. Which vehicle spends more time along the surface of the lot? (The faster or slower one?)

   TOWN CAR (SLOWER DUE TO LESS ACCELERATION).

3. Which vehicle has the larger impulse imparted to it by the applied force? (Think Impulse = Ft.) Defend your answer.

   TOWN CAR. SAME FORCE IS APPLIED OVER A LONGER TIME.

4. Which vehicle has the greater momentum at the cliff's edge? (Think $Ft = \Delta mv$.) Defend your answer.

   TOWN CAR. MORE IMPULSE PRODUCES MORE MOMENTUM CHANGE.

5. Which vehicle has the greater work done on it by the applied force? (Think $W = Fd$.) Defend your answer in terms of the distance traveled.

   SAME ON EACH. FORCE × DISTANCE IS SAME FOR EACH.

6. Which vehicle has the greater kinetic energy at the edge of the cliff? (Think $W = \Delta KE$.) Does your answer follow from your explanation of Question 5? Does it contradict your answer to Question 3? Why or why not?

   SAME, BECAUSE OF SAME WORK.
   YES.
   NO CONTRADICTION BECAUSE GREATER IMPULSE MEAN GREATER WORK.

7. Which vehicle spends more time in the air, from the edge of the cliff to the ground below?

   BOTH THE SAME!

8. Which vehicle lands farther horizontally from the edge of the cliff onto the ground below?

   THE MINI COOPER, WHICH IS MOVING FASTER.

Challenge: Suppose the slower vehicle crashes a horizontal distance of 10 m from the ledge. Then at what horizontal distance does the faster car hit? **14 m (MINI COOPER MOVES √2 TIMES FASTER DUE TO EQUAL KE AT CLIFF EDGE. 1/2(2 m)v^2 = 1/2MV^2, WHERE V = √2v. SO √2 FASTER MEANS √2 FARTHER IN SAME TIME!**

Name ______________________ Date ________

CONCEPTUAL Physics FUNDAMENTALS PRACTICE PAGE

Chapter 6 Gravity, Projectiles, and Satellites

Inverse-Square Law

1. Paint spray travels radially away from the nozzle of the can in straight lines. Like gravity, the strength (intensity) of the spray obeys an inverse-square law. Complete the diagram by filling in the blank spaces.

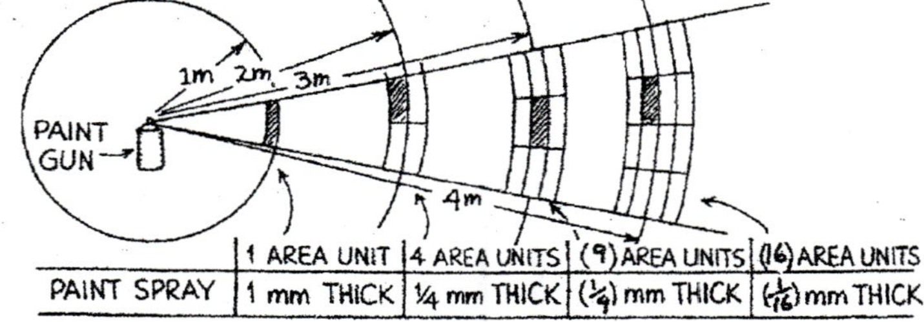

	1 AREA UNIT	4 AREA UNITS	(9) AREA UNITS	(16) AREA UNITS
PAINT SPRAY	1 mm THICK	¼ mm THICK	(1/9) mm THICK	(1/16) mm THICK

2. A small light source located 1 m in front of an opening of area 1 m² illuminates a wall behind. If the wall is 1 m behind the opening (2 m from the light source), the illuminated area covers 4 m². How many square meters will be illuminated if the wall is

5 m from the source? __25 m²__

10 m from the source? __100 m²__

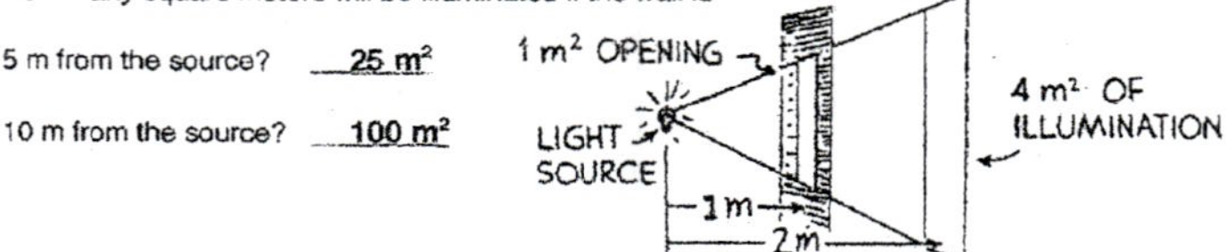

3. If you stand at rest on a weighing scale and find that you are pulled toward Earth with a force of 500 N, then the normal force on the scale is also __500__ N and you weigh __500__ N. How much does Earth weigh? If you tip the scale upside down and repeat the weighing process, you and Earth are still pulled together with a force of __500__ N, and therefore, relative to you, the whole 6,000,000,000,000,000,000,000,000-kg Earth weighs __500__ N! Weight, unlike mass, is a relative quantity.

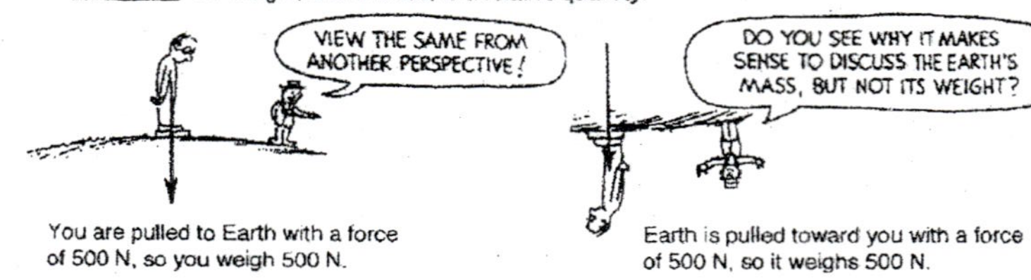

VIEW THE SAME FROM ANOTHER PERSPECTIVE!
DO YOU SEE WHY IT MAKES SENSE TO DISCUSS THE EARTH'S MASS, BUT NOT ITS WEIGHT?

You are pulled to Earth with a force of 500 N, so you weigh 500 N.

Earth is pulled toward you with a force of 500 N, so it weighs 500 N.

CONCEPTUAL Physics FUNDAMENTALS PRACTICE PAGE

Chapter 6 Gravity, Projectiles, and Satellites

Inverse-Square Law—continued

4. The spaceship is attracted to both the planet and the planet's moon. The planet has four times the mass of its moon. The force of attraction of the spaceship to the planet is shown by the vector.

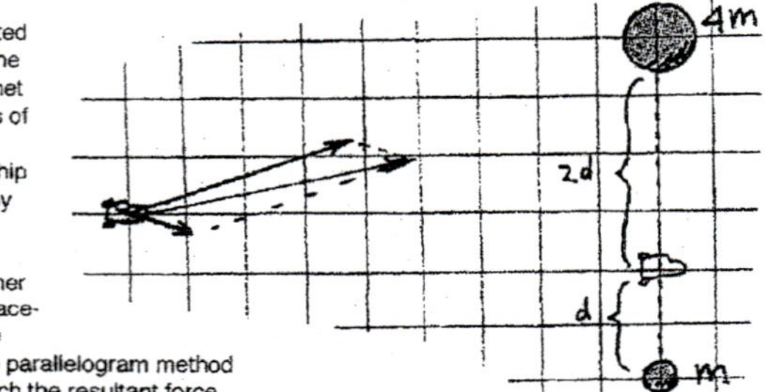

a. Carefully sketch another vector to show the spaceship's attraction to the moon. Then apply the parallelogram method of Chapter 3 and sketch the resultant force.

b. Determine the location between the planet and its moon (along the dotted line) where gravitational forces cancel. Make a sketch of the spaceship there.

5. Consider a planet of uniform density that has a straight tunnel from the North Pole through the center to the South Pole. At the surface of the planet, an object weighs 1 ton.

a. Fill in the gravitational force on the object when it is halfway to the center, then at the center.

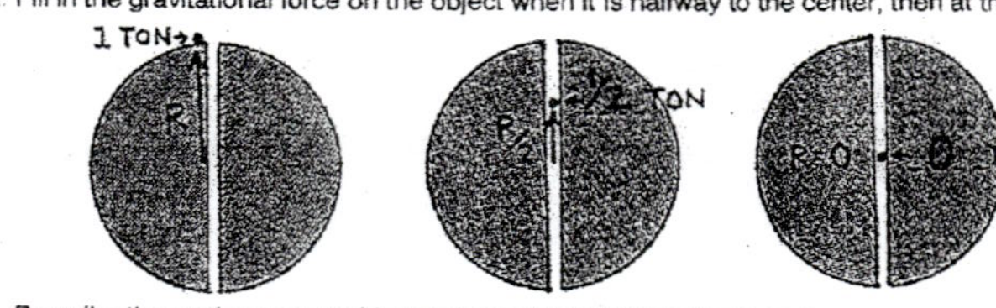

b. Describe the motion you would experience if you fell into the tunnel.

TO AND FRO (IN SIMPLE HARMONIC MOTION).

6. Consider an object that weighs 1 ton at the surface of a planet, just before the planet gravitationally collapses.

a. Fill in the weights of the object on the planet's shrinking surface at the radial values shown.

b. When the planet has collapsed to 1/10 of its initial radius, a ladder is erected that puts the object as far from its center as the object was originally. Fill in its weight at this position.

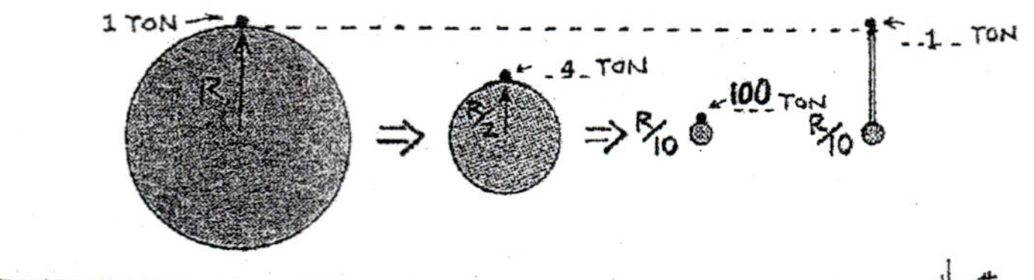

Name ______________________ Date __________

CONCEPTUAL Physics FUNDAMENTALS PRACTICE PAGE

Chapter 6 Gravity, Projectiles, and Satellites

Our Ocean Tides

1. Consider two equal-mass blobs of water, A and B, initially at rest in the Moon's gravitational field. The vector shows the gravitational force of the Moon on A.

a. Draw a force vector on B due to the Moon's gravity.

b. Is the force on B more or less than the force on A? LESS

c. Why? FARTHER AWAY

d. The blobs accelerate toward the Moon. Which has the greater acceleration? [A] [B]

e. Because of the different accelerations, with time

[A gets farther ahead of B] [A and B gain identical speeds] and the distance between A and B

[increases] [stays the same] [decreases].

f. If A and B were connected by a rubber band, with time the rubber band would

[stretch] [not stretch].

g. This [stretching] [nonstretching] is due to the [difference] [nondifference] in the Moon's gravitational pulls.

h. The two blobs will eventually crash into the Moon. To orbit around the Moon instead of crashing into it, the blobs should move

[away from the Moon] [tangentially]. Then their accelerations will consist of changes in

[speed] [direction].

2. Now consider the same two blobs located on opposite sides of Earth.

a. Because of difference in the Moon's pull on the blobs,

they tend to [spread away from each other] [approach each other].

b. Does this spreading produce ocean tides? [Yes] [No]

c. If Earth and Moon were closer, gravitational force between them would be

[more] [the same] [less], and the difference in gravitational forces on the near and far parts

of the ocean would be [more] [the same] [less].

d. Because Earth's orbit about the Sun is slightly elliptical, Earth and Sun are closer in December than in June. Taking the Sun's tidal force into account, on a world average, ocean tides are

greater in [December] [June] [no difference].

Name ______________________ Date __________

CONCEPTUAL Physics FUNDAMENTALS PRACTICE PAGE

Chapter 6 Gravity, Projectiles, and Satellites

Independence of Horizontal and Vertical Components of Motion

1. Above left: Use the scale 1 cm: 5 m and draw the positions of the dropped ball at 1-second intervals. Neglect air resistance and assume $g = 10\ m/s^2$.
   Estimate the number of seconds the ball is in the air. 4 seconds

2. Above right: The four positions of the thrown ball with no gravity are at 1-second intervals. At 1 cm: 5 m, carefully draw the positions of the ball with gravity. Connect your positions with a smooth curve to show the path of the ball.
   How is the motion in the vertical direction affected by motion in the horizontal direction?
   ONLY VERTICAL MOTION AFFECTED BY GRAVITY; HORIZONTAL MOTION IS INDEPENDENT.

Chapter 6 Gravity, Projectiles, and Satellites

***Independence of Horizontal and Vertical Components of Motion**—continued*

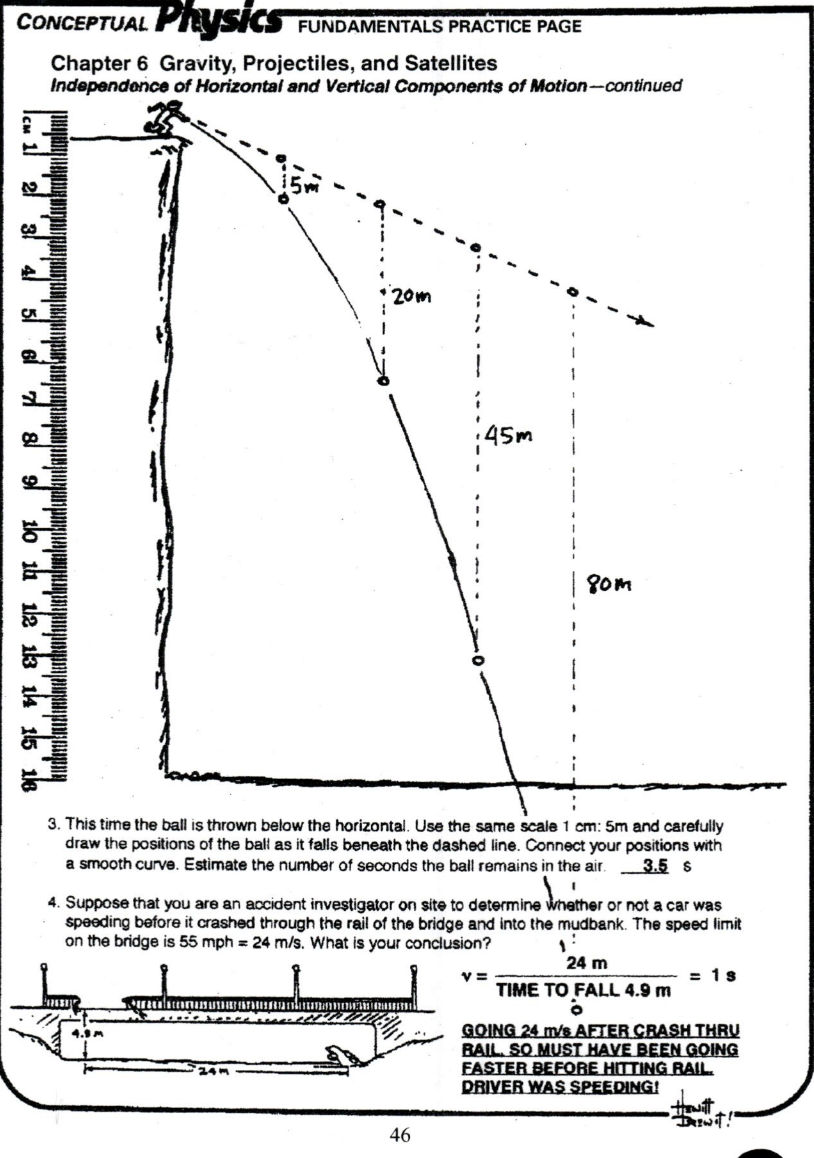

3. This time the ball is thrown below the horizontal. Use the same scale 1 cm: 5m and carefully draw the positions of the ball as it falls beneath the dashed line. Connect your positions with a smooth curve. Estimate the number of seconds the ball remains in the air. __3.5__ s

4. Suppose that you are an accident investigator on site to determine whether or not a car was speeding before it crashed through the rail of the bridge and into the mudbank. The speed limit on the bridge is 55 mph = 24 m/s. What is your conclusion?

$$v = \frac{24\text{ m}}{\text{TIME TO FALL 4.9 m}} = 1\text{ s}$$

GOING 24 m/s AFTER CRASH THRU RAIL, SO MUST HAVE BEEN GOING FASTER BEFORE HITTING RAIL. DRIVER WAS SPEEDING!

Name ______________________ Date ________

Chapter 6 Gravity, Projectiles, and Satellites

Tossed Ball

A ball tossed upward has initial velocity components 30 m/s vertical, and 5 m/s horizontal. The location of the ball is shown at 1-second intervals. Air resistance is negligible, and $g = 10\text{ m/s}^2$. Write the values in the boxes for ascending velocity components, and your calculated resultant descending velocities.

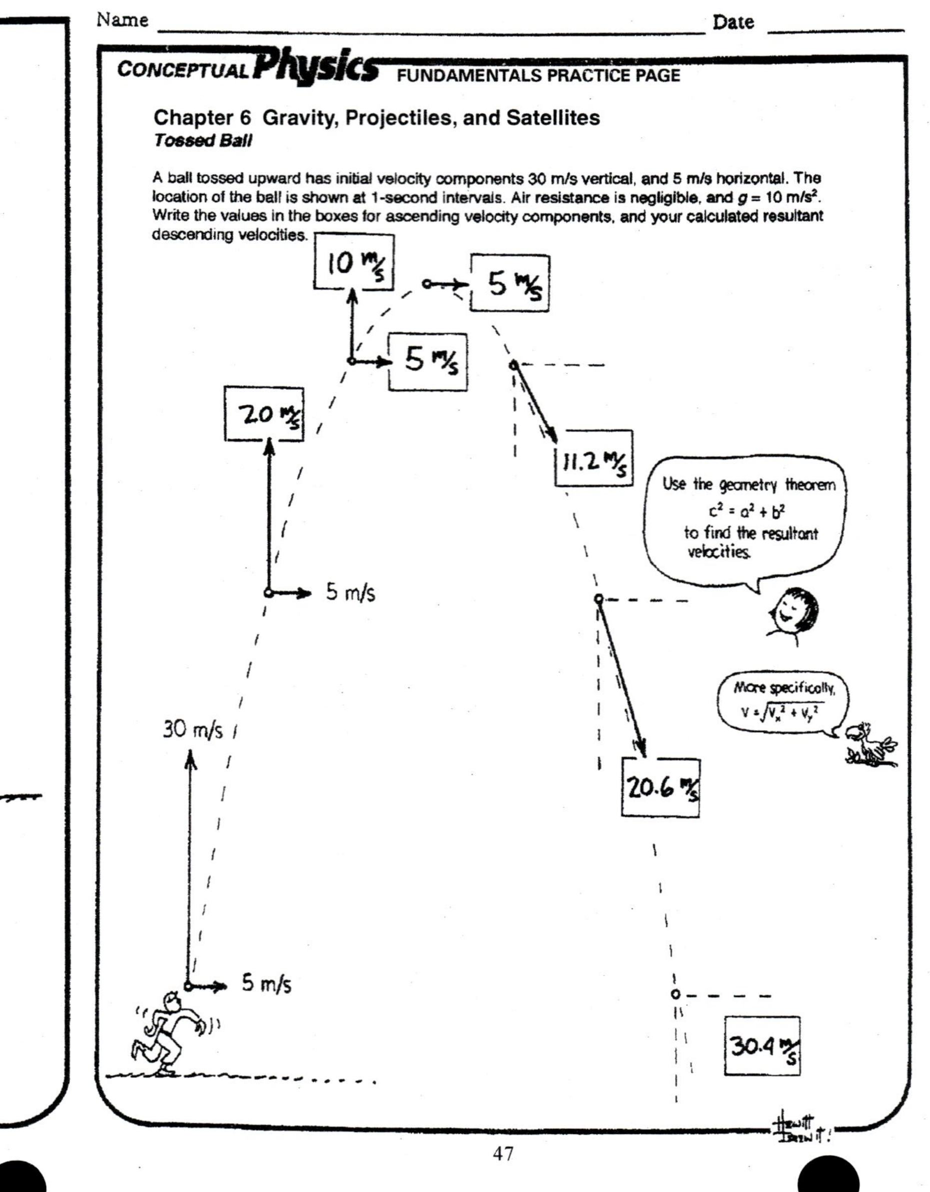

Name ______________________ Date ________

CONCEPTUAL **Physics** FUNDAMENTALS PRACTICE PAGE

Chapter 6 Gravity, Projectiles, and Satellites
Satellite in Circular Orbit

1. Figure A shows "Newton's Mountain," so high that its top is above the drag of the atmosphere. The cannonball is fired and hits the ground.

   a. Draw a likely path that the cannonball might take if it were fired a little bit faster.

   b. Repeat for a still greater speed, but less than 8 km/s.

   c. Then draw its orbital path for a speed of 8 km/s.

   d. What is the shape of the 8-km/s curve?

   CIRCLE

   Figure A

   e. What would be the shape of the orbital path if the cannonball were fired at a speed of 9 km/s?

   ELLIPSE

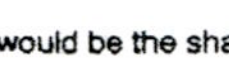

2. Figure B shows a satellite in circular orbit.

   a. At each of the four positions, draw a vector that represents the gravitational *force* exerted on the satellite.

   b. Label the force vectors ***F***.

   c. Then draw a vector at each location to represent the *velocity* of the satellite, and label it ***V***.

   Figure B

   d. Are all four ***F*** vectors the same length? Why or why not?

   YES; SATELLITE IS AT SAME DISTANCE, SAME FORCE.

   e. Are all four ***V*** vectors the same length? Why or why not?

   YES; IN CIRCULAR ORBIT F ⊥ v SO THERE'S NO COMPONENT OF FORCE ALONG v TO CHANGE SPEED v.

   f. What is the angle between your ***F*** and ***V*** vectors? **90°**

   g. Is there any component of ***F*** parallel to ***V***? **NO (F ⊥ v)**

   h. What does this indicate about the work the force of gravity can do on the satellite?

   NO WORK BECAUSE THERE'S NO COMPONENT OF FORCE ALONG PATH.

   i. Does the KE of the satellite in Figure B remain constant or does it vary? **CONSTANT**

   j. Does the PE of the satellite remain constant or does it vary? **REMAINS CONSTANT**

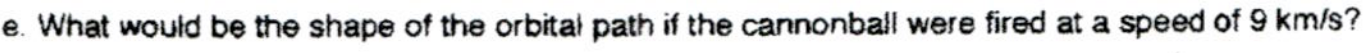

CONCEPTUAL **Physics** FUNDAMENTALS PRACTICE PAGE

Chapter 6 Gravity, Projectiles, and Satellites
Satellite in Elliptical Orbit

3. Repeat the procedure you used for the circular orbit, drawing vectors ***F*** and ***V*** for each position in Figure C, including proper labeling. Show greater magnitudes with greater lengths. Don't bother making the scale accurate.

   a. Are your vectors ***F*** all the same magnitude? Why or why not?

   NO, FORCE DECREASES WHEN DISTANCE FROM EARTH INCREASES.

   b. Are your vectors ***V*** all the same magnitude? Why or why not?

   NO, WHEN KE DECREASES (AS SATELLITE MOVES FARTHER FROM EARTH) SPEED DECREASES, WHEN KE INCREASES (CLOSER TO EARTH) SPEED INCREASES.

   c. Is the angle between vectors ***F*** and ***V*** everywhere the same, or does it vary?

   IT VARIES.

   d. Are there places where there is a component of ***F*** parallel to ***V***?

   YES (EVERYWHERE EXCEPT AT THE APOGEE AND PERIGEE).

   Figure C

   e. Is work done on the satellite where there is a component of ***F*** parallel to ***V***? If so, does this change the KE of the satellite?

   YES. THIS INCREASES KINETIC ENERGY OF SATELLITE.

   f. Where there is a component of ***F*** parallel to or in the direction of ***V***, does this increase or decrease the KE of the satellite?

   THIS DECREASES KE OF SATELLITE.

   g. What can you say about the sum of KE + PE along the orbit?

   CONSTANT (IN ACCORD WITH CONSERVATION OF ENERGY).

Be very very careful when placing both velocity and force vectors on the same diagram. Not a good practice, for one may construct the resultant of the vectors—ouch!

Name ______________________ Date ________

CONCEPTUAL Physics FUNDAMENTALS PRACTICE PAGE

Chapter 7 Fluid Mechanics

Archimedes' Principle I

1. Consider a balloon filled with 1 liter of water (1000 cm^3) in equilibrium in a container of water, as shown in Figure 1.

   a. What is the mass of the 1 liter of water?

   1 kg

   b. What is the weight of the 1 liter of water?

   9.8 N (OR 10 N)

   c. What is the weight of water displaced by the balloon?

   9.8 N

   d. What is the buoyant force on the balloon?

   9.8 N

   e. Sketch a pair of vectors in Figure 1: one for the weight of the balloon and the other for the buoyant force that acts on it. How do the size and directions of your vectors compare?

   VECTORS EQUAL IN MAGNITUDE, OPPOSITE IN DIRECTION

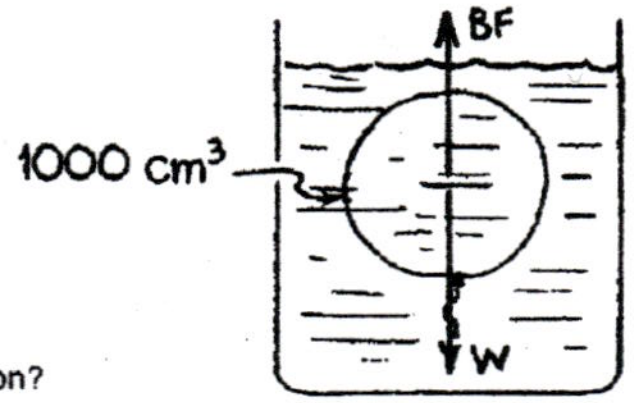

BF
1000 cm^3
W

Figure 1

WATER DOES NOT SINK IN WATER!

2. As a thought experiment, pretend we could remove the water from the balloon but still retain the same size of 1 liter. Then inside the balloon is a vacuum.

   a. What is the mass of the liter of nothing?

   0 kg

   b. What is the weight of the liter of nothing?

   0 N

   c. What is the weight of water displaced by the nearly massless 1-liter balloon?

   9.8 N

   d. What is the buoyant force on the nearly massless balloon?

   9.8 N

   e. In which direction would the nearly massless balloon accelerate?

   UPWARD

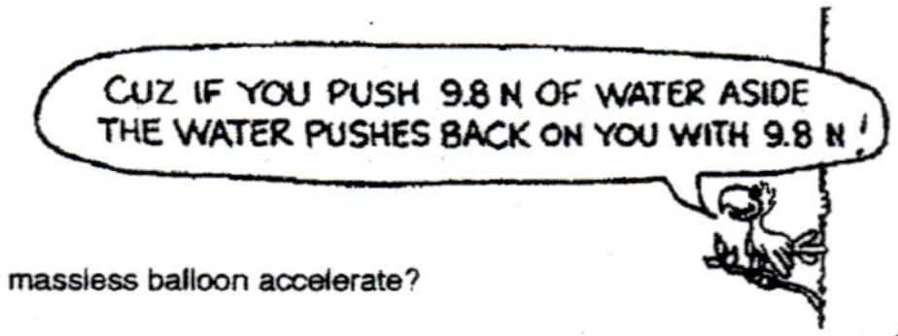

ANYTHING THAT DISPLACES 9.8 N OF WATER EXPERIENCES 9.8 N OF BUOYANT FORCE.
CUZ IF YOU PUSH 9.8 N OF WATER ASIDE THE WATER PUSHES BACK ON YOU WITH 9.8 N!

Chapter 7 Fluid Mechanics

Archimedes' Principle I—continued

3. Assume the balloon is replaced by a 0.5-kilogram piece of wood that has exactly the same volume (1000 cm³), as shown in Figure 2. The wood is held in the same submerged position beneath the surface of the water.

Figure 2

a. What volume of water is displaced by the wood? **1000 cm³ = 1 L**

b. What is the mass of the water displaced by the wood? **1 kg**

c. What is the weight of the water displaced by the wood? **9.8 N**

d. How much buoyant force does the surrounding water exert on the wood? **9.8 N**

e. When the hand is removed, what is the net force on the wood?
NET FORCE = BUOYANT FORCE − WEIGHT OF WOOD = 9.8 N − 4.9 N = 4.9 N (UPWARD)

f. In which direction does the wood accelerate when released? **UPWARD**

THE BUOYANT FORCE ON A SUBMERGED OBJECT EQUALS THE WEIGHT OF WATER DISPLACED

... NOT THE WEIGHT OF THE OBJECT ITSELF!

... UNLESS IT IS FLOATING!

4. Repeat parts *a* through *f* in the previous question for a 5-kg rock that has the same volume (1000 cm³), as shown in Figure 3. Assume the rock is suspended in the container of water by a string.

WHEN THE WEIGHT OF AN OBJECT IS GREATER THAN THE BUOYANT FORCE EXERTED ON IT, IT SINKS!

Figure 3

a. **1000 cm³ (SAME)**

b. **1 kg (SAME)**

c. **9.8 N (SAME)**

d. **9.8 N (SAME)**

e. **39 N DOWNWARD***

f. **DOWNWARD**

* **NET FORCE = BUOYANT FORCE − WT ROCK = 9.8 N − 49 N = −39 N**

Name ______ Date ______

Chapter 7 Fluid Mechanics

Archimedes' Principle II

1. The water lines for the first three cases are shown. Sketch in the appropriate water lines for cases *d* and *e*, and make up your own for case *f*.

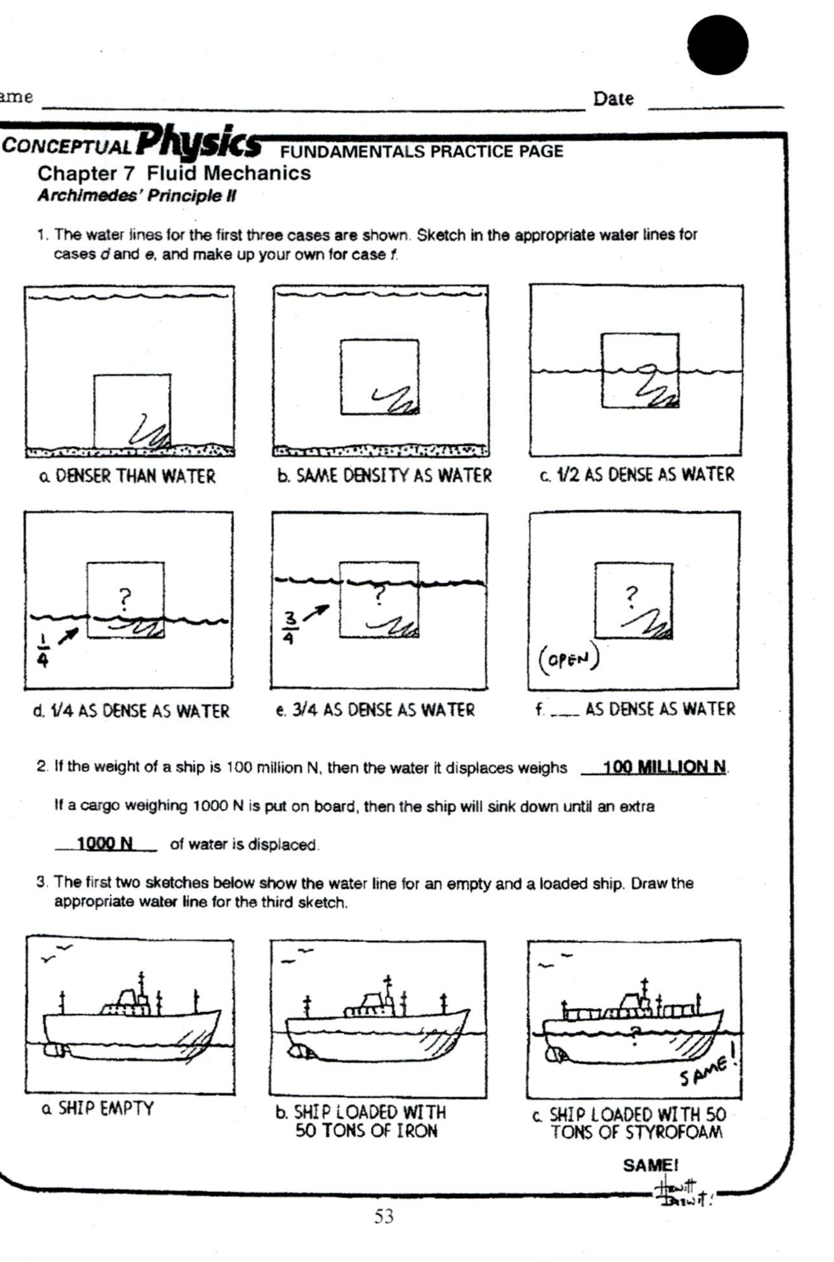

2. If the weight of a ship is 100 million N, then the water it displaces weighs **100 MILLION N**.

If a cargo weighing 1000 N is put on board, then the ship will sink down until an extra **1000 N** of water is displaced.

3. The first two sketches below show the water line for an empty and a loaded ship. Draw the appropriate water line for the third sketch.

a. SHIP EMPTY

b. SHIP LOADED WITH 50 TONS OF IRON

c. SHIP LOADED WITH 50 TONS OF STYROFOAM

SAME!

SAME!

Chapter 7 Fluid Mechanics

Archimedes' Principles II—continued

4. Here is an ice cube floating in a glass of ice water. Draw the water line after the ice cube melts. (Will the water line rise, fall, or remain the same?)

REMAINS THE SAME. VOLUME OF WATER WITH SAME WEIGHT OF ICE CUBE EQUALS VOLUME OF SUBMERGED PORTION OF ICE CUBE. THIS IS ALSO VOLUME OF WATER FROM MELTED ICE.

SAME!

5. The air-filled balloon is weighted so it sinks in water. Near the surface, the balloon has a certain volume. Draw the balloon at the bottom (inside the dashed square) and show whether it is bigger, smaller, or the same size.

   a. Since the weighted balloon sinks, how does its overall density compare to the density of water?

   THE DENSITY OF BALLOON IS GREATER.

   b. As the weighted balloon sinks, does its density increase, decease, or remain the same?

   DENSITY INCREASES (BECAUSE VOL DECREASES).

   c. Since the weighted balloon sinks, how does the buoyant force on it compare to its weight?

   BUOYANT FORCE IS LESS THAN ITS WEIGHT.

   d. As the weighted balloon sinks deeper, does the buoyant force on it increase, decrease, or remain the same?

   BUOYANT FORCE DECREASES (BECAUSE VOLUME DECREASES).

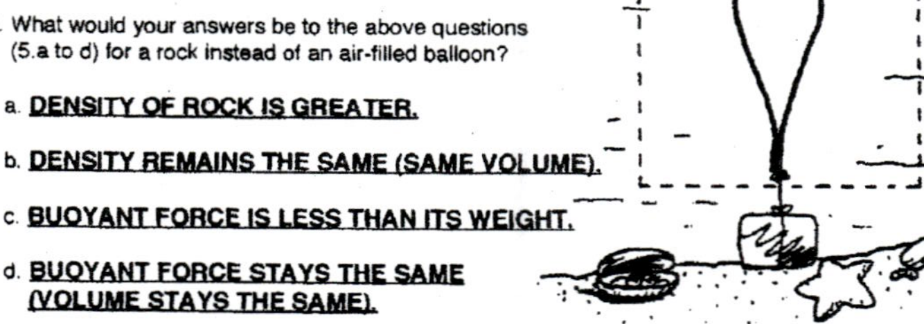

6. What would your answers be to the above questions (5.a to d) for a rock instead of an air-filled balloon?

   a. **DENSITY OF ROCK IS GREATER.**

   b. **DENSITY REMAINS THE SAME (SAME VOLUME).**

   c. **BUOYANT FORCE IS LESS THAN ITS WEIGHT.**

   d. **BUOYANT FORCE STAYS THE SAME (VOLUME STAYS THE SAME).**

Name ______________________ Date ________

Chapter 7 Fluid Mechanics

Gas Pressure

1. A principle difference between a liquid and a gas is that when a liquid is under pressure, its volume
   [increases] [decreases] **[doesn't change noticeably]**
   and its density
   [increases] [decrease] **[doesn't change noticeably]**.
   When a gas is under pressure, its volume
   [increases] **[decreases]** [doesn't change noticeably]
   and its density
   [increases] [decreases] [doesn't change noticeably].

2. The sketch above shows the launching of a weather balloon at sea level. Make a sketch of the same weather balloon when it is high in the atmosphere. In words, what is different about its size and why?

BALLOON GROWS AS IT RISES. ATMOSPHERIC PRESSURE TENDS TO COMPRESS THINGS—EVEN BALLOONS. MORE PRESSURE AT GROUND LEVEL, DECREASES AND MORE COMPRESSION. LESS COMPRESSION AT HIGH ALTITUDES, AND BIGGER BALLOONS.

3. A hydrogen-filled balloon that weighs 10 N must displace **10** N of air in order to float in air. If it displaces less than **10** N it will be buoyed up with less than **10** N and sink. If it displaces more than **10** N of air it will move upward.

4. Why is the cartoon more humorous to physics types than nonphysics types? What physics concept has occurred?

IN ACCORD WITH BERNOULLI'S PRINCIPLE, MOVEMENT OF AIR OVER CURVED TOP OF UMBRELLA CAUSES A REDUCTION OF AIR PRESSURE (LIKE AIRPLANE WING). THIS LIKELY PRODUCED AN UPWARD FORCE THAT TURNED THE UMBRELLA INSIDE OUT.

Name ______________________ Date __________

CONCEPTUAL Physics FUNDAMENTALS PRACTICE PAGE

Mechanics Overview—Chapters 1 to 7

1. The sketch shows the elliptical path described by a satellite about Earth. In which of the labeled positions, A - D, (place an "S" for "same everywhere") does the satellite experience the maximum

   a. gravitational force? __A__

   b. speed? __A__

   c. momentum? __A__

   d. kinetic energy? __A__

   e. gravitational potential energy? __C__

   f. total energy (KE + PE)? __S__

   g. acceleration? __A__ ← $a = \frac{F}{m}$

   h. angular momentum? __S__

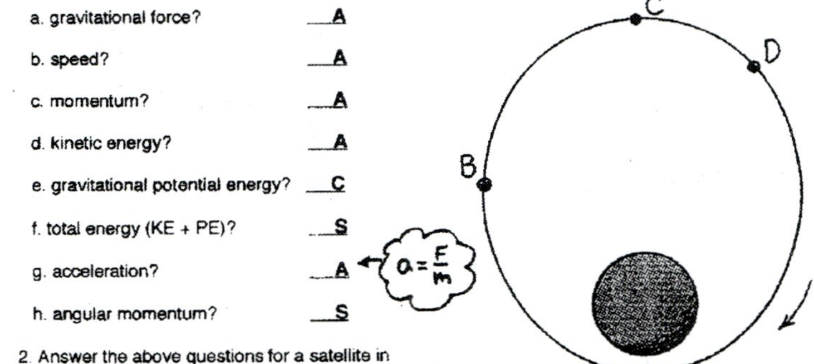

2. Answer the above questions for a satellite in circular orbit.

   a. __S__ b. __S__ c. __S__ d. __S__

   e. __S__ f. __S__ g. __S__ h. __S__

3. In which position(s) is there momentarily no work being done on the satellite by the force of gravity? Why?

   __A AND C BECAUSE OF NO FORCE COMPONENTS IN DIRECTION OF MOTION.__

4. Work changes energy. Let the equation for work, $W = Fd$, guide your thinking on the following: Defend your answers in terms of $W = Fd$.

   a. In which position will a several-minutes thrust of rocket engines pushing the satellite forward do the most work on the satellite and give it the greatest change in kinetic energy? (Hint: Think about where the most distance will be traveled during the application of a several-minutes thrust?)

   __A. WHERE FORCE ACTS OVER LONGEST DISTANCE.__

   b. In which position will a several-minutes thrust of rocket engines pushing the satellite forward do the least work on the satellite and give it the least boost in kinetic energy?

   __C. WHERE FORCE ACTS OVER SHORTEST DISTANCE.__

   c. In which position will a several-minutes thrust of a retro-rocket (pushing opposite to the satellite's direction of motion) do the most work on the satellite and change its kinetic energy the most?

   __A. MOST "NEGATIVE WORK" DECREASES AND MOST ΔKE OCCURS WHERE FORCE ACTS OVER THE LONGEST DISTANCE.__

Hewitt Drew it!

Name ____________________ Date ________

CONCEPTUAL Physics FUNDAMENTALS PRACTICE PAGE

Chapter 8 Temperature, Heat, and Thermodynamics

Measuring Temperatures

1. Complete the table:

TEMPERATURE OF MELTING ICE	°C	32 °F	273K
TEMPERATURE OF BOILING WATER	°C	212°F	373K

2. Suppose you apply a flame and warm one liter of water, raising its temperature 10°C. If you transfer the same heat energy to two liters, how much will the temperature rise? For three liters? Record your answers on the blanks in the drawing at the right.

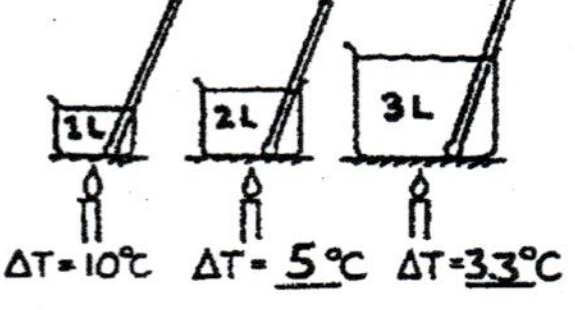

3. A thermometer is in a container half-filled with 20°C water.

   a. When an equal volume of 20°C water is added, the temperature of the mixture will be

   [10°C] [20°C] (circled) [40°C].

   b. When instead an equal volume of 40°C water is added, the temperature of the mixture will be

   [20°C] [30°C] (circled) [40°C].

   c. When instead a small amount of 40°C water is added, the temperature of the mixture will be

   [20°C] [between 20°C and 30°C] (circled) [30°C] [more than 30°C].

Circle one:

4. A small red-hot piece of iron is placed into a large bucket of cool water. (Ignore the heat transfer to the bucket.)

   a. [True] [False] (circled) The decrease in iron temperature equals the increase in the water temperature.

   b. [True] (circled) [False] The quantity of heat lost by the iron equals the quantity of heat gained by the water.

   c. [True] (circled) [False] The iron and water both will eventually reach the same temperature.

   d. [True] [False] (circled) The final temperature of the iron and water is halfway between the initial temperatures of each.

CAN COMMON ICE BE COLDER THAN 0°C?
YES!

CONCEPTUAL Physics FUNDAMENTALS PRACTICE PAGE

Chapter 8 Temperature, Heat, and Thermodynamics

Thermal Expansion

1. The weight hangs above the floor from the copper wire. When a candle is moved along the wire and warms it, what happens to the height of the weight above the floor? Why?

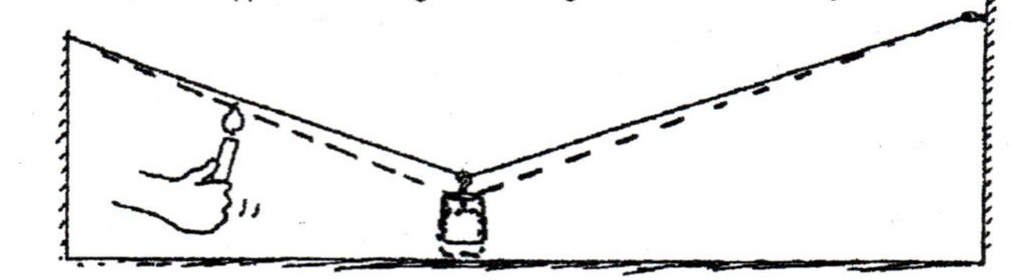

HEIGHT DECREASES AS WIRE LENGTHENS

2. The levels of water at 0°C and 1°C are shown below in the first two flasks. At these temperatures there is microscopic slush in the water. There is slightly more slush at 0°C than at 1°C. As the water is warmed, some of the slush collapses as it melts, and the level of the water falls in the tube. That's why the level of water is slightly lower in the 1°C-tube. Make rough estimates and sketch in the appropriate levels of water at the other temperatures shown. What is important about the level when the water reaches 4°C?

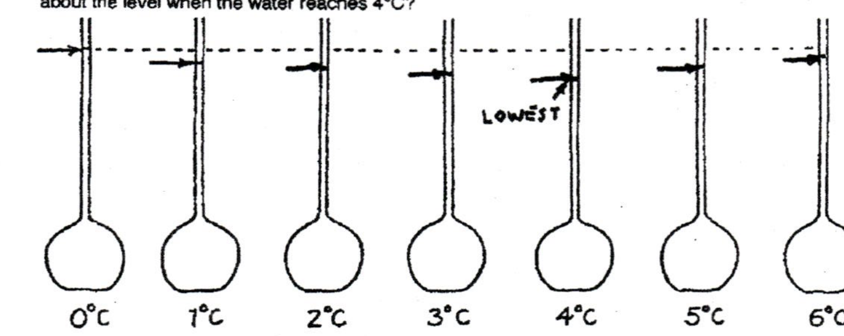

SINCE WATER IS MOST DENSE AT 4°C, WATER LEVEL IS LOWEST AT 4°C.

3. The diagram to the left shows an ice-covered pond. Fill in the blanks for likely temperatures of the water at the top and bottom of the pond.

ICE

0 °C

4 °C

WHICH WILL WEIGH MORE, 1 LITER OF ICE OR 1 LITER OF WATER?

WATER (MORE DENSE)

I CAN'T GET THIS METAL LID OFF THE JAR··· SHOULD I HEAT THE LID OR COOL IT? WHY? HEAT IT SO IT WILL EXPAND.

Name ____________________ Date __________

CONCEPTUAL **Physics** FUNDAMENTALS PRACTICE PAGE

Chapter 8 Temperature, Heat, and Thermodynamics

Absolute Zero

A mass of air is contained so that the volume can change but the pressure remains constant. Table I shows air volumes at various temperatures when the air is warmed slowly.

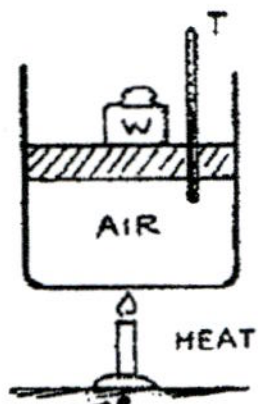

1. Plot the data in Table I on the graph and connect the points.

TABLE I

TEMP. (°C)	VOLUME (mL)
0	50
25	55
50	60
75	65
100	70

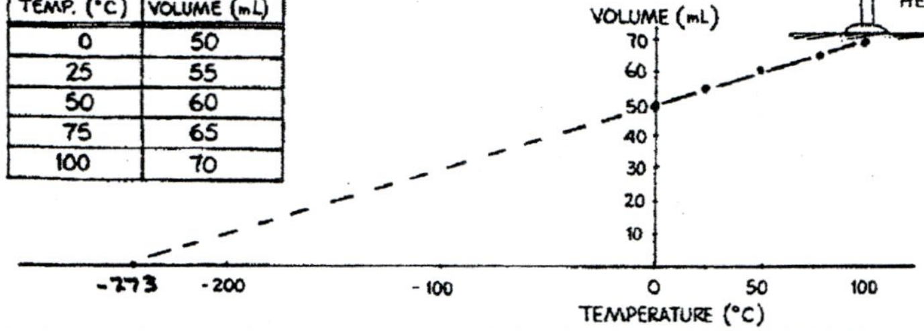

VOLUME (mL): 70, 60, 50, 40, 30, 20, 10

-273, -200, -100, 0, 50, 100

TEMPERATURE (°C)

AIR, T, W, HEAT

2. The graph shows how the volume of air varies with temperature at constant pressure. The straightness of the line means that the air expands uniformly with temperature. From your graph, you can predict what will happen to the volume of air when it is cooled.

   Extrapolate (extend) the straight line of your graph to find the temperature at which the volume of the air would become zero. Mark this point on your graph. Estimate this temperature: **−273°C**

3. Although air would liquefy before cooling to this temperature, the procedure suggests that there is a lower limit to how cold something can be. This is the absolute zero of temperature.

   Careful experiments show that absolute zero is **−273** °C.

4. Scientists measure temperature in *kelvins* instead of degrees Celsius, where the absolute zero of temperature is 0 kelvins. If you relabeled the temperature axis on the graph in Question 1 so that it shows temperature in kelvins, would your graph look like the one below?

   YES

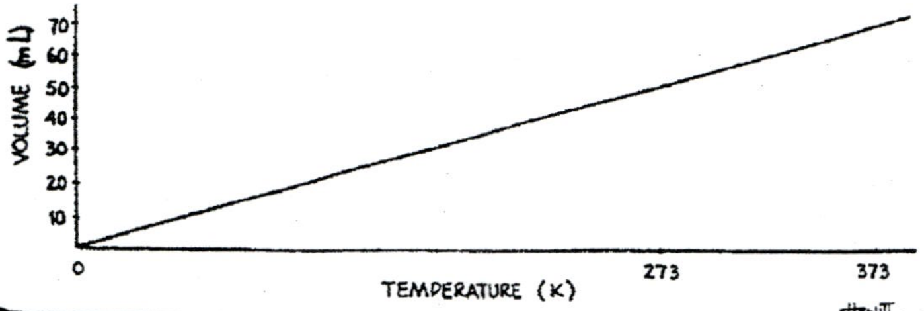

VOLUME (mL): 70, 60, 50, 40, 30, 20, 10

0, 273, 373

TEMPERATURE (K)

Name ______________________ Date __________

CONCEPTUAL **Physics** FUNDAMENTALS PRACTICE PAGE

Chapter 9 Heat Transfer and Change of Phase
Transmission of Heat

Circle one:

1. The tips of both brass rods are held in the gas flame.

   a. [True] ([False]) Heat is conducted only along Rod A.

   b. [True] ([False]) Heat is conducted only along Rod B.

   c. ([True]) [False] Heat is conducted equally along both Rod A and Rod B.

   d. ([True]) [False] The idea that "heat rises" applies to heat transfer by *convection*, not by *conduction*.

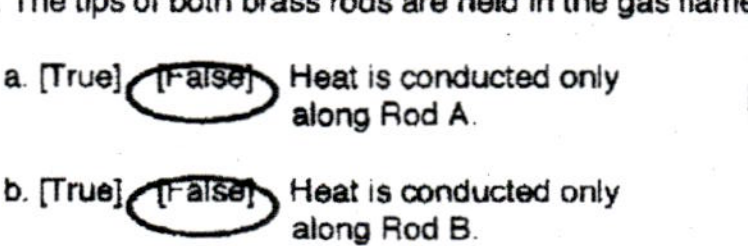

2. Why does a bird fluff its feathers to keep warm on a cold day?

   FLUFFED FEATHERS TRAP AIR THAT ACTS AS AN INSULATOR.

3. Why does a down-filled sleeping bag keep you warm on a cold night? Why is it useless if the down is wet?

   AS IN #2, WHEN WATER TAKES PLACE OF TRAPPED AIR, INSULATION IS REDUCED.

4. What does *convection* have to do with the holes in the shade of the desk lamp?

   WARMED AIR RISES THROUGH HOLES INSTEAD OF BEING TRAPPED.

5. The warmth of equatorial regions and coldness of polar regions on Earth can be understood by considering light from a flashlight striking a surface. If it strikes perpendicularly, light energy is more concentrated as it covers a smaller area; if it strikes at an angle, the energy spreads over a larger area. So the energy per unit area is less.

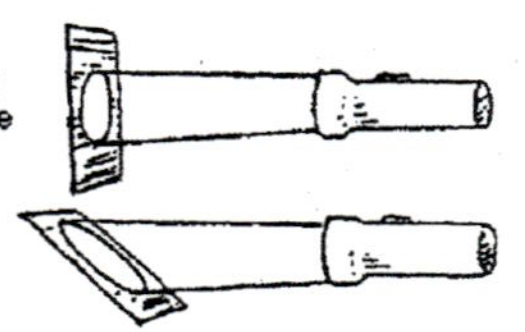

The arrows represent rays of light from the distant Sun incident upon Earth. Two areas of equal size are shown, Area A near the North Pole and Area B near the equator. Count the rays that reach each area, and explain why B is warmer than A.

3 ON A; 6 ON B.

AREA B GETS TWICE THE SOLAR ENERGY AS AREA A SO IT IS WARMER.

CONCEPTUAL **Physics** FUNDAMENTALS PRACTICE PAGE

Chapter 9 Heat Transfer and Change of Phase
Transmission of Heat

6. The Earth's seasons arise from the 23.5-degree tilt of Earth's daily spin axis as it orbits the Sun. When Earth is at the position shown on the right in the sketch below (not to scale), the Northern Hemisphere tilts toward the Sun, and sunlight striking it is strong (more rays per area). Sunlight striking Southern Hemisphere is weak (fewer rays per area). Days in the north are warmer, and daylight is longer. You can see this by imagining Earth making its complete daily 24-hour spin.

   Do two things on the sketch:
   (i) Shade the part of Earth in nighttime darkness for all positions, as is already done in the left position.
   (ii) Label each position with the proper month—March, June, September, or December.

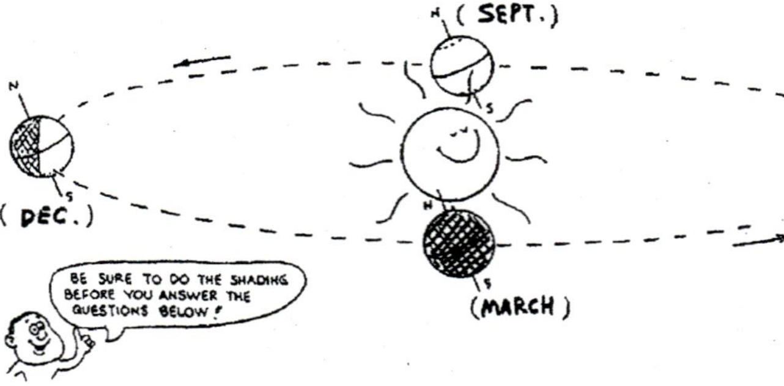

   a. When Earth is in any of the four positions shown, during one 24-hour spin, a location at the equator receives sunlight half the time and is in darkness the other half the time. This means that regions at the equator always receive about **12** hours of sunlight and **12** hours of darkness.

   b. Can you see that in the June position regions farther north have longer daylight hours and shorter nights? Locations north of the Arctic Circle (dotted line in Northern Hemisphere) are continually in view of the Sun as Earth spins, so they get daylight **24** hours a day.

   c. How many hours of light and darkness are there in June at regions south of the Antarctic Circle (dotted line in Southern Hemisphere)?

   ZERO HOURS OF LIGHT, OR 24 HOURS OF DARKNESS PER DAY

   d. Six months later, when Earth is at the December position, is the situation in the Antarctic Circle the same or is it the reverse?

   REVERSE; MORE SUNLIGHT PER AREA IN DECEMBER IN SOUTHERN HEMISPHERE

   e. Why do South America and Australia enjoy warm weather in December instead of June?

   IN DECEMBER THE SOUTHERN HEMISPHERE IS TILTED TOWARD THE SUN AND GETS MORE SUNLIGHT PER AREA THAN IN JUNE.

Name ______________________ Date ________

CONCEPTUAL **Physics** FUNDAMENTALS PRACTICE PAGE

Chapter 9 Heat Transfer and Change of Phase

Ice, Water, and Steam

All matter can exist in the solid, liquid, or gaseous phases. The solid phase normally exists at relatively low temperatures, the liquid phase at higher temperatures, and the gaseous phase at still higher temperatures. Water is the most common example, not only because of its abundance but also because the temperatures for all three phases are common. Study "Energy and Changes of Phase" in your textbook and then answer the following:

1. How many calories are needed to change 1 gram of 0°C ice to water?

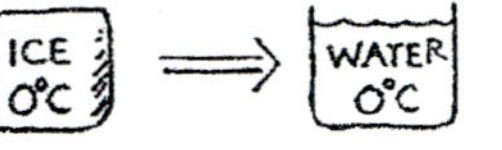

__80__

2. How many calories are needed to change the temperature of 1 gram of water by 1°C?

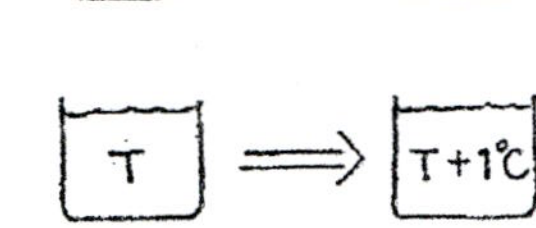

__1__

3. How many calories are needed to melt 1 gram of 0°C ice and turn it to water at a room temperature of 23°C?

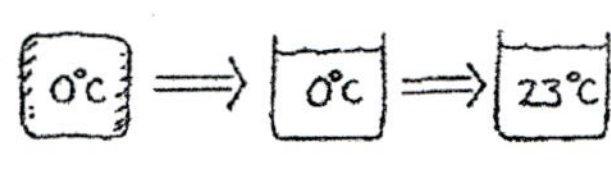

__80 CAL + 23 CAL = 103 CALORIES__

4. A 50-gram sample of ice at 0°C is placed in a glass beaker that contains 200 g of water at 20°C.

   a. How much heat is needed to melt the ice? __4000 CALORIES__
   SINCE THERE'S 50 g OF ICE, AND 80 CAL IS REQUIRED PER GRAM, HEAT REQUIRED IS 50 g × (80 CAL/g) = 4000 CAL

   b. By how much would the temperature of the water change if it released this much heat to the ice? __BY 20°C__

   c. What will be the final temperature of the mixture? (Disregard any heat absorbed by the glass or given off by the surrounding air.)

   __0°C__

5. How many calories are needed to change 1 gram of 100°C boiling water to 100°C steam?

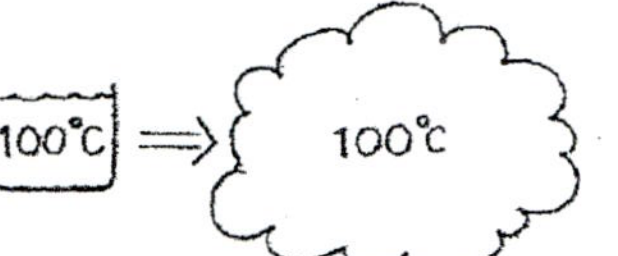

__540 CALORIES__

6. Fill in the number of calories at each step below for changing the phase of 1 gram of 0°C ice to 100°C steam.

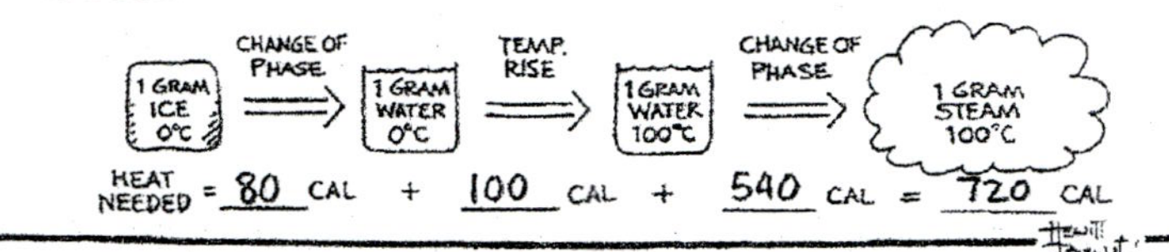

CONCEPTUAL **Physics** FUNDAMENTALS PRACTICE PAGE

Chapter 9 Heat Transfer and Change of Phase

***Ice, Water, and Steam**—continued*

7. One gram of steam at 100°C condenses, and the water cools to 22°C.

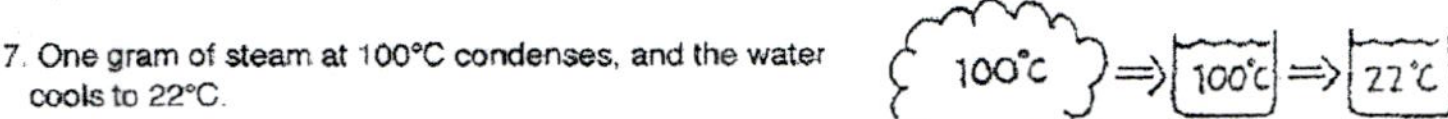

   a. How much heat is released when the steam condenses? __540 CALORIES__

   b. How much heat is released when the water cools from 100°C to 22°C? __78 CALORIES__
   (SINCE WATER COOLS BY 100°C − 22°C = 78°C)

   c. How much heat is released altogether? __618 CALORIES__

8. In a household radiator 1000 g of steam at 100°C condenses, and the water cools to 90°C.

   a. How much heat is released when the steam condenses? __540,000 CAL__

   b. How much heat is released when the water cools from 100°C to 90°C?

   __10,000 CALORIES__

   c. How much heat is released altogether? __550,000 CALORIES__

9. Why is it difficult to brew tea on the top of a high mountain?

   __WATER BOILS AT A LOWER TEMP. AND GETS NO HOTTER THAN THIS TEMP__

10. How many calories are given up by 1 gram of 100°C steam that condenses to 100°C water?

   __540 CALORIES__

11. How many calories are given up by 1 gram of 100°C steam that condenses and drops in temperature to 22°C water?

   __540 CALORIES + (100°C − 22°C) = 618 CALORIES__

12. How many calories are given to a household radiator when 1000 grams of 100°C steam condenses, and drops in temperature to 90°C water?

   __1000 CALORIES (540 CALORIES + [100°C − 90°C]) = 550,000 CALORIES__

13. To get water from the ground, even in the hot desert, dig a hole about a half meter wide and a half meter deep. Place a cup at the bottom. Spread a sheet of plastic wrap over the hole and place stones along the edge to hold it secure. Weight the center of the plastic with a stone so it forms a cone shape. Why will water collect in the cup? (Physics can save your life if you're ever stranded in a desert!) __EVAPORATED WATER FROM GROUND IS TRAPPED, AND CONDENSES ON UNDERSIDE OF PLASTIC AND RUNS INTO THE CUP. (AT NIGHT, CONDENSATION FROM AIR COLLECTS ON *TOP* OF THE PLASTIC.)__

Name ______________________ Date ____________

CONCEPTUAL **Physics** FUNDAMENTALS PRACTICE PAGE

Chapter 9 Heat Transfer and Change of Phase

Evaporation

1. Why do you feel colder when you swim in a pool on a windy day?

WATER EVAPORATES FROM YOUR BODY FASTER AND COOLS YOU.

2. Why does your skin feel cold when a little rubbing alcohol is applied to it?

ALCOHOL RAPIDLY EVAPORATES AND COOLS YOU IN PROCESS.

3. Briefly explain from a molecular point of view why evaporation is a cooling process.

THE MORE ENERGETIC AND FASTER MOLECULES ESCAPE INTO THE AIR.

ENERGY TAKEN WITH THEM REDUCES AVERAGE KE OF REMAINING MOLECULES.

4. When hot water rapidly evaporates, the result can be dramatic. Consider 4 g of boiling water spread over a large surface so that 1 g rapidly evaporates. Suppose further that the surface and surroundings are very cold so that all 540 calories for evaporation come from the remaining 3 g of water.

a. How many calories are taken from each gram of water that remains?

$$\frac{\text{540 CALORIES}}{3} = \text{180 CALORIES}$$

b. How many calories are released when 1 g of 100°C water cools to 0°C?

100 CALORIES

c. How many calories are released when 1 g of 0°C water changes to 0°C ice?

80 CALORIES

d. What happens in this case to the remaining 3 g of boiling water when 1 g rapidly evaporates?

THE REMAINING WATER FREEZES!

(EACH GRAM OF WATER RELEASES

180 CALORIES IN COOLING AND FREEZING.)

CONCEPTUAL **Physics** FUNDAMENTALS PRACTICE PAGE

Chapter 9 Heat Transfer and Change of Phase

Our Earth's Hot Interior

A major puzzle faced scientists in the 19th century. Volcanoes showed that Earth is molten beneath its crust. Penetration into the crust by bore holes and mines showed that Earth's temperature increases with depth. Scientists found that heat flows from the interior to the surface. They assumed that the source of Earth's internal heat was primordial, the afterglow of its fiery birth. Measurements of cooling rates indicated a relatively young Earth—some 25 to 30 millions years in age. But geological evidence indicated an older Earth. This puzzle wasn't solved until the discovery of radioactivity. Then it was learned that the interior is kept hot by the energy of radioactive decay. We now know the age of Earth is some 4.5 billions years—a much older Earth.

All rock contains trace amounts of radioactive minerals. Those in common granite release energy at the rate 0.03 Joule/kilogram•year. Granite at Earth's surface transfers this energy to the surroundings as fast as it is generated, so we don't find granite warm to the touch. But what if a sample of granite were thermally insulated? That is, suppose the increase of internal energy due to radioactivity were contained. Then it would get hotter. How much? Let's figure it out, using 790 joule/kilogram kelvin as the specific heat of granite.

Calculations to make:

1. How many joules are required to increase the temperature of 1 kg of granite by 1000 K?

$Q = cm\Delta T = (790\ J/kg{\bullet}C°)(1\ kg)(1000\ C°) = 790{,}000\ J$

2. How many years would it take radioactive decay in a kilogram of granite to produce this many joules?

790,000 J/0.03 J/kg•yr × 1 kg = 26.3 MILLION YEARS

Questions to answer:

1. How many years would it take a thermally insulated 1-kg chunk of granite to undergo a 1000 K increase in temperature?

SAME 26.3 MILLION YEARS

2. How many years would it take a thermally insulated one-million-kilogram chunk of granite to undergo a 1000 K increase in temperature?

SAME
(DUE TO CORRESPONDINGLY MORE RADIATION)

3. Why are your answers to the above the same (or different)?
THE ENERGY RELEASED PER KG IS THE SAME FOR BOTH. A BIGGER CHUNK GIVES MORE AND REQUIRES THE SAME AMOUNT MORE FOR A CHANGE IN TEMP.

An electric toaster stays hot while electric energy is supplied, and doesn't cool until switched off. Similarly, do you think the energy source now keeping the Earth hot will one day suddenly switch off like a disconnected toaster — or gradually decrease over a long time?

Circle one:
4. [True] (circled) [False] The energy produced by Earth radioactivity ultimately becomes terrestrial radiation.

Name ______________________ Date __________

CONCEPTUAL *Physics* FUNDAMENTALS PRACTICE PAGE

Chapter 10 Static and Current Electricity

Static Charge

1. Consider the diagram below.
   a. A pair of insulated metal spheres, A and B, touch each other, so in effect they form a single uncharged conductor.
   b. A positively charged rod is brought near A, but not touching, and electrons in the metal sphere are attracted toward the rod. Charges in the spheres have redistributed, and the negative charge is labeled. Draw the appropriate + signs that are repelled to the far side of B.
   c. Draw the signs of charge when the spheres are separated while the rod is still present, and
   d. after the rod has been removed. Your completed work should be similar to Figure 22.7 in the textbook. The spheres have been charged by *induction*.

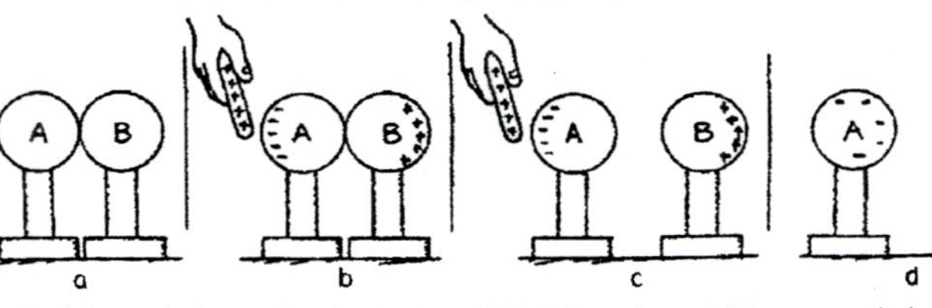

2. Consider below a single metal insulated sphere, (*a*) initially uncharged. When a negatively charged rod is nearby, (*b*), charges in the metal are separated. Electrons are repelled to the far side. When the sphere is touched with your finger, (*c*), electrons flow out of the sphere to Earth through your hand. The sphere is "grounded." Note the positive charge remaining (*d*) while the rod is still present and your finger removed, and (*e*) when the rod is removed. This is an example of *charge induction by grounding*. In this procedure the negative rod "gives" a positive charge to the sphere.

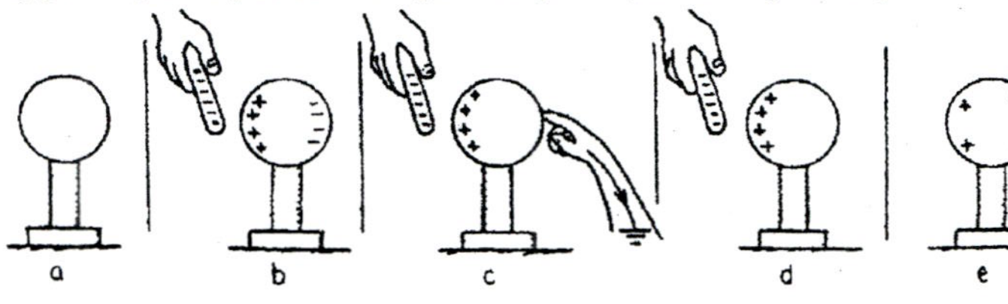

The diagrams below show a similar procedure with a positive rod. Draw the correct charges for *a* through *e*.

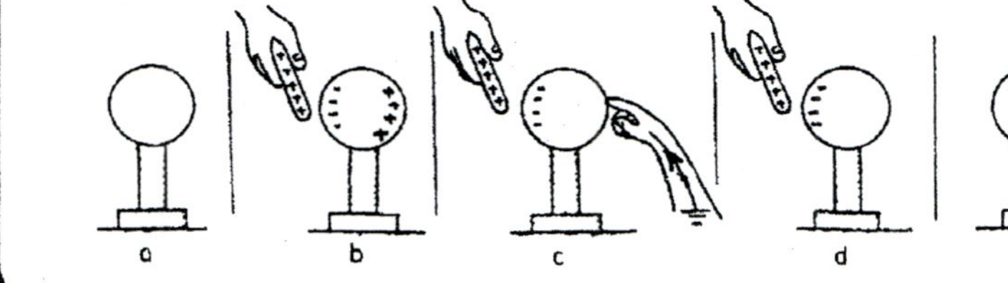

CONCEPTUAL Physics FUNDAMENTALS PRACTICE PAGE

Chapter 10 Static and Current Electricity

Electric Potential

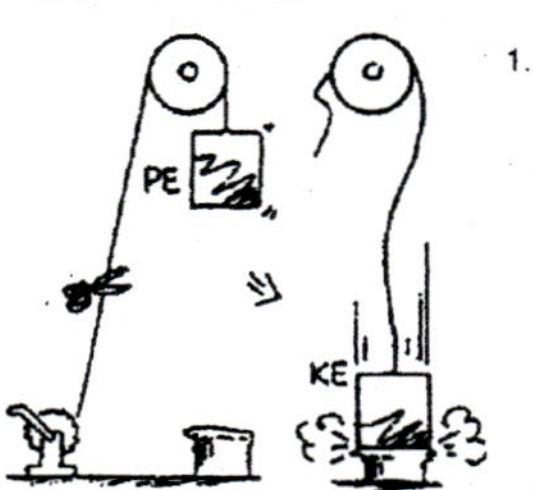

1. Just as PE (potential energy) transforms to KE (kinetic energy) for a mass lifted against the gravitational field (left), the electric PE of an electric charge transforms to other forms of energy when it changes location in an electric field (right). When released, how does the KE acquired by each compare to the decrease in PE?

   KE = DECREASE IN PE

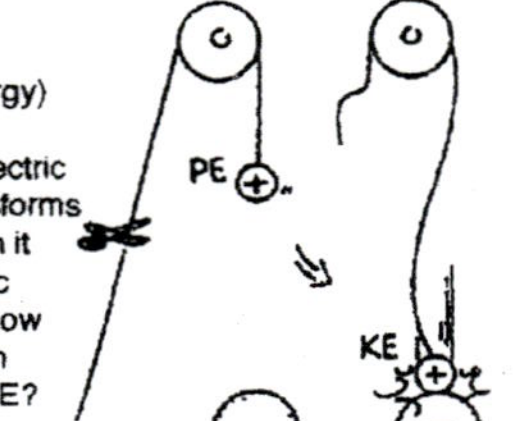

Complete the statements:

2. A force compresses the spring. The work done in compression is the product of the average force and the distance moved. $W = Fd$. This work increases the PE of the spring.

   Similarly, a force pushes the charge (call it a test charge) closer to the charged sphere. The work done in moving the test charge

   is the product of the average **FORCE** and the **DISTANCE**

   moved. $W =$ **Fd**. This work **INCREASES** the PE of the test charge.

   At any point, a greater quantity of test charge means a greater amount of PE, but not a greater amount of PE *per quantity* of charge. The quantities PE (measured in joules) and PE/charge (measured in volts) are different concepts.

   By definition: **Electric Potential** = $\dfrac{\textbf{PE}}{\textbf{charge}}$. 1 volt = 1 joule/coulomb.

3. Complete the statements:

ELECTRIC PE/CHARGE HAS THE SPECIAL NAME ELECTRIC POTENTIAL
SINCE IT IS MEASURED IN VOLTS IT IS COMMONLY CALLED VOLTAGE

4. If a conductor connected to the terminal of a battery has a potential of 12 volts, then each coulomb of charge on the conductor has a PE of **12** J.

5. Some people are confused between force and pressure. Recall that pressure is force *per area*. Similarly, some people get mixed up between electric PE and voltage. According to this chapter, voltage is electric PE per **CHARGE**.

Name ______________________ Date __________

CONCEPTUAL Physics FUNDAMENTALS PRACTICE PAGE

Chapter 10 Static and Current Electricity

Flow of Charge

1. Water doesn't flow in the pipe when both ends (*a*) are at the same level. Another way of saying this is that water will not flow in the pipe when both ends have the same potential energy (PE). Similarly, charge will not flow in a conductor if both ends of the conductor are the same electric potential. But tip the water pipe, as in (*b*), and water will flow. Similarly, charge will flow when you increase the electric potential of an electric conductor so there is a potential difference across the ends.

   a. The unit of electric potential difference is

   ([volt]) [ampere] [ohm] [watt].

   b. It is common to call electric potential difference

   ([voltage]) [amperage] [wattage].

   c. The flow of electric charge is called electric

   [voltage] ([current]) [power]

   and is measured in

   [volts] ([amperes]) [ohms] [watts].

A VOLT IS A UNIT OF POTENTIAL AND AN AMPERE IS A UNIT OF CURRENT

DOES VOLTAGE CAUSE CURRENT, OR DOES CURRENT CAUSE VOLTAGE? WHICH IS THE CAUSE AND WHICH IS THE EFFECT?

VOLTAGE (THE CAUSE) PRODUCES CURRENT (THE EFFECT).

Complete the statements:

2. a. A current of 1 ampere is a flow of charge at the rate of **ONE** coulomb per second.

   b. When a charge of 15 C flows through any area in a circuit each second, the current is **15** A.

   c. One volt is the potential difference between two points if 1 joule of energy is needed to move **ONE** coulomb of charge between the two points.

   d. When a lamp is plugged into a 120-V socket, each coulomb of charge that flows in the circuit is raised to a potential energy of **120** joules.

   e. Which offers more resistance to water flow, a wide pipe or a narrow pipe? **NARROW PIPE**

   Similarly, which offers more resistance to the flow of charge, a thick wire or a thin wire?

   THIN WIRE

CONCEPTUAL Physics FUNDAMENTALS PRACTICE PAGE

Chapter 10 Static and Current Electricity
Ohm's Law

MATH CRUTCH

CURRENT = $\frac{\text{VOLTAGE}}{\text{RESISTANCE}}$ OR $I = \frac{V}{R}$

USE OHM'S LAW IN THE TRIANGLE TO FIND THE QUANTITY YOU WANT, COVER THE LETTER WITH YOUR FINGER AND THE REMAINING TWO SHOW YOU THE FORMULA!

$\frac{V}{I \times R}$

CONDUCTORS AND RESISTORS HAVE RESISTANCE TO THE CURRENT IN THEM.

1. How much current flows in a 1000-ohm resistor when 1.5 volts are impressed across it?

   0.0015 A

2. If the filament resistance in an automobile headlamp is 3 ohms, how many amps does it draw when connected to a 12-volt battery?

   4 A

3. The resistance of the side lights on an automobile are 10 ohms. How much current flows in them when connected to 12 volts?

   1.2 A

4. What is the current in the 30-ohm heating coil of a coffee maker that operates on a 120-volt circuit?

   4 A

5. During a lie detector test, a voltage of 6 V is impressed across two fingers. When a certain question is asked, the resistance between the fingers drops from 400,000 ohms to 200,000 ohms.

   a. What is the current initially through the fingers? **0.000015 A (15 μA)**

   b. What is the current through the fingers when the resistance between them drops? **0.000030 A (30 μA)**

6. How much resistance allows an impressed voltage of 6 V to produce a current of 0.006 A?

   1000 Ω

7. What is the resistance of a clothes iron that draws a current of 12 A at 120 V?

   10 Ω

8. What is the voltage across a 100-ohm circuit element that draws a current of 1 A?

   100 V

OHM MY GOODNESS!

9. What voltage will produce 3 A through a 15-ohm resistor?

   45 V

10. The current in an incandescent lamp is 0.5 A when connected to a 120-V circuit, and 0.2 A when connected to a 10-V source. Does the resistance of the lamp change in these cases? Explain your answer and defend it with numerical values.

    YES. RESISTANCE INCREASES WITH HIGHER TEMP OR GREATER CURRENT. AT 0.2 A, R = 10 V/0.2 A = 50 Ω; AT 0.5 A, R = 120 V/0.5 A = 240 Ω (APPRECIABLY GREATER).

Hewitt Drew It!

Name ______________________ Date ________

CONCEPTUAL Physics FUNDAMENTALS PRACTICE PAGE

Chapter 10 Static and Current Electricity
Electric Power

Recall that the rate at which energy is converted from one form to another is *power.*

$$\text{Power} = \frac{\text{energy converted}}{\text{time}} = \frac{\text{voltage} \times \text{charge}}{\text{time}} = \text{voltage} \times \frac{\text{charge}}{\text{time}} = \text{voltage} \times \text{current}$$

The unit of power is the *watt* (or *kilowatt*), so in units form,

Electric power (*watts*) = current (*amperes*) × voltage (*volts*), where 1 *watt* = 1 *ampere* × 1 *volt.*

THAT'S RIGHT··· VOLTAGE = $\frac{\text{ENERGY}}{\text{CHARGE}}$, SO ENERGY = VOLTAGE × CHARGE··· AND $\frac{\text{CHARGE}}{\text{TIME}}$ = CURRENT ; NEAT ;

A 100-WATT BULB CONVERTS ELECTRIC ENERGY INTO HEAT AND LIGHT MORE QUICKLY THAN A 25-WATT BULB. THAT'S WHY FOR THE SAME VOLTAGE A 100-WATT BULB GLOWS BRIGHTER THAN A 25-WATT BULB!

1. What is the power when a voltage of 120 V drives a 2-A current through a device?

   240 W

2. What is the current when a 60-W lamp is connected to 120 V?

   0.5 A

3. How much current does a 100-W lamp draw when connected to 120 V?

   0.83 A

WHICH DRAWS MORE CURRENT··· THE 100-WATT OR THE 25-WATT BULB?

4. If part of an electric circuit dissipates energy at 6 W when it draws a current of 3 A, what voltage is impressed across it?

   2 V

5. The equation

$$\text{power} = \frac{\text{energy converted}}{\text{time}}$$

rearranged gives energy converted = **POWER × TIME**

WATT'S HAPPENING?

6. Explain the difference between a kilowatt and a kilowatt-hour.

   A KILOWATT IS A UNIT OF POWER; Kw-HOUR IS UNIT OF ENERGY (POWER × TIME)

7. One deterrent to burglary is to leave your front porch light constantly on. If your fixture contains a 60-W bulb at 120 V, and your local power utility sells energy at 10 cents per kilowatt-hour, how much will it cost to leave the light on for the entire month? Show your work on the other side of this page.

   E = P x t = 60 W x 1 mo x 30 day/1 mo x 24 h/1 day x 1 kW/1000 W = 43.2 kWh MULTIPLY BY $0.10/kWh = $4.32

Hewitt Drew It!

Name ________________ Date ________

CONCEPTUAL **Physics** FUNDAMENTALS PRACTICE PAGE

Chapter 10 Static and Current Electricity

Series Circuits

1. In the circuit shown at the right, a voltage of 6 V pushes charge through a single resistor of 2 Ω. According to Ohm's law, the current in the resistor (and therefore in the whole circuit) is

 __3__ A.

2. Two 3-Ω resistors and a 6-V battery comprise the circuit on the right. The total resistance of the circuit is __6__ Ω.

 The current in the circuit is then __1__ A.

3. The equivalent resistance of three 4-Ω resistors in series would be

 __12__ Ω.

4. Does current flow *through* a resistor, or *across* a resistor? __THROUGH__

 Is voltage established *through* a resistor, or *across* a resistor? __ACROSS__

5. Does current in the lamps of a circuit occur simultaneously, or does charge flow first through one lamp, then the other, and finally the last in turn?

 __SIMULTANEOUSLY (SPEED OF LIGHT)__

6. Circuits *a* and *b* below are identical with all bulbs rated at equal wattage (therefore equal resistance). The only difference between the circuits is that Bulb 5 has a short circuit, as shown.

 a. In which circuit is the current greater? __b__

 b. In which circuit are all three bulbs equally bright? __a__

 c. Which bulbs are the brightest? __4 AND 6__

 d. Which bulb is the dimmest? __5 (NOT LIT)__

 e. Which bulbs have the largest voltage drops across them? __4 AND 6 (2.25 V EACH)__

 f. Which circuit dissipates more power? __b (GREATER CURRENT, SAME VOLTAGE)__

 g. Which circuit produces more light? __b (MORE POWER)__

CONCEPTUAL **Physics** FUNDAMENTALS PRACTICE PAGE

Chapter 10 Static and Current Electricity

Parallel Circuits

1. In the circuit shown below, there is a voltage drop of 6 V across *each* 2 Ω resistors.

 a. By Ohm's law, the current in *each* resistor is __3__ A.

 b. The current through the battery is the sum of the currents in the resistors, __6__ A.

 c. Fill in the current in the eight blank spaces in the diagram above of the same circuit.

2. Cross out the circuit below that is *not* equivalent to the circuit above.

3. Consider the parallel circuit at the right.

 a. The voltage drop across each resistor is

 __6__ V.

 b. The current in each branch is:

 2-Ω resistor __3__ A.

 2-Ω resistor __3__ A.

 1-Ω resistor __6__ A.

 c. The current through the battery equals the sum of the currents which equals __12__ A.

 d. The equivalent resistance of the circuit equals __0.5__ Ω.

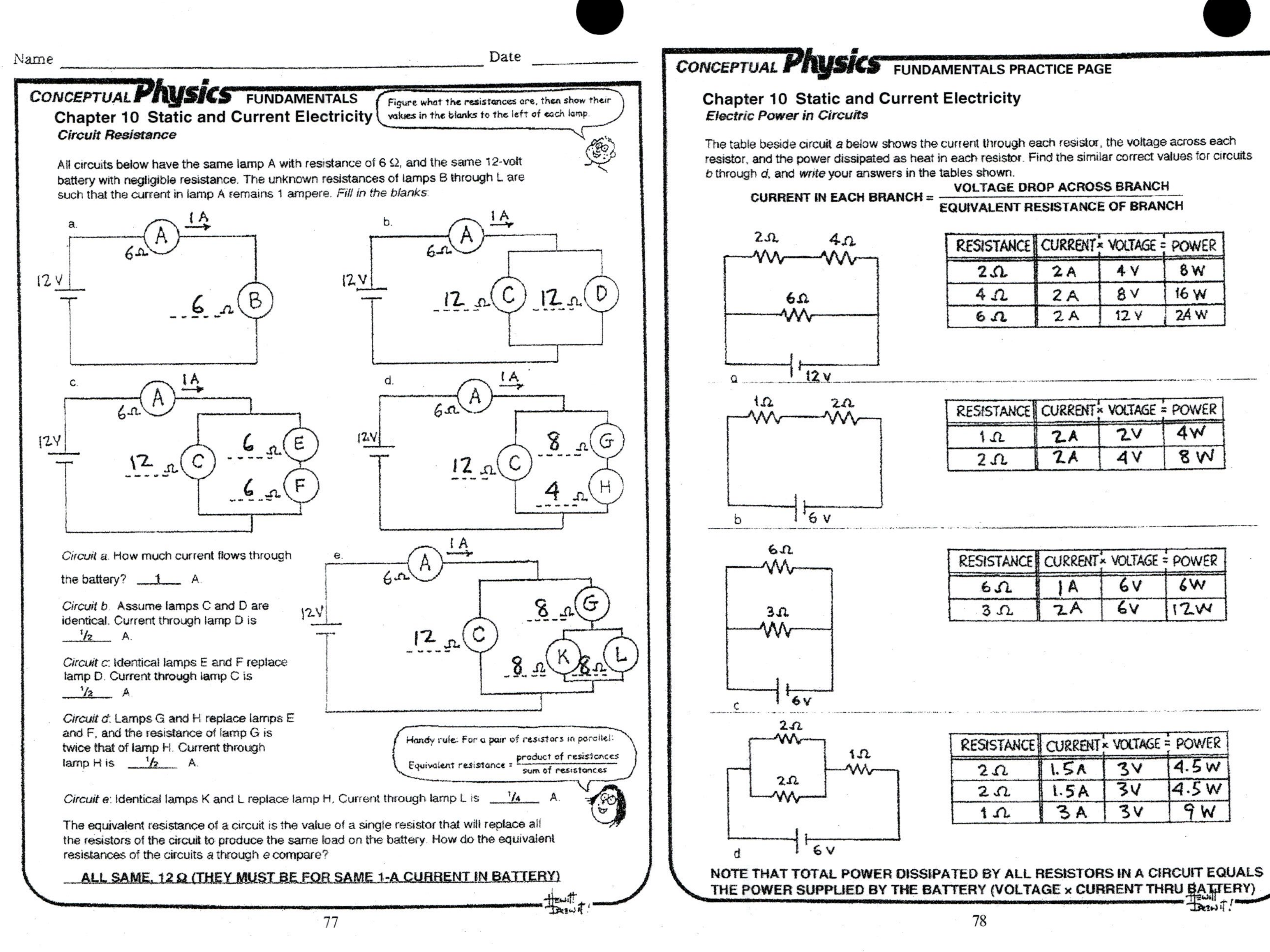

Name ______________ Date ______

CONCEPTUAL Physics FUNDAMENTALS

Chapter 10 Static and Current Electricity

Circuit Resistance

Figure what the resistances are, then show their values in the blanks to the left of each lamp.

All circuits below have the same lamp A with resistance of 6 Ω, and the same 12-volt battery with negligible resistance. The unknown resistances of lamps B through L are such that the current in lamp A remains 1 ampere. *Fill in the blanks.*

Circuit a. How much current flows through the battery? __1__ A.

Circuit b. Assume lamps C and D are identical. Current through lamp D is __1/2__ A.

Circuit c: Identical lamps E and F replace lamp D. Current through lamp C is __1/2__ A.

Circuit d: Lamps G and H replace lamps E and F, and the resistance of lamp G is twice that of lamp H. Current through lamp H is __1/2__ A.

Handy rule: For a pair of resistors in parallel:
Equivalent resistance = product of resistances / sum of resistances

Circuit e: Identical lamps K and L replace lamp H. Current through lamp L is __1/4__ A.

The equivalent resistance of a circuit is the value of a single resistor that will replace all the resistors of the circuit to produce the same load on the battery. How do the equivalent resistances of the circuits a through *e* compare?

__ALL SAME, 12 Ω (THEY MUST BE FOR SAME 1-A CURRENT IN BATTERY)__

CONCEPTUAL Physics FUNDAMENTALS PRACTICE PAGE

Chapter 10 Static and Current Electricity

Electric Power in Circuits

The table beside circuit *a* below shows the current through each resistor, the voltage across each resistor, and the power dissipated as heat in each resistor. Find the similar correct values for circuits *b* through *d*, and *write* your answers in the tables shown.

$$\text{CURRENT IN EACH BRANCH} = \frac{\text{VOLTAGE DROP ACROSS BRANCH}}{\text{EQUIVALENT RESISTANCE OF BRANCH}}$$

RESISTANCE	CURRENT ×	VOLTAGE =	POWER
2 Ω	2 A	4 V	8 W
4 Ω	2 A	8 V	16 W
6 Ω	2 A	12 V	24 W

RESISTANCE	CURRENT ×	VOLTAGE =	POWER
1 Ω	2 A	2 V	4 W
2 Ω	2 A	4 V	8 W

RESISTANCE	CURRENT ×	VOLTAGE =	POWER
6 Ω	1 A	6 V	6 W
3 Ω	2 A	6 V	12 W

RESISTANCE	CURRENT ×	VOLTAGE =	POWER
2 Ω	1.5 A	3 V	4.5 W
2 Ω	1.5 A	3 V	4.5 W
1 Ω	3 A	3 V	9 W

NOTE THAT TOTAL POWER DISSIPATED BY ALL RESISTORS IN A CIRCUIT EQUALS THE POWER SUPPLIED BY THE BATTERY (VOLTAGE × CURRENT THRU BATTERY)

Name ____________________ Date __________

CONCEPTUAL **Physics** FUNDAMENTALS PRACTICE PAGE

Chapter 11 Magnetism and Electromagnetic Induction

Magnetic Fundamentals

Fill in the blanks:

1. Attraction or repulsion of charges depends on their *signs*, positives or negatives. Attraction or repulsion of magnets depends on their magnetic __POLES__ : __NORTH__ or __SOUTH__.
2. Opposite poles attract; like poles __REPEL__.
3. A magnetic field is produced by the __MOTION__ of electric charge.
4. Clusters of magnetically aligned atoms are magnetic __DOMAINS__.
5. A magnetic __FIELD__ surrounds a current-carrying wire.
6. When a current-carrying wire is made to form a coil around a piece of iron, the result is an __ELECTROMAGNET__.
7. A charged particle moving in a magnetic field experiences a deflecting __FORCE__ that is maximum when the charge moves __PERPENDICULAR__ to the field.
8. A current-carrying wire experiences a deflecting __FORCE__ that is maximum when the wire and magnetic field are __PERPENDICULAR__ to one another.

YOU HAVE A MAGNETIC PERSONALITY!

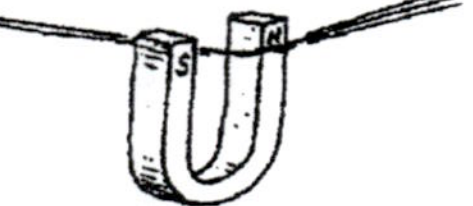

9. A simple instrument designed to detect electric current is the __GALVANOMETER__; when calibrated to measure current, it is an __AMMETER__; when calibrated to measure voltage, it is a __VOLTMETER__.
10. The largest size magnet in the world is the __WORLD (OR EARTH)__ itself.

THEN TO REALLY MAKE THINGS "SIMPLE," THERE'S THE RIGHT-HAND RULE!
F
B
I

CONCEPTUAL **Physics** FUNDAMENTALS PRACTICE PAGE

Chapter 11 Magnetism and Electromagnetic Induction

Magnetic Fundamentals—continued

11. The illustration below is similar to Figure 24.2 in your textbook. Iron filings trace out patterns of magnetic field lines about a bar magnet. In the field are some magnetic compasses. The compass needle in only one compass is shown. Draw in the needles with proper orientation in the other compasses.

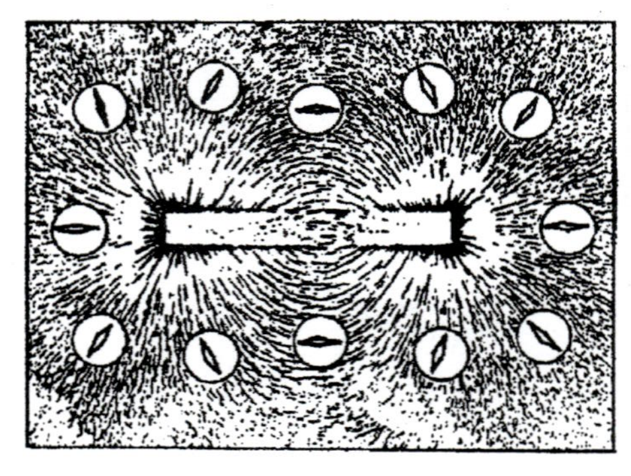

12. The illustration below is similar to Figure 24.10b in your textbook. Iron filings trace out magnetic field pattern about the loop of current-carrying wire. Draw in the compass needle orientations for all the compasses.

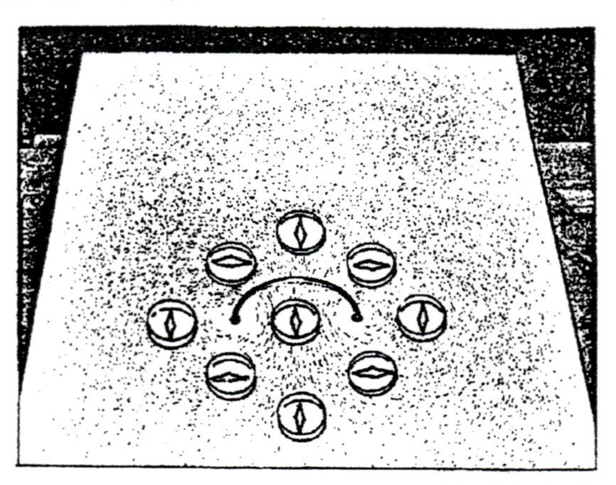

Name ______________________ Date __________

CONCEPTUAL Physics FUNDAMENTALS PRACTICE PAGE

Chapter 11 Magnetism and Electromagnetic Induction

Faraday's Law

Circle the correct answers:

1. Hans Christian Oersted discovered that magnetism and and electricity are

   [related] [independent of each other].

   Magnetism is produced by

   [batteries] **[motion of electric charges]**.

Faraday and Henry discovered that electric current can be produced by

[batteries] **[motion of a magnet]**.

More specifically, voltage is induced in a loop of wire if there is a change in

[batteries] **[magnetic field in the loop]**.

This phenomenon is called

[electromagnetism] **[electromagnetic induction]**.

2. When a magnet is plunged in and out of a coil of wire, voltage is induced in the coil. If the rate of the in-and-out motion of the magnet is doubled, the induced voltage

   [doubles] [halves] [remains the same].

   If instead the number of loops in the coil is doubled, the induced voltage

   [doubles] [halves] [remains the same].

3. A rapidly changing magnetic field in any region of space induces a rapidly changing

   [electric field] [magnetic field] [gravitational field]

   which in turn induces a rapidly changing

   [magnetic field] [electric field] [baseball field].

   This generation and regeneration of electric and magnetic fields make up

   [electromagnetic waves] [sound waves] [both of these].

PHYSICS SIGH

CONCEPTUAL Physics FUNDAMENTALS PRACTICE PAGE

Chapter 11 Magnetism and Electromagnetic Induction

Transformers

Consider a simple transformer that has a 100-turn primary coil and a 1000-turn secondary coil. The primary is connected to a 120-V AC source and the secondary is connected to an electrical device with a resistance of 1000 ohms.

1. What will be the voltage output of the secondary?

   __1200__ V.

2. What current flows in the secondary circuit?

   __1.2__ A.

3. Now that you know the voltage and the current, what is the power in the secondary coil?

   __1440__ W.

4. Neglecting small heating losses, and knowing that energy is conserved, what is the power in the primary coil?

   __1440__ W.

5. Now that you know the power and the voltage across the primary coil, what is the current drawn by the primary coil?

   __12__ A.

$I = V/R$ $P = IV$ 100 TURNS 1000 TURNS PHYSICS PHYSICS

Circle the answers:

6. The results show voltage is stepped **[up]** [down] from primary to secondary, and that current is correspondingly stepped [up] **[down]**.
7. For a step-up transformer, there are **[more]** [fewer] turns in the secondary coil than in the primary. For such a transformer, there is [more] **[less]** current in the secondary than in the primary.
8. A transformer can step up **[voltage]** [energy and power], but in no way can it step up [voltage] **[energy and power]**.
9. If 120 V is used to power a toy electric train that operates on 6 V, then a [step up] **[step down]** transformer should be used that has a primary to secondary turns ratio of [1/20] **[20/1]**.
10. A transformer operates on [dc] **[ac]** because the magnetic field within the iron core must **[continually change]** [remain steady].

Electricity and magnetism connect to become light!

Name ______________________ Date ________

CONCEPTUAL *Physics* FUNDAMENTALS PRACTICE PAGE

Chapter 12 Waves and Sound

Vibration and Wave Fundamentals

1. A sine curve that represents a transverse wave is drawn below. With a ruler, measure the wavelength and amplitude of the wave.

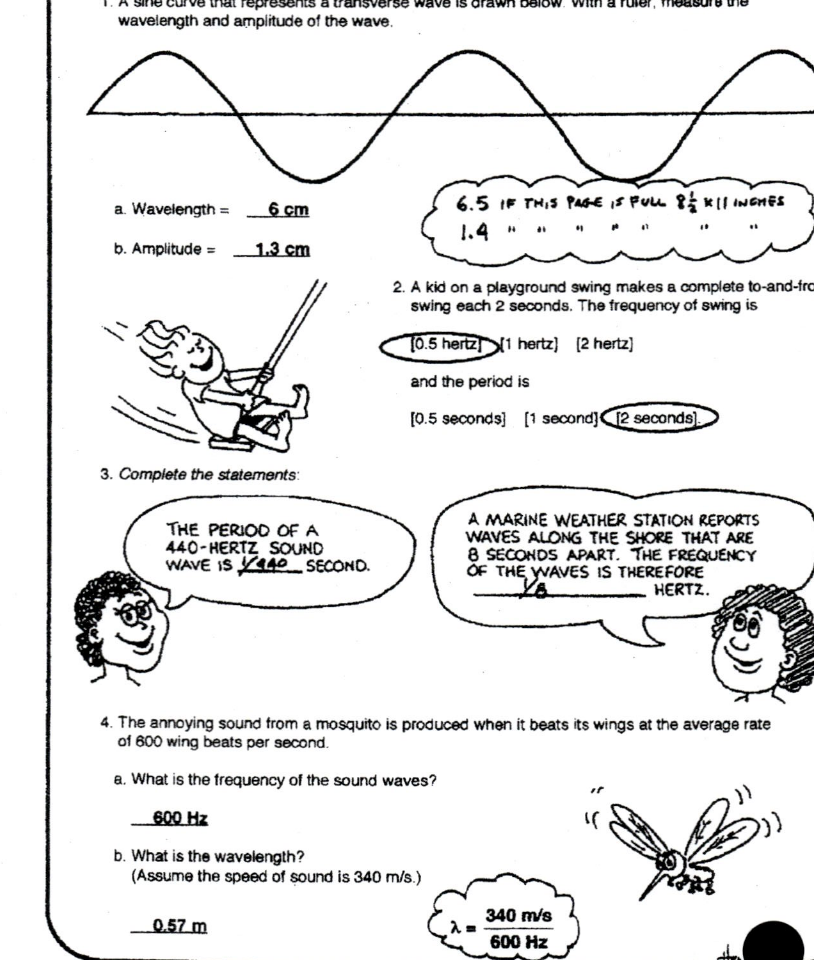

6.5 IF THIS PAGE IS FULL 8½ x 11 INCHES
1.4 " " " " " " "
THE PERIOD OF A 440-HERTZ SOUND WAVE IS 1/440 SECOND.
A MARINE WEATHER STATION REPORTS WAVES ALONG THE SHORE THAT ARE 8 SECONDS APART. THE FREQUENCY OF THE WAVES IS THEREFORE 1/8 HERTZ.

a. Wavelength = 6 cm

b. Amplitude = 1.3 cm

2. A kid on a playground swing makes a complete to-and-fro swing each 2 seconds. The frequency of swing is

[0.5 hertz] [1 hertz] [2 hertz]

and the period is

[0.5 seconds] [1 second] [2 seconds].

3. *Complete the statements:*

4. The annoying sound from a mosquito is produced when it beats its wings at the average rate of 600 wing beats per second.

a. What is the frequency of the sound waves?

600 Hz

b. What is the wavelength?
(Assume the speed of sound is 340 m/s.)

0.57 m

$$\lambda = \frac{340 \text{ m/s}}{600 \text{ Hz}}$$

CONCEPTUAL Physics FUNDAMENTALS PRACTICE PAGE

Chapter 12 Waves and Sound

Vibration and Wave Fundamentals—continued

5. A machine gun fires 10 rounds per second. The speed of the bullets is 300 m/s.

   a. What is the distance in the air between the flying bullets? __30 m__

   b. What happens to the distance between the bullets if the rate of fire is increased?

   __DISTANCE BETWEEN BULLETS DECREASES__

6. Consider a wave generator that produces 10 pulses per second. The speed of the waves is 300 cm/s.

   a. What is the wavelength of the waves? __30 cm__

   b. What happens to the wavelength if the frequency of pulses is increased?

   __λ DECREASES, JUST AS DISTANCE BETWEEN BULLETS IN #5 DECREASES.__

7. The bird at the right watches the waves. If the portion of a wave between 2 crests passes the pole each second,

   a. what is the speed of the waves? __$v = f\lambda = 2 \times 1\text{ m} = 2\text{ m/s}$__

   b. what is the period of wave motion? __$T = \frac{1}{f} = \frac{1}{2} = 0.5\text{ s}$__

   1 m

   c. If the distance between crests were 1.5 meters apart, and 2 crests pass the pole each second, what would be the speed of the wave?

   __$v = f\lambda = 2 \times 1.5 = 3\text{ m/s}$__

   d. What would the period of wave motion be for 7.c ?

   __SAME (0.5 s)__

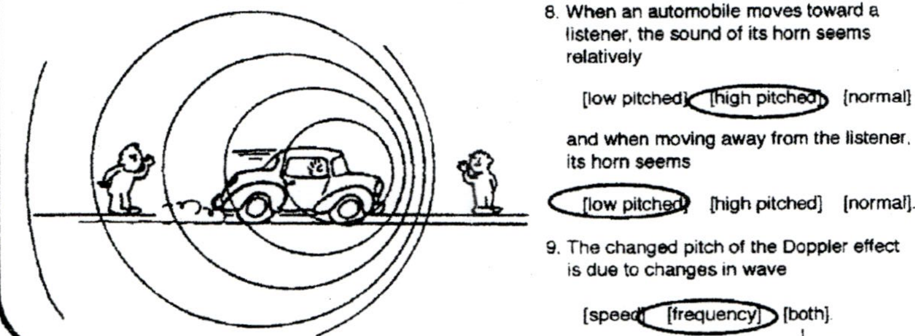

8. When an automobile moves toward a listener, the sound of its horn seems relatively

   [low pitched] **([high pitched])** [normal]

   and when moving away from the listener, its horn seems

   ([low pitched]) [high pitched] [normal].

9. The changed pitch of the Doppler effect is due to changes in wave

   [speed] **([frequency])** [both].

Name ______________________ Date ________

CONCEPTUAL Physics FUNDAMENTALS PRACTICE PAGE

Chapter 12 Waves and Sound

Shock Waves

The cone-shaped shock wave produced by a supersonic aircraft is actually the result of overlapping spherical waves of sound, as indicated by the overlapping circles in Figure 19.19 in your textbook. Sketches *a* through *e* below show the "animated" growth of only one of the many spherical sound waves (shown as an expanding circle in the two-dimensional drawing).

a

The circle originates when the aircraft is in the position shown in *a*.

b

Sketch *b* shows both the growth of the circle and position of the aircraft at a later time.

c

Still later times are shown in *c*, *d*, and *e*. Note that the circle grows and the aircraft moves farther to the right. Note also that the aircraft is moving farther than the sound wave. This is because the aircraft is moving faster than sound.

Careful examination will reveal how fast the aircraft is moving compared to the speed of sound. Sketch *e* shows that in the same time the sound travels from O to A, the aircraft has traveled from O to B—twice as far. You can check with a ruler.

d

Circle the answer.

1. Inspect sketches *b* and *d*. Has the aircraft traveled twice as far as sound in the same time in these positions also?

   ([Yes]) [No]

2. For greater speeds, the angle of the shock wave would be

   [wider] [the same] **([narrower])**.

e

O A B

DURING THE TIME THAT SOUND TRAVELS FROM O TO A, THE PLANE TRAVELS TWICE AS FAR --- FROM O TO B.

SO IT'S FLYING AT TWICE THE SPEED OF SOUND!

CONCEPTUAL Physics FUNDAMENTALS PRACTICE PAGE

Chapter 12 Waves and Sound

Shock Waves — continued

3. Use a ruler to estimate the speeds of the aircraft that produce the shock waves in the two sketches below.

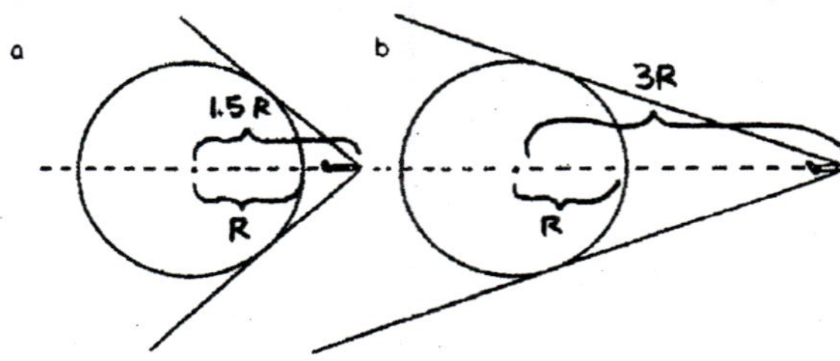

Aircraft *a* is traveling about __1.5__ times the speed of sound.

Aircraft *b* is traveling about __3.0__ times the speed of sound.

4. Draw your own circle (anywhere) and estimate the speed of the aircraft to produce the shock wave shown below:

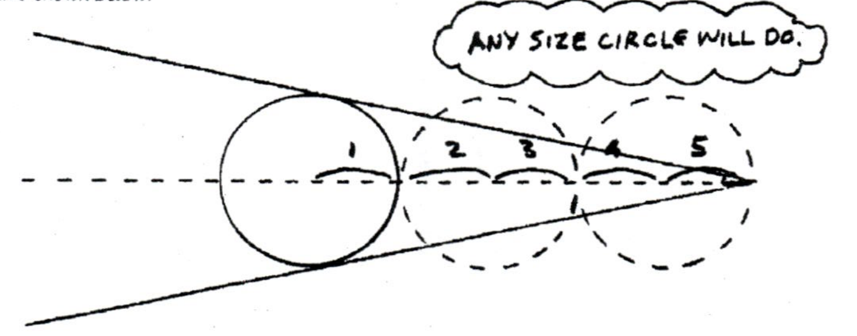

The speed is about __5__ times the speed of sound.

5. In the space below, draw the shock wave made by a supersonic missile that travels at four times the speed of sound.

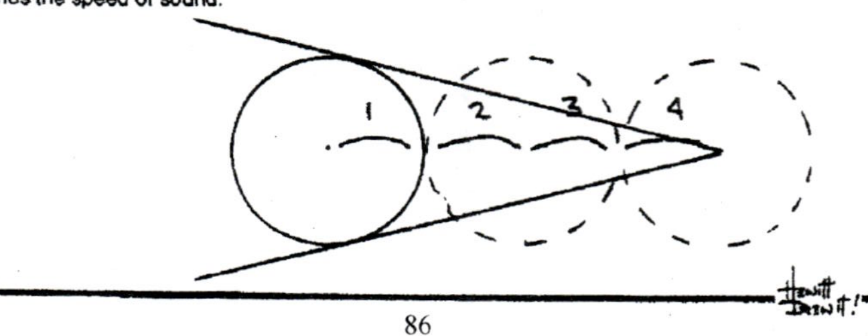

Name ______________________ Date __________

CONCEPTUAL Physics FUNDAMENTALS PRACTICE PAGE

Chapter 12 Waves and Sound

Wave Superposition

A pair of pulses travel toward each at equal speeds. The composite waveforms, as they pass through each other and interfere, are shown at 1-second intervals. In the left column note how the pulses interfere to produce the composite waveform (solid line). Make a similar construction for the two wave pulses in the right column. Like the pulses in the first column, they each travel at 1 space per second.

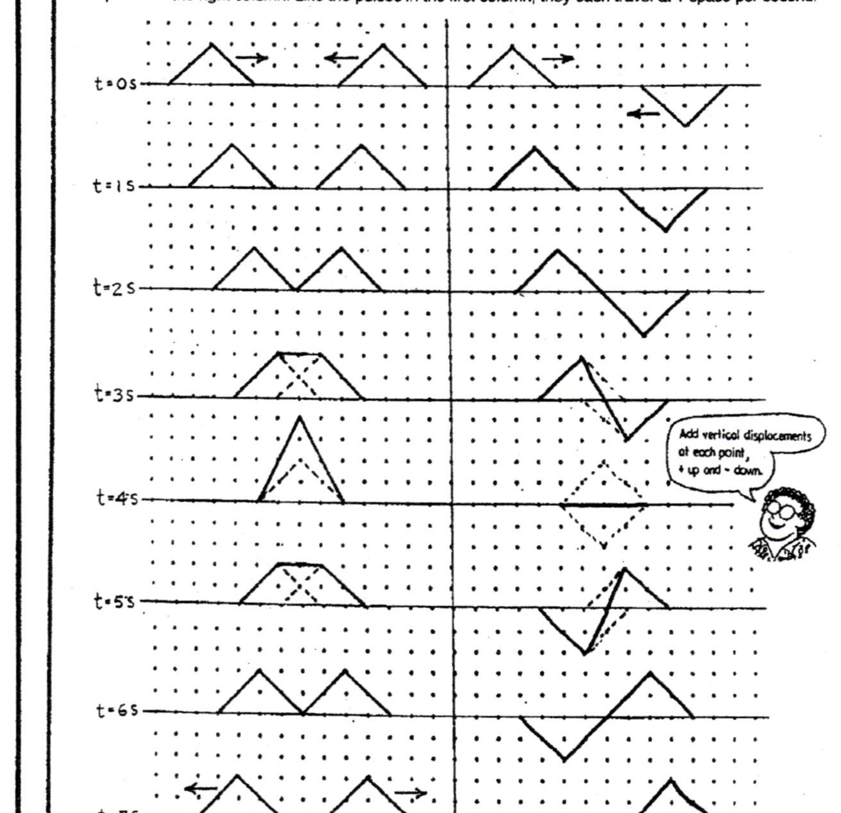

Chapter 12 Waves and Sound
***Wave Superposition** — continued*

Construct the composite waveforms at 1-second intervals for the two waves traveling toward each other at equal speed.

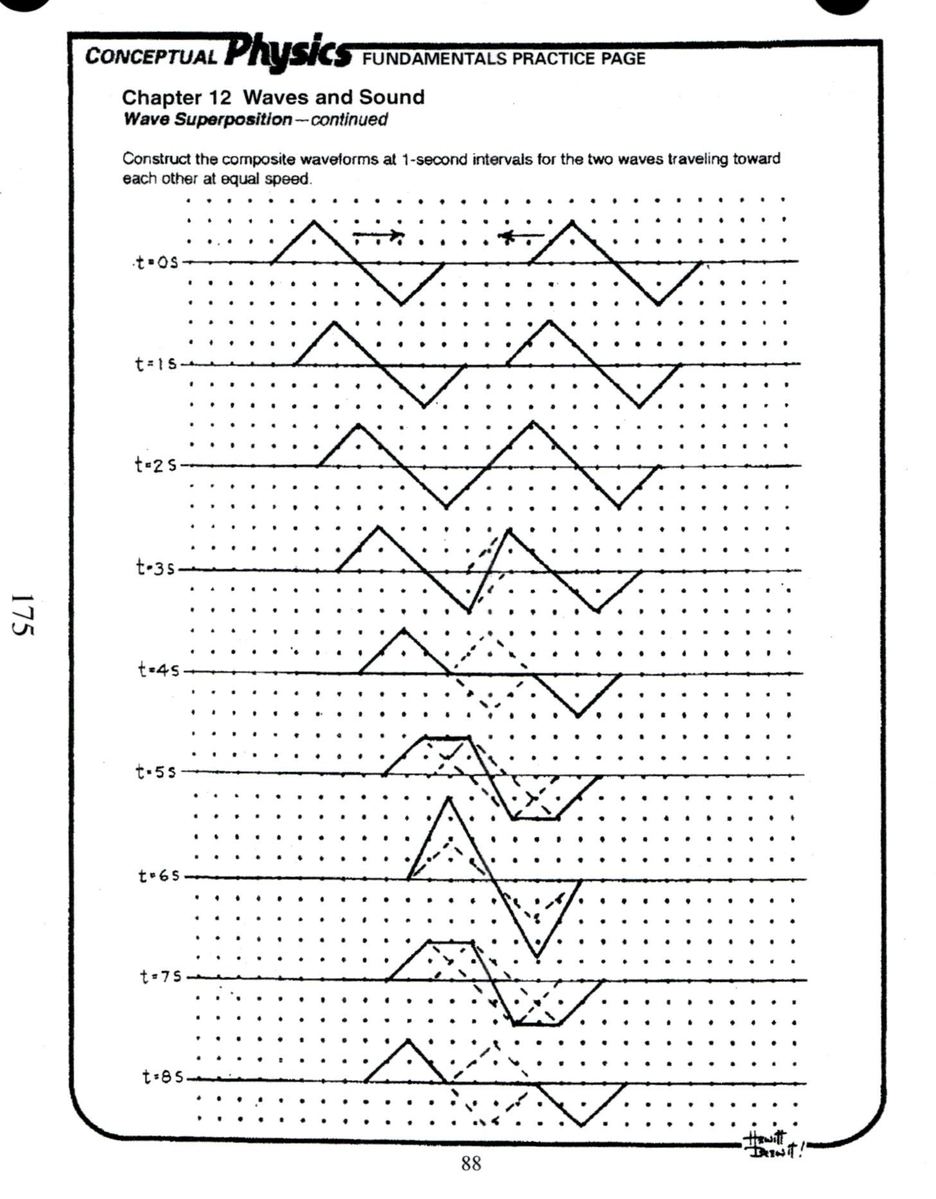

Name ______________________ Date ________

CONCEPTUAL Physics FUNDAMENTALS PRACTICE PAGE

Chapter 13 Light Waves

Speed, Wavelength, and Frequency

1. The first investigation that led to a determination of the speed of light was performed in about 1675 by the Danish astronomer Olaus Roemer. He made careful measurements of the period of Io, a moon about the planet Jupiter, and was surprised to find an irregularity in Io's observed period. While Earth was moving away from Jupiter, the measured periods were slightly longer than average. While Earth approached Jupiter, they were shorter than average. Roemer estimated that the cumulative discrepancy amounted to about 16.5 minutes. Later interpretations showed that what occurs is that light takes about 16.5 minutes to travel the extra 300,000,000-km distance across Earth's orbit. Aha! We have enough information to calculate the speed of light!

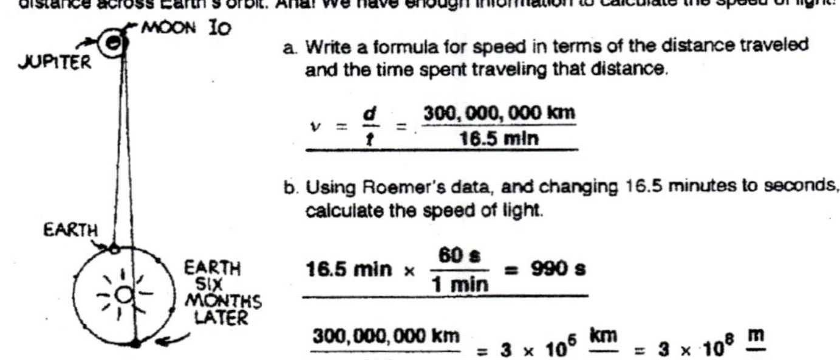

a. Write a formula for speed in terms of the distance traveled and the time spent traveling that distance.

$v = \frac{d}{t} = \frac{300,000,000 \text{ km}}{16.5 \text{ min}}$

b. Using Roemer's data, and changing 16.5 minutes to seconds, calculate the speed of light.

$16.5 \text{ min} \times \frac{60 \text{ s}}{1 \text{ min}} = 990 \text{ s}$

$\frac{300,000,000 \text{ km}}{990 \text{ s}} = 3 \times 10^5 \frac{\text{km}}{\text{s}} = 3 \times 10^8 \frac{\text{m}}{\text{s}}$

Study Figure 26.3 in your textbook and answer the following:

2. a. Which has the longer *wavelengths*? [radio waves] [light waves].

b. Which has the longer *wavelengths*? [light waves] [gamma waves].

c. Which has the higher *frequencies*? [ultraviolet waves] [infrared waves].

d. Which has the higher *frequencies*? [ultraviolet waves] [gamma rays].

LIGHT IS THE ONLY THING WE SEE!

Carefully study the section "Transparent Materials" in your textbook and answer the following:

3. a. Exactly what do vibrating electrons emit?

ELECTROMAGNETIC WAVES

b. When ultraviolet light shines on glass, what does it do to electrons in the glass structure?

UV CAUSES ELECTRONS TO VIBRATE IN RESONANCE WITH THE INCIDENT UV

c. When energetic electrons in the glass structure vibrate against neighboring atoms, what happens to the energy of vibration?

BECOMES THERMAL ENERGY (HEAT)

PHYSICS YEA!

d. What happens to the energy of a vibrating electron that does not collide with neighboring atoms?

EMITTED AS LIGHT

CONCEPTUAL Physics FUNDAMENTALS PRACTICE PAGE

Chapter 13 Light Waves

Speed, Wavelength, and Frequency—continued

e. Light in which range of frequencies is absorbed in glass? [visible] [ultraviolet].

f. Light in which range of frequencies is transmitted through glass? [visible] [ultraviolet].

g. How is the speed of light in glass affected by the succession of time delays that accompany the absorption and re-emission of light from atom to atom in the glass?

THE AVERAGE SPEED OF LIGHT IS LESS IN GLASS THAN IN AIR

h. How does the speed of light compare in water, glass, and diamond?

SPEED OF LIGHT IS 0.75c IN WATER; 0.41c IN A DIAMOND

4. The Sun normally shines on both Earth and Moon. Both cast shadows. Sometimes the Moon's shadow falls on Earth, and at other times Earth's shadow falls on the Moon.

a. The sketch shows the Sun and Earth. Draw the Moon at a position for a solar eclipse.

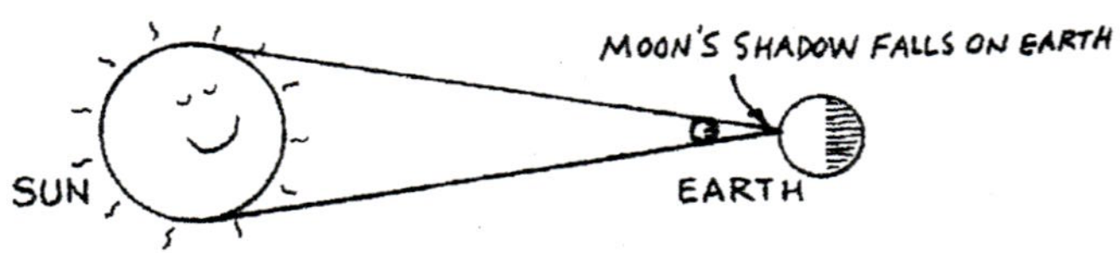

b. This sketch also shows the Sun and Earth. Draw the Moon at a position for a lunar eclipse.

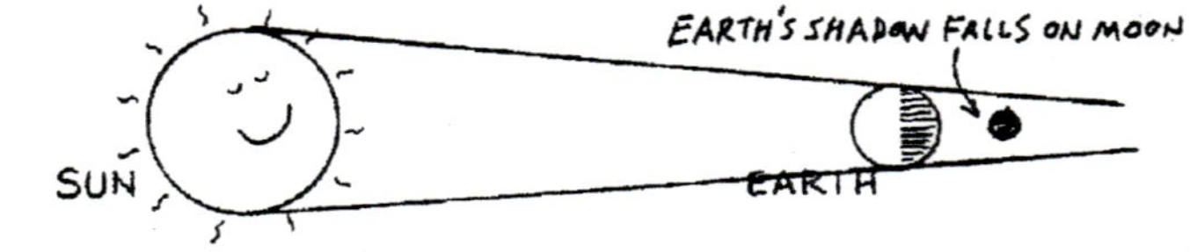

5. The diagram shows the limits of light rays when a large lamp makes a shadow of a small object on a screen. Make a sketch of the shadow on the screen, shading the umbra darker than the penumbra. In what part of the shadow could an ant on the screen see part of the lamp?

PENUMBRA

LAMP APPLE DRAW COMPLETE SHADOW OF APPLE ON SCREEN

Name ______________________ Date __________

CONCEPTUAL Physics FUNDAMENTALS PRACTICE PAGE

Chapter 13 Light Waves

Color Addition

The sketch to the right shows the shadow of an instructor in front of a white screen in a dark room. The light source is red, so the screen looks red and the shadow looks black. Color the sketch, or label the colors with pen or pencil.

RED
BLACK
RED

A green lamp is added and makes a second shadow. The shadow cast by the red lamp is no longer black, but is illuminated by green light. So it is green. Color or mark it green. The shadow cast by the green lamp is not black because it is illuminated by the red lamp. Indicate its color. Do the same for the background, which receives a mixture of red and green light.

RED
YELLOW
GREEN
RED
GREEN

A blue lamp is added and three shadows appear. Indicate the appropriate colors of the shadows and the background.

WHITE
MAGENTA
YELLOW
CYAN
RED
GREEN
BLUE

The lamps are placed closer together so the shadows overlap. Indicate the colors of all screen areas.

MAGENTA
YELLOW
CYAN
WHITE
RED
GREEN
RED
GREEN
BLUE

Hewitt Drew it!

CONCEPTUAL Physics FUNDAMENTALS PRACTICE PAGE

Chapter 13 Light Waves

Color Addition—continued

If you have colored markers or pencils, have a try at these.

COLOR ME RED
COLOR ME MAGENTA
COLOR ME YELLOW
PLEASE DON'T COLOR ME!
COLOR ME BLUE
COLOR ME GREENISH BLUE
COLOR ME GREEN

Hewitt Drew it!

Name ______________________ Date __________

CONCEPTUAL **Physics** FUNDAMENTALS PRACTICE PAGE

Chapter 13 Light Waves

Diffraction and Interference

1. Shown are concentric solid and dashed circles, each different in radius by 1 cm. Consider the circular pattern a top view of water waves, where the solid circles are crests and the dashed circles are troughs.

   a. Draw another set of the same concentric circles with a compass. Choose any part of the paper for your center (except the present central point). Let the circles run off the edge of the paper.

   b. Find where a dashed line crosses a solid line and draw a large dot at the intersection. Do this for ALL places where a solid and dashed line intersect.

   c. With a wide felt marker, connect the dots with the solid lines. These *nodal lines* lie in regions where the waves have cancelled—where the crest of one wave overlaps the trough of another (see Figures 29.15 and 29.16 in your textbook).

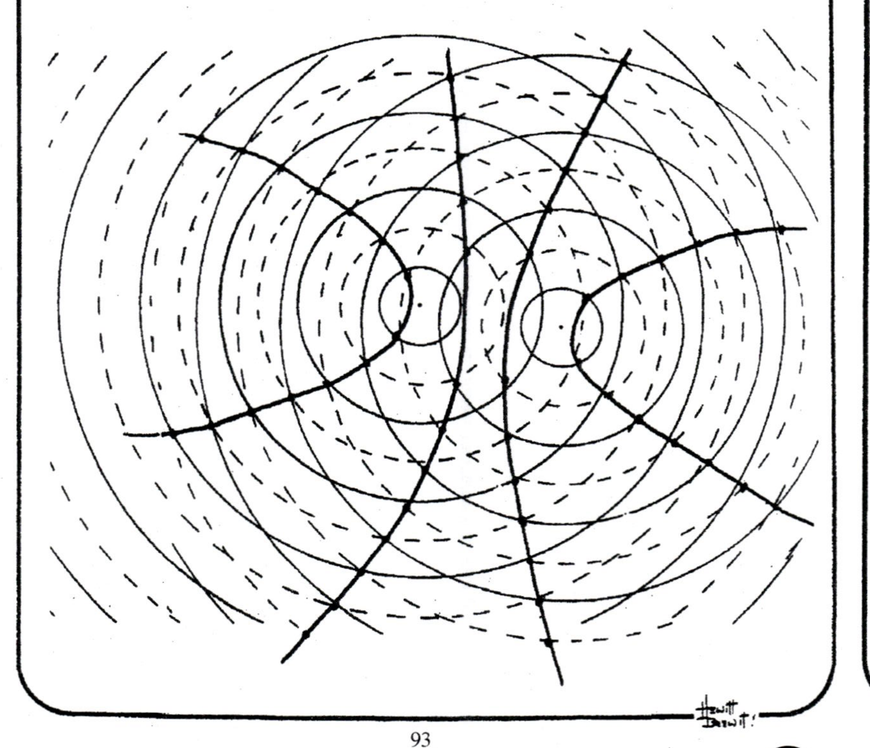

CONCEPTUAL **Physics** FUNDAMENTALS PRACTICE PAGE

Chapter 13 Light Waves

Diffraction and Interference—continued

2. Look at the construction of overlapping circles on your classmates' papers. Some will have more nodal lines than others, due to different starting points. How does the number of nodal lines in a pattern relate to the distance between centers of circles, (or sources of waves)?

   THE FARTHER APART THE CENTERS (OR WAVE SOURCES) THE MORE NODAL LINES. (NOTE IN FIGURE 29.15 IN YOUR TEXT MORE NODAL LINES IN THE RIGHT PATTERN COMPARED WITH CLOSER SOURCES IN THE CENTRAL PATTERN.)

3. Figure 29.19 from your textbook is repeated below. Carefully count the number of wavelengths (same as the number of wave crests) along the following paths between the slits and the screen.

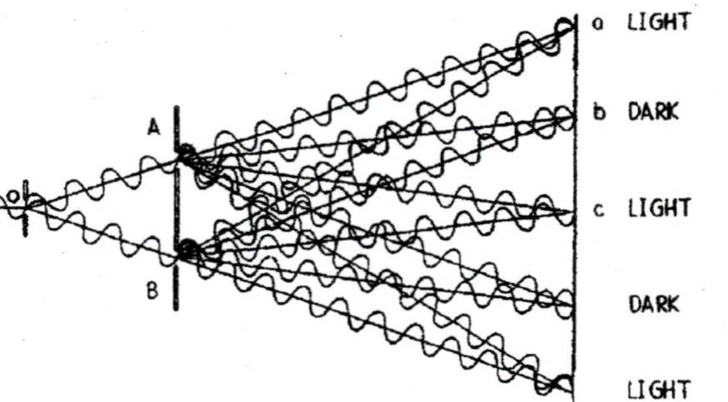

   a. Number of wavelengths between slit A and point a is __10.5__

   b. Number of wavelengths between slit B and point a is __11.5__

   c. Number of wavelengths between slit A and point b is __10.0__

   d. Number of wavelengths between slit B and point b is __10.5__

   e. Number of wavelengths between slit A and point c is __10.0__

   f. Number of wave crests between slit B and point c is __10.0__

4. When the number of wavelengths along each path is the same or differs by one or more whole wavelengths, interference is

   [constructive] (circled) [destructive]

   and when the number of wavelengths differ by a half-wavelength (or odd multiples of a half-wavelength), interference is

   [constructive] [destructive] (circled)

"It's nice how knowing some physics really changes the way we see things!"

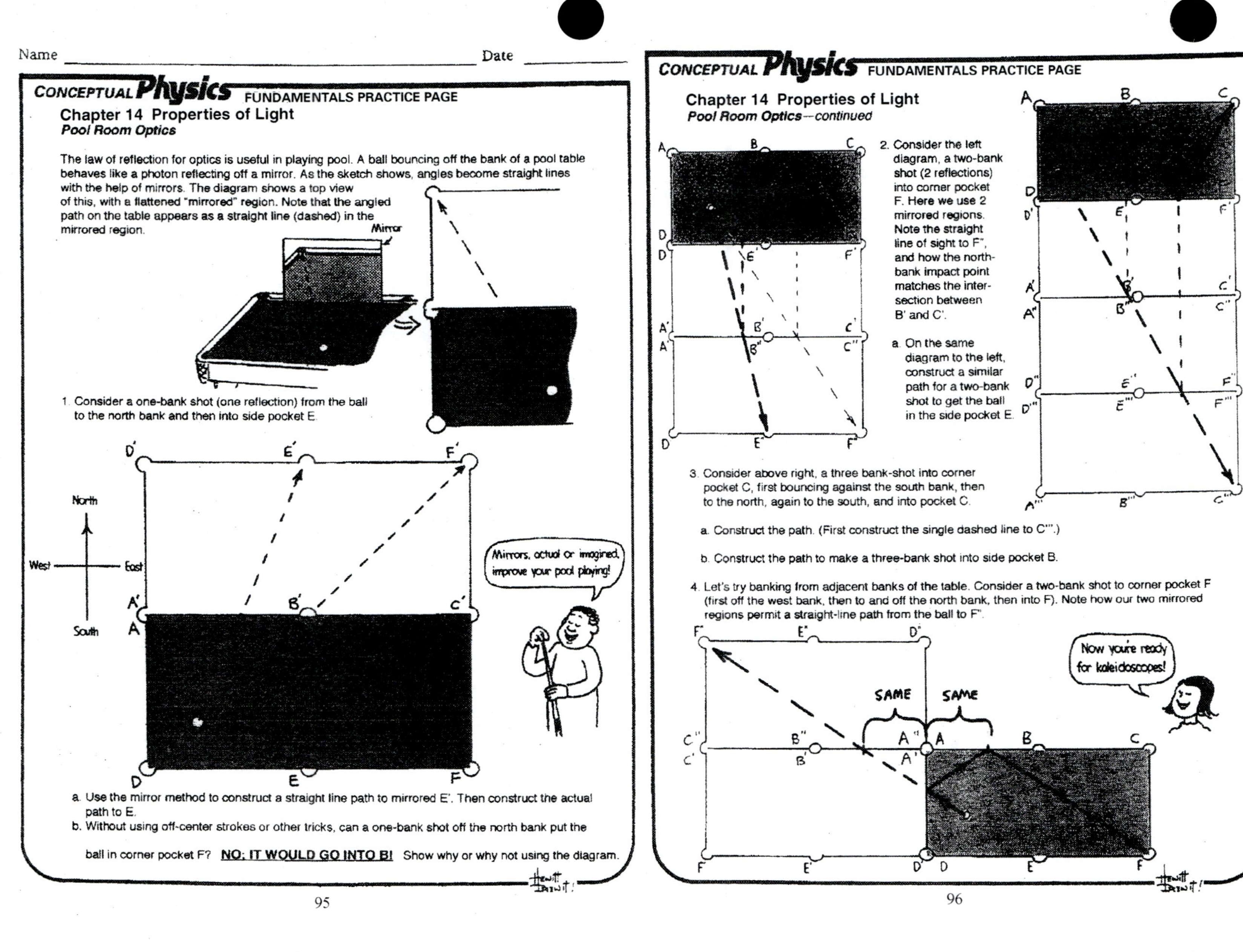

Name ______________________ Date ________

CONCEPTUAL **Physics** FUNDAMENTALS PRACTICE PAGE

Chapter 14 Properties of Light
Pool Room Optics

The law of reflection for optics is useful in playing pool. A ball bouncing off the bank of a pool table behaves like a photon reflecting off a mirror. As the sketch shows, angles become straight lines with the help of mirrors. The diagram shows a top view of this, with a flattened "mirrored" region. Note that the angled path on the table appears as a straight line (dashed) in the mirrored region.

1. Consider a one-bank shot (one reflection) from the ball to the north bank and then into side pocket E.

a. Use the mirror method to construct a straight line path to mirrored E'. Then construct the actual path to E.

b. Without using off-center strokes or other tricks, can a one-bank shot off the north bank put the ball in corner pocket F? **NO; IT WOULD GO INTO B!** Show why or why not using the diagram.

CONCEPTUAL **Physics** FUNDAMENTALS PRACTICE PAGE

Chapter 14 Properties of Light
Pool Room Optics*—continued*

2. Consider the left diagram, a two-bank shot (2 reflections) into corner pocket F. Here we use 2 mirrored regions. Note the straight line of sight to F'', and how the north-bank impact point matches the intersection between B' and C'.

a. On the same diagram to the left, construct a similar path for a two-bank shot to get the ball in the side pocket E.

3. Consider above right, a three bank-shot into corner pocket C, first bouncing against the south bank, then to the north, again to the south, and into pocket C.

a. Construct the path. (First construct the single dashed line to C'''.)

b. Construct the path to make a three-bank shot into side pocket B.

4. Let's try banking from adjacent banks of the table. Consider a two-bank shot to corner pocket F (first off the west bank, then to and off the north bank, then into F). Note how our two mirrored regions permit a straight-line path from the ball to F''.

Name ____________________ Date ________

CONCEPTUAL **Physics** FUNDAMENTALS PRACTICE PAGE

Chapter 14 Properties of Light

Reflection

Abe and Bev both look in a plane mirror directly in front of Abe (left view). Abe can see himself while Bev cannot see herself—but can Abe see Bev, and can Bev see Abe?

MIRROR

ABE BEV ABE BEV

C

To find the answer, we construct their artificial locations "through" the mirror, the same distance behind as Abe and Bev are in front (right view). If straight-line connections intersect the mirror, as at point C, then each sees the other. The mouse, for example, cannot see or be seen by Abe and Bev (because there's no mirror in its line of sight).

Here we have eight students in front of a small plane mirror. Their positions are shown in the diagram below. Make appropriate straight-line constructions to answer the following:

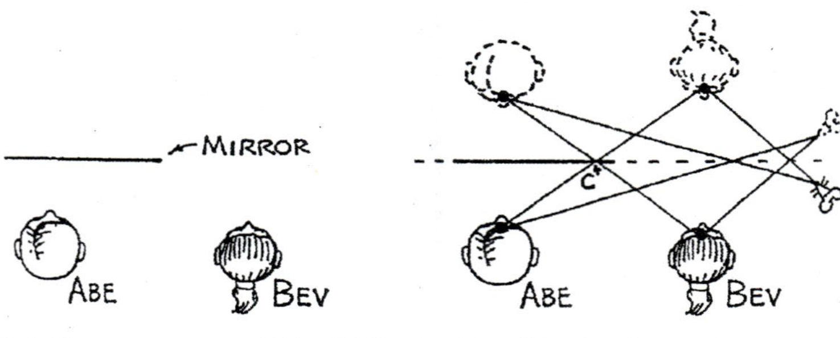

Abe can see	EVA, FLO, GUY, HAN	Abe cannot see	ABE –> DON
Bev can see	DON –> HAN	Bev cannot see	ABE –> CIS
Cis can see	CIS –> HAN	Cis cannot see	ABE, BEV
Don can see	BEV –> GUY	Don cannot see	ABE, HAN
Eva can see	ABE –> FLO	Eva cannot see	GUY, HAN
Flo can see	ABE –> EVA	Flo cannot see	FLO, GUY, HAN
Guy can see	ABE –> DON	Guy cannot see	EVA –> HAN
Han can see	ABE, BEV, CIS	Han cannot see	DON –> HAN

thanx to Marshall Ellenstein

Hewitt Drew it!

CONCEPTUAL **Physics** FUNDAMENTALS PRACTICE PAGE

Chapter 14 Properties of Light

Reflection—continued

Six of our group are now arranged differently in front of the same plane mirror. Their positions are shown below. Make appropriate constructions for this interesting arrangement, and answer the questions provided below:

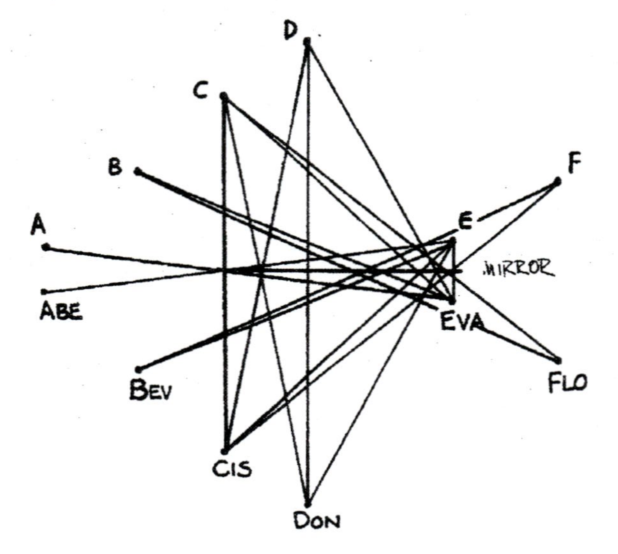

Who can Abe see?	EVA	Who can Abe not see?	EVERYONE ELSE
Who can Bev see?	EVA, FLO	Who can Bev not see?	EVERYONE ELSE
Who can Cis see?	CIS, DON, EVA, FLO	Who can Cis not see?	ABE, BEV
Who can Don see?	CIS, DON, EVA	Who can Don not see?	ABE, BEV, FLO
Who can Eva see?	ABE,BEV,CIS,DON,EVA	Who can Eva not see?	FLO
Who can Flo see?	BEV, CIS	Who can Flo not see?	ABE, DON, EVA, FLO

Harry Hostshot views himself in a full-length mirror (right). Construct straight lines from Harry's eyes to the image of his feet, and to the top of his head. Mark the mirror to indicate the minimum area Harry uses to see a full view of himself.

Does this region of the mirror depend on Harry's distance from the mirror? NO

Hewitt Drew it!

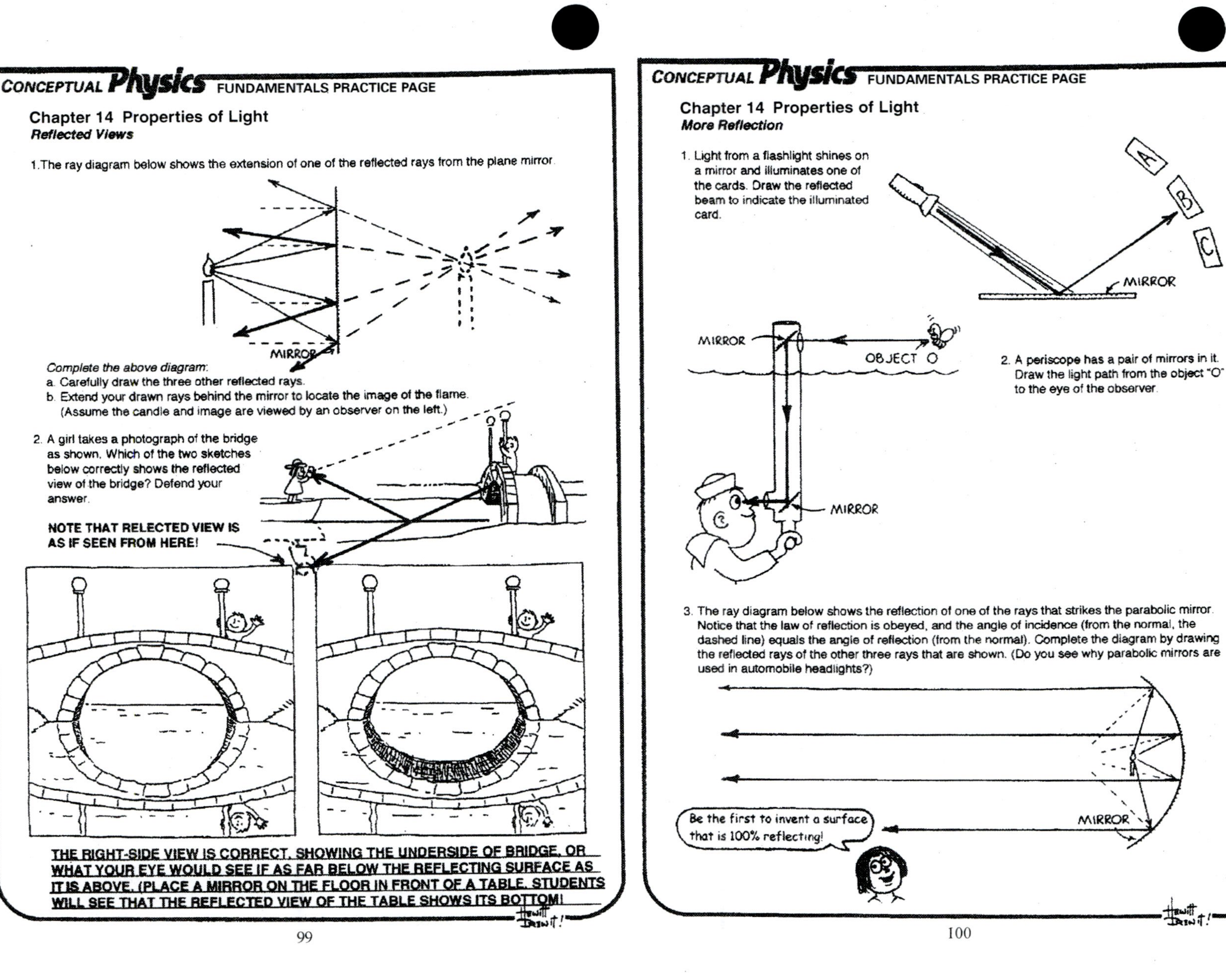

CONCEPTUAL **Physics** FUNDAMENTALS PRACTICE PAGE

Chapter 14 Properties of Light
Reflected Views

1. The ray diagram below shows the extension of one of the reflected rays from the plane mirror.

Complete the above diagram:
a. Carefully draw the three other reflected rays.
b. Extend your drawn rays behind the mirror to locate the image of the flame. (Assume the candle and image are viewed by an observer on the left.)

2. A girl takes a photograph of the bridge as shown. Which of the two sketches below correctly shows the reflected view of the bridge? Defend your answer.

NOTE THAT RELECTED VIEW IS AS IF SEEN FROM HERE!

THE RIGHT-SIDE VIEW IS CORRECT, SHOWING THE UNDERSIDE OF BRIDGE, OR WHAT YOUR EYE WOULD SEE IF AS FAR BELOW THE REFLECTING SURFACE AS IT IS ABOVE. (PLACE A MIRROR ON THE FLOOR IN FRONT OF A TABLE. STUDENTS WILL SEE THAT THE REFLECTED VIEW OF THE TABLE SHOWS ITS BOTTOM!

CONCEPTUAL **Physics** FUNDAMENTALS PRACTICE PAGE

Chapter 14 Properties of Light
More Reflection

1. Light from a flashlight shines on a mirror and illuminates one of the cards. Draw the reflected beam to indicate the illuminated card.

2. A periscope has a pair of mirrors in it. Draw the light path from the object "O" to the eye of the observer.

3. The ray diagram below shows the reflection of one of the rays that strikes the parabolic mirror. Notice that the law of reflection is obeyed, and the angle of incidence (from the normal, the dashed line) equals the angle of reflection (from the normal). Complete the diagram by drawing the reflected rays of the other three rays that are shown. (Do you see why parabolic mirrors are used in automobile headlights?)

Name ______________________ Date __________

CONCEPTUAL Physics FUNDAMENTALS PRACTICE PAGE

Chapter 14 Properties of Light

Refraction

1. A pair of toy cart wheels are rolled obliquely from a smooth surface onto two plots of grass—a rectangular plot on the left, and a triangular plot on the right. The ground is on a slight incline so that after slowing down in the grass, the wheels speed up again when emerging on the smooth surface. Finish each sketch and show some positions of the wheels inside the plots and on the other side. Clearly indicate their paths and directions of travel.

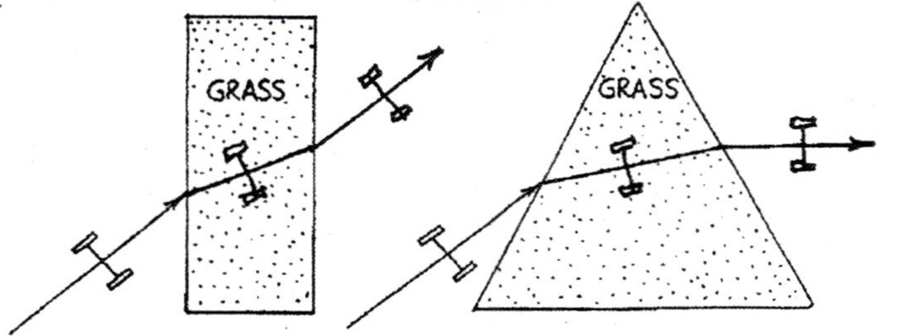

2. Red, green, and blue rays of light are incident upon a glass prism as shown below. The average speed of red light in the glass is less than in air, so the red ray is refracted. When it emerges into the air it regains its original speed and travels in the direction shown. Green light takes longer to get through the glass. Because of its slower speed it is refracted as shown. Blue light travels even slower in glass. Complete the diagram by estimating the path of the blue ray.

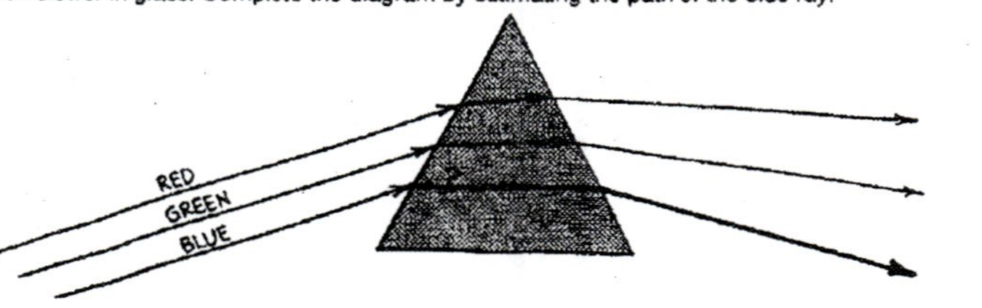

3. Below we consider a prism-shaped hole in a piece of glass—that is, an "air prism." Complete the diagram, showing likely paths of the beams of red, green, and blue light as they pass through this "prism" and then into glass.

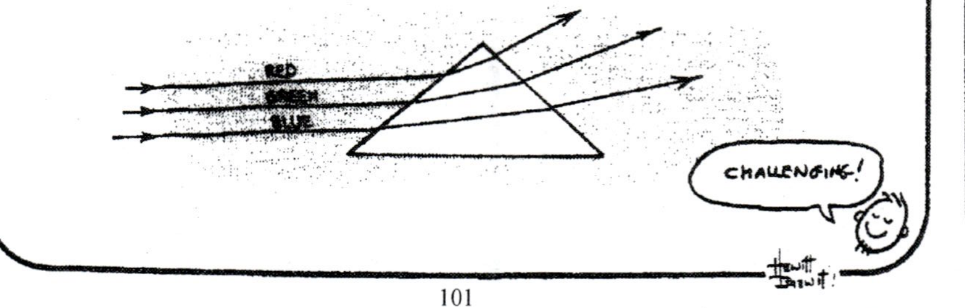

CONCEPTUAL Physics FUNDAMENTALS PRACTICE PAGE

Chapter 14 Properties of Light

Refraction—continued

4. Light of different colors diverges when emerging from a prism. Newton showed that with a second prism he could make the diverging beams become parallel again. Which placement of the second prism will do this?

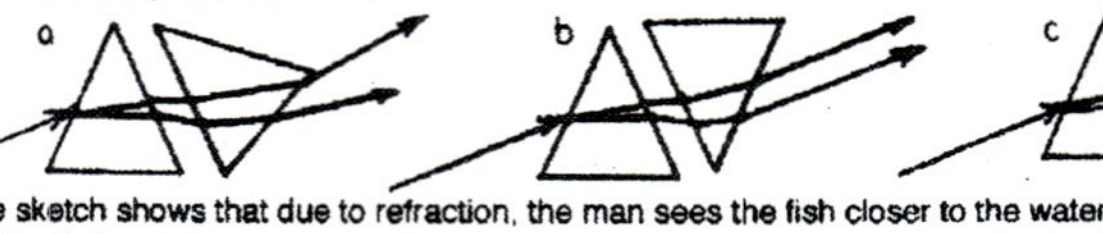

5. The sketch shows that due to refraction, the man sees the fish closer to the water surface than it actually is.

OBSERVED FISH

50° 50°

ACTUAL FISH

a. Draw a ray beginning at the fish's eye to show the line of sight of the fish when it looks upward at 50° to the normal at the water surface. Draw the direction of the ray after it meets the surface of water.

b. At the 50° angle, does the fish see the man, or does it see the reflected view of the starfish at the bottom of the pond? Explain.
FISH SEES REFLECTED VIEW OF STARFISH (50° > 48° CRITICAL ANGLE, SO THERE IS TOTAL INTERNAL REFLECTION).

c. To see the man, should the fish look higher or lower than the 50° path? **HIGHER, SO LINE OF SIGHT TO THE WATER IS LESS THAN 48° WITH NORMAL**

d. If the fish's eye were barely above the water surface, it would see the world above in a 180° view, horizon to horizon. The fisheye view of the world above as seen beneath the water, however, is very different. Due to the 48° critical angle of water, the fish sees a normally 180° horizon-to-horizon view compressed within an angle of **96°**.

Name ______________________ Date ________

CONCEPTUAL **Physics** FUNDAMENTALS PRACTICE PAGE

Chapter 14 Properties of Light
More Refraction

1. The sketch to the right shows a light ray moving from air into water, at 45° to the normal. Which of the three rays indicated with capital letters is most likely the light ray that continues inside the water?

__C__

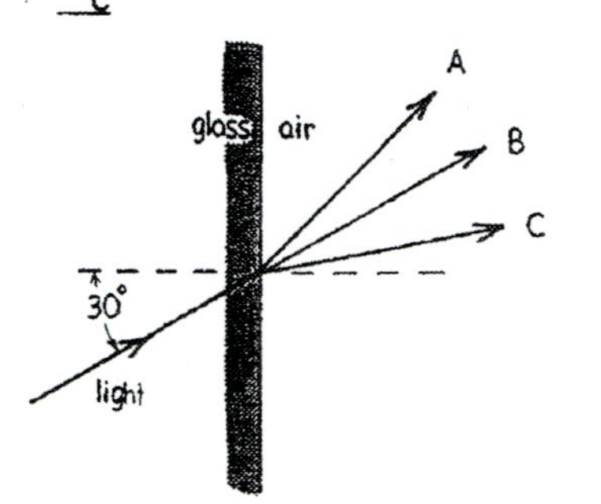

2. The sketch on the left shows a light ray moving from glass into air, at 30° to the normal. Which of the three is most likely the light ray that continues in the air?

__A__

3. To the right, a light ray is shown moving from air into a glass block, at 40° to the normal. Which of the three rays is most likely the light ray that travels in the air after emerging from the opposite side of the block? (Sketch the path the light would take inside the glass.)

__A__

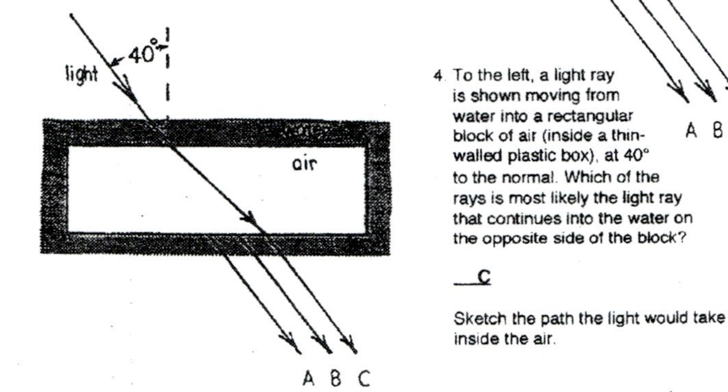

4. To the left, a light ray is shown moving from water into a rectangular block of air (inside a thin-walled plastic box), at 40° to the normal. Which of the rays is most likely the light ray that continues into the water on the opposite side of the block?

__C__

Sketch the path the light would take inside the air.

thanx to Clarence Bakken

CONCEPTUAL **Physics** FUNDAMENTALS PRACTICE PAGE

Chapter 14 Properties of Light
More Refraction*—continued*

5. The two transparent blocks (right) are made of different materials. The speed of light in the left block is greater than the speed of light in the right block. Draw an appropriate light path through and beyond the right block. Is the light that emerges displaced more or less than light emerging from the left block?

__MORE__

6. Light from the air passes through plates of glass and plastic below. The speeds of light in the different materials are shown to the right (these different speeds are often implied by the "index of refraction" of the material). Construct a rough sketch showing an appropriate path through the system of four plates.

Compared to the 50° incident ray at the top, what can you say about the angles of the ray in the air between and below the block pairs?

__SAME 50°__

7. Parallel rays of light are refracted as they change speed in passing from air into the eye (left below). Construct a rough sketch showing appropriate light paths when parallel light under water meets the same eye (right below).

If a fish out of water wishes to clearly view objects in air, should it wear goggles filled with water or with air

WATER!

8. Why do we need to wear a face mask or goggles to see clearly when under water?
__SO THAT LIGHT GOES FROM AIR TO EYE FOR PROPER REFRACTION__

Name ______________________ Date __________

CONCEPTUAL Physics FUNDAMENTALS PRACTICE PAGE

Chapter 14 Properties of Light
Lenses

Rays of light bend as shown when passing through the glass blocks.

1. Show how light rays bend when they pass through the arrangement of glass blocks below.

2. Show how light rays bend when they pass through the lens below. Is the lens a converging or a diverging lens? What is your evidence?

   CONVERGING, AS EVIDENT IN THE CONVERGING RAYS

3. Show how light rays bend when they pass through the arrangement of glass blocks below.

4. Show how light rays bend when they pass through the lens shown below. Is the lens a converging or diverging lens? What is your evidence?

   DIVERGING, AS EVIDENT IN THE DIVERGING RAYS

Hewitt Drew it!

CONCEPTUAL Physics FUNDAMENTALS PRACTICE PAGE

Chapter 14 Properties of Light
***Lenses**—continued*

5. Which type of lens is used to corrected farsightedness? CONVERGING
   Nearsightedness? DIVERGING

6. Construct rays to find the location and relative size of the arrow's image for each of the lenses. Rays that pass through the middle of a lens continue undeviated. In a converging lens, rays from the tip of the arrow that are parallel to the optic axis extend through the far focal point after going through the lens. Rays that go through the near focal point travel parallel to the axis after going through the lens. In a diverging lens, rays parallel to the axis diverge and appear to originate from the near focal point after passing through the lens. Have fun!

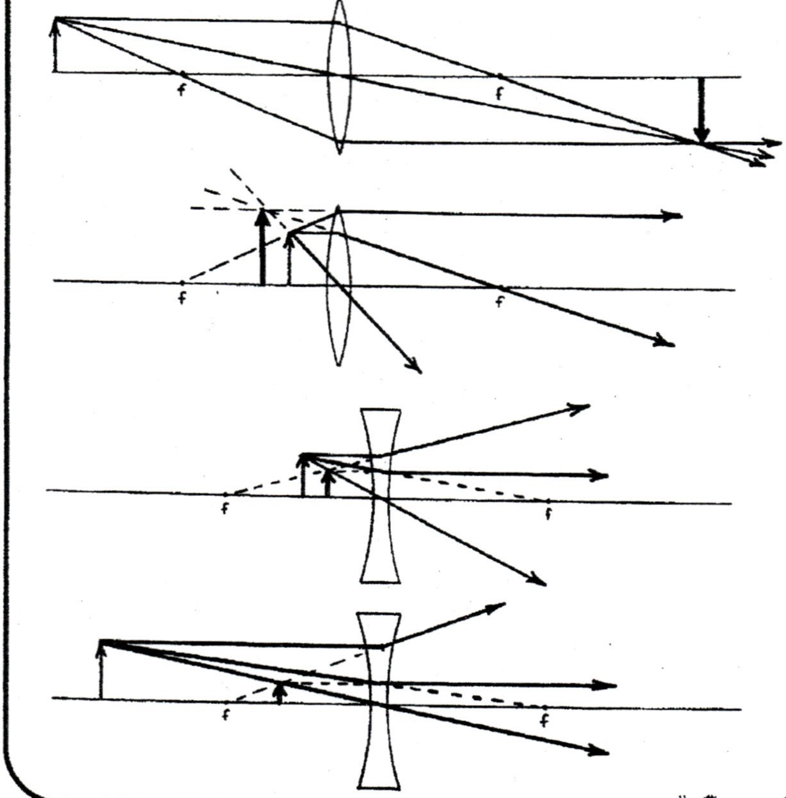

Hewitt Drew it!

Name ______________________ Date ________

CONCEPTUAL **Physics** FUNDAMENTALS PRACTICE PAGE

Chapter 14 Properties of Light

Polarization

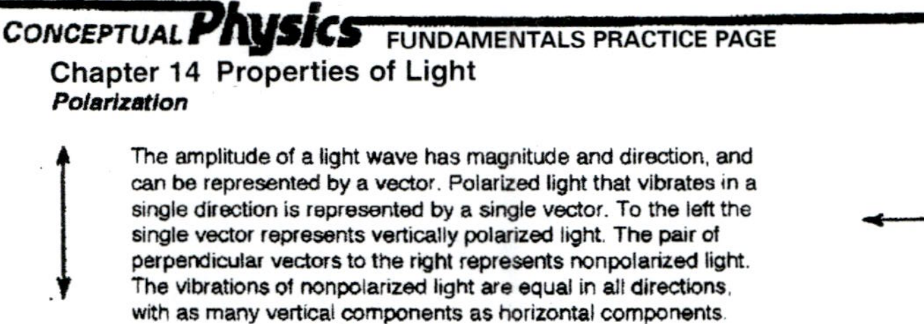

The amplitude of a light wave has magnitude and direction, and can be represented by a vector. Polarized light that vibrates in a single direction is represented by a single vector. To the left the single vector represents vertically polarized light. The pair of perpendicular vectors to the right represents nonpolarized light. The vibrations of nonpolarized light are equal in all directions, with as many vertical components as horizontal components.

1. In the sketch below, nonpolarized light from a flashlight strikes a pair of Polaroid filters.

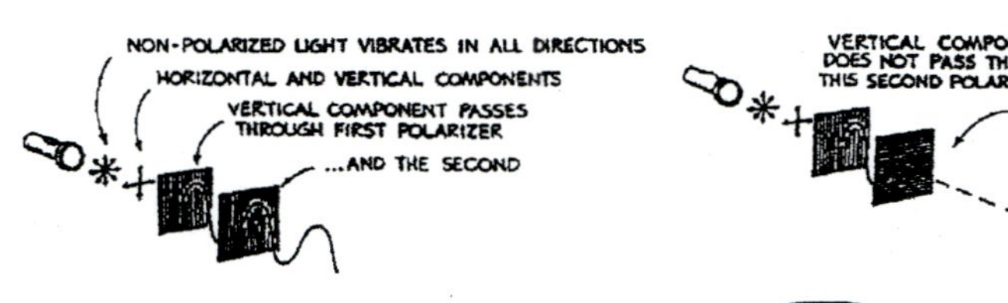

a. Light is transmitted by a pair of Polaroids when their axes are [aligned] [crossed at right angles]

and light is blocked when their axes are [aligned] [crossed at right angles].

b. Transmitted light is polarized in a direction [the same as] [different than] the polarization axis of the filter.

2. Consider the transmission of light through a pair of Polaroids with polarization axes at 45° to each other. Although in practice the Polaroids are one atop the other, we show them spread out side by side below. From left to right:
(a) Nonpolarized light is represented by its horizontal and vertical components.
(b) These components strike filter A.
(c) The vertical component is transmitted, and
(d) falls upon filter B. This vertical component is not aligned with the polarization axis of filter B, but it has a component that is aligned—component *t*,
(e) which is transmitted.

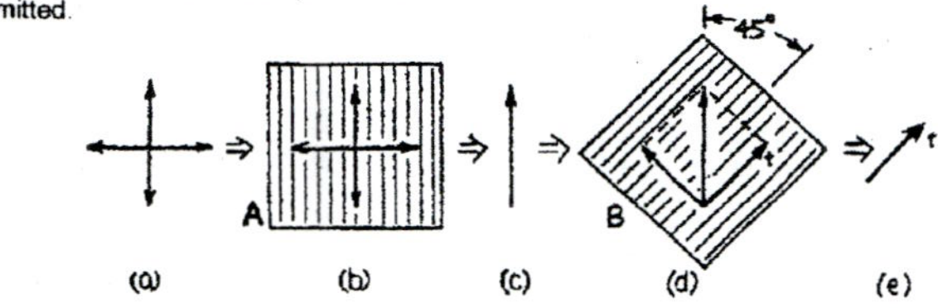

a. The amount of light that gets through Filter B, compared to the amount that gets through

Filter A is [more] [less] [the same].

b. The component perpendicular to *t* that falls on Filter B is [also transmitted] [absorbed].

Hewitt Drew it!

CONCEPTUAL **Physics** FUNDAMENTALS PRACTICE PAGE

Chapter 14 Properties of Light

Polarization—continued

3. Below are a pair of Polaroids with polarization axes at 30° to each other. Carefully draw vectors and appropriate components (as in Question 2) to show the vector that emerges at *e*.

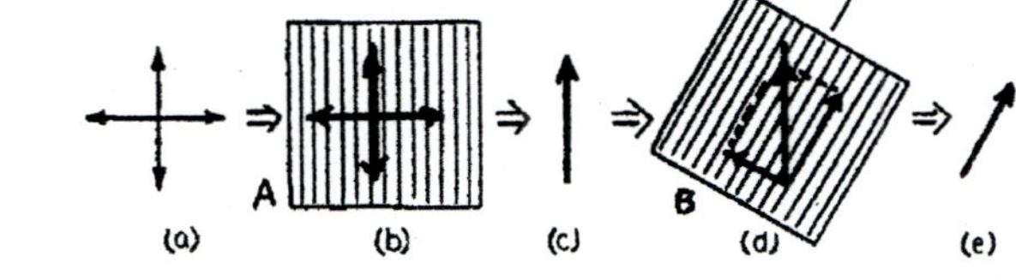

a. The amount of light that gets through the Polaroids at 30°, compared to the amount that gets

through the 45° Polaroids is [less] [more] [the same].

4. Figure 29.35 in your textbook shows the smile of Ludmila Hewitt emerging through three Polaroids. Use vector diagrams to complete steps *b* through *g* below to show how light gets through the three-Polaroid system.

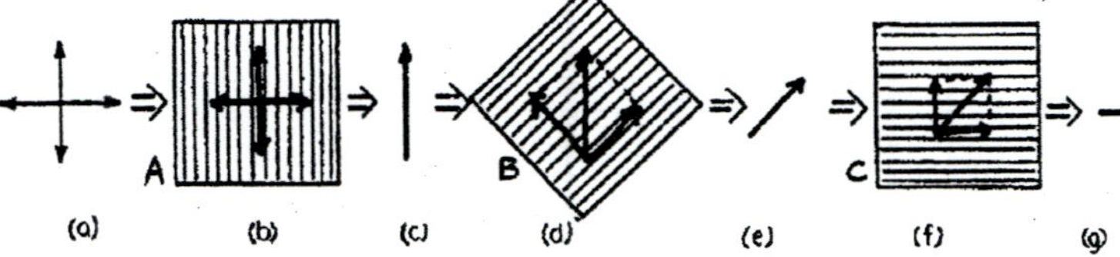

5. A novel use of polarization is shown below. How do the polarized side windows in these next-to-each-other houses provide privacy for the occupants? (Who can see what?)

OCCUPANTS CAN SEE OUTSIDE VIEWS NORMALLY, BUT IF SIDE WINDOWS ARE POLARIZED WITH AXES AT 90° TO EACH OTHER, THEN FROM INSIDE THEIR HOMES THEY CANNOT SEE THROUGH THE SIDE WINDOWS OF THEIR NEIGHBORS.

Hewitt Drew it!

Name ______________________ Date __________

CONCEPTUAL Physics FUNDAMENTALS PRACTICE PAGE

Chapter 15 Quantum Theory

Light Quanta

1. To say that light is quantized means that light is made up of

   [elemental units] [waves].

2. Compared to photons of low-frequency light, photons of higher-frequency light have more

   [energy] [speed] [quanta].

3. The photoelectric effect supports the

   [wave model of light] **[particle model of light]**.

4. The photoelectric effect is evident when light shone on certain

   photosensitive materials ejects [photon] **[electrons]**.

5. The photoelectric effect is more effective with violet light than with red light because the photons

   [resonate with the atoms in the material]

   [deliver more energy to the material]

   [are more numerous].

6. According to De Broglie's wave model of matter, a beam of light

   and a beam of electrons [are fundamentally different] **[are similar]**.

7. According to De Broglie, the greater the speed of an electron beam, the

   [longer is its wavelength] **[shorter is its wavelength]**.

8. The discreteness of the energy levels of electrons about the atomic nucleus is best understood

   by considering the electron to be a **[wave]** [particle].

9. Heavier atoms are not appreciably larger in size than lighter atoms. The main reason for this is that the greater nuclear charge

   [pulls surrounding electrons into tighter orbits]

   [holds more electrons about the atomic nucleus]

   [produces a denser atomic structure].

10. Whereas in the everyday macroworld the study of motion is called *mechanics* in the microworld the study of quanta is called

    [Newtonian mechanics] **[quantum mechanics]**.

A QUANTUM MECHANIC!

Name ______________________ Date __________

CONCEPTUAL Physics FUNDAMENTALS PRACTICE PAGE

Chapter 16 The Atomic Nucleus and Radioactivity

Radioactivity

Complete the following statements:

1.a. A lone neutron spontaneously decays into a proton plus an __ELECTRON__.

b. Alpha and beta rays are made of streams of particles, whereas gamma rays are streams of

   __PHOTONS__.

c. An electrically charged atom is called an __ION__.

d. Different __ISOTOPES__ of an element are chemically identical but differ in the number of neutrons in the nucleus.

e. Transuranic elements are those beyond atomic number __92__.

f. If the amount of a certain radioactive sample decreases by half in four weeks,

   in four more weeks the amount remaining should be __ONE-FOURTH__ the original amount.

g. Water from a natural hot spring is warmed by __RADIOACTIVITY__ inside Earth.

2. The gas in the little girl's balloon is made up of former alpha and beta particles produced by radioactive decay.

a. If the mixture is electrically neutral, how many more beta particles than alpha particles are in the balloon?

   __THERE ARE TWICE AS MANY BETA PARTICLES AS ALPHA PARTICLES.__

b. Why is your answer to the above not the "same"?

   __AN ALPHA HAS DOUBLE CHARGE; THE CHARGE OF 2 BETAS = MAGNITUDE OF CHARGE OF 1 ALPHA PARTICLE.__

c. Why are the alpha and beta particles no longer harmful to the child?

   __THEY HAVE LONG LOST THEIR HIGH KE, WHICH BECOMES THE THERMAL ENERGY ENERGY OF RANDOM MOLECULAR MOTION.__

d. What element does this mixture make?

   __HELIUM__

CONCEPTUAL Physics FUNDAMENTALS PRACTICE PAGE

Chapter 16 The Atomic Nucleus and Radioactivity

Nuclear Reactions

Complete these nuclear reactions:

1. $^{238}_{92}U \rightarrow ^{234}_{90}Th + ^{4}_{2}\underline{He}$

THORIUM LATE, I OVERTHLEPT !

2. $^{234}_{90}Th \rightarrow ^{234}_{91}Pa + ^{0}_{-1}\underline{e}$

3. $^{234}_{91}Pa \rightarrow \underline{^{230}_{89}Ac} + ^{4}_{2}He$

4. $^{220}_{86}Rn \rightarrow \underline{^{216}_{84}Po} + ^{4}_{2}He$

5. $^{216}_{84}Po \rightarrow \underline{^{216}_{85}At} + ^{0}_{-1}e$

6. $^{216}_{84}Po \rightarrow \underline{^{212}_{82}Pb} + ^{4}_{2}He$

7. $^{210}_{83}Bi \rightarrow \underline{^{210}_{84}Po} + ^{0}_{-1}e$

NUCLEAR PHYSICS --- IT'S THE SAME TO ME WITH THE FIRST TWO LETTERS INTERCHANGED?

8. $^{1}_{0}n + ^{10}_{5}B \rightarrow \underline{^{7}_{3}Li} + ^{4}_{2}He$

Name ______________________ Date __________

CONCEPTUAL Physics FUNDAMENTALS PRACTICE PAGE

Chapter 16 The Atomic Nucleus and Radioactivity

Natural Transmutation

Fill in the decay-scheme diagram below, similar to that shown in Figure 33.14 in your textbook, but beginning with U-235 and ending with an isotope of lead. Use the table at the left, and identify each element in the series with its chemical symbol.

Step	Particle emitted
1	Alpha
2	Beta
3	Alpha
4	Alpha
5	Beta
6	Alpha
7	Alpha
8	Alpha
9	Beta
10	Alpha
11	Beta
12	Stable

MASS NUMBER: 235, 231, 227, 223, 219, 215, 211, 207, 203

ATOMIC NUMBER: 81, 82, 83, 84, 85, 86, 87, 88, 89, 90, 91, 92

Diagram entries: U (235, 92); Th → Pa (231, 90–91); Ac (227, 89); Fr → Ra (223, 87–88); Rn (219, 86); Po (215, 84); Pb → Bi (211, 82–83); Tl → Pb (207, 81–82)

What is the final-product isotope?

$^{207}_{82}Pb$ (LEAD – 207)

Name ______________________ Date ________

CONCEPTUAL **Physics** FUNDAMENTALS PRACTICE PAGE

Chapter 16 The Atomic Nucleus and Radioactivity

Nuclear Reactions

1. Complete the table for a chain reaction in which two neutrons from each step individually cause a new reaction.

EVENT	1	2	3	4	5	6	7
NO. OF REACTIONS	1	2	4	8	16	32	64

2. Complete the table for a chain reaction where three neutrons from each reaction cause a new reaction.

EVENT	1	2	3	4	5	6	7
NO. OF REACTIONS	1	3	9	27	81	243	729

3. Complete these beta reactions, which occur in a breeder reactor.

$^{239}_{92}U \rightarrow \underline{^{239}_{93}Np} + ^{0}_{-1}e$

$^{239}_{93}Np \rightarrow \underline{^{239}_{94}Pu} + ^{0}_{-1}e$

4. Complete the following fission reactions.

$^{1}_{0}n + ^{235}_{92}U \rightarrow ^{143}_{54}Xe + ^{90}_{38}Sr + \underline{3}\,(^{1}_{0}n)$

$^{1}_{0}n + ^{235}_{92}U \rightarrow ^{152}_{60}Nd + \underline{^{80}_{32}Ge} + 4\,(^{1}_{0}n)$

$^{1}_{0}n + ^{239}_{94}Pu \rightarrow \underline{^{141}_{54}Xe} + ^{97}_{40}Zr + 2\,(^{1}_{0}n)$

5. Complete the following fusion reactions.

$^{2}_{1}H + ^{2}_{1}H \rightarrow ^{3}_{2}He + \underline{^{1}_{0}n}$

$^{2}_{1}H + ^{3}_{1}H \rightarrow ^{4}_{2}He + \underline{^{1}_{0}n}$

KNOW NUKES!

Name ______________________ Date ________

CONCEPTUAL Physics FUNDAMENTALS PRACTICE PAGE

Appendix B Linear and Rotational Motion

Torques

1. Apply what you know about torques by making a mobile. Shown below are five horizontal arms with fixed 1- and 2-kg masses attached, and four hangers with ends that fit in the loops of the arms, lettered *A* through *R*. You are to determine where the loops should be attached so that when the whole system is suspended from the spring scale at the top, it will hang as a proper mobile, with its arms suspended horizontally. This is best done by working from the bottom upward. Circle the loops where the hangers should be attached. When the mobile is complete, how many kilograms will be indicated on the scale? (Assume the horizontal struts and connecting hooks are practically massless compared with the 1- and 2-kg masses.) On a separate sheet of paper, make a sketch of your completed mobile.

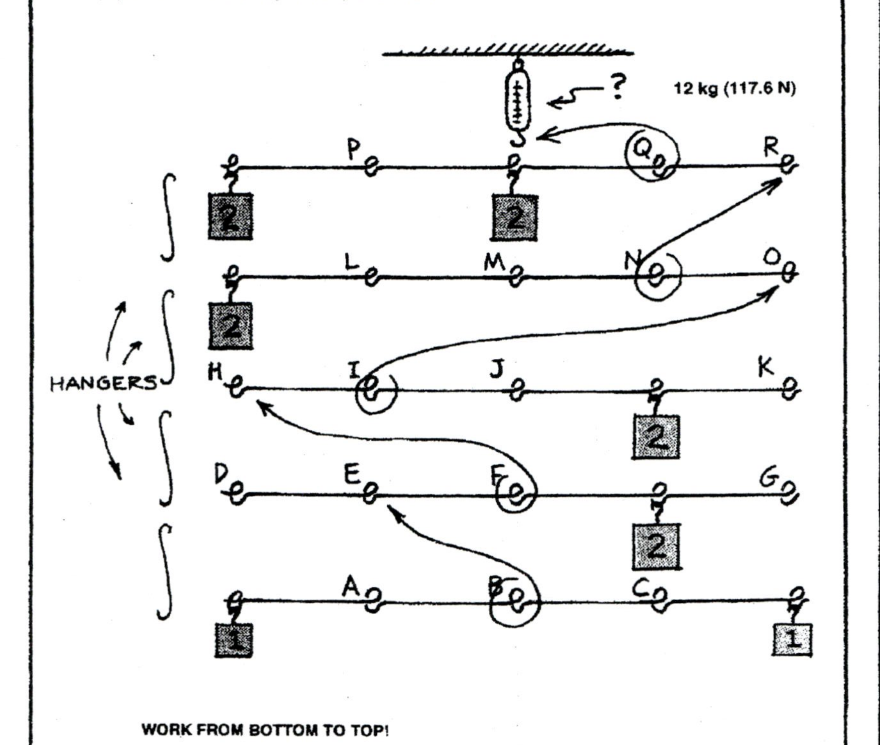

WORK FROM BOTTOM TO TOP!

CONCEPTUAL Physics FUNDAMENTALS PRACTICE PAGE

Appendix B Linear and Rotational Motion

Torques—continued

2. Complete the data for the three seesaws in equilibrium.

4 m | 2 m — W = 250 N; W = 500 N

4 m | ___ m — W = 300 N; W = 400 N

8 m; 3 m | 4 m — W = 600 N; W OF BOARD = 200 N

$600 \times 1\text{ m} = W \times 3\text{ m}$

$W = \dfrac{600\text{ N} \times 1\text{ m}}{300\text{ m}} = 200\text{ N}$

3. The broom balances at its CG. If you cut the broom in half at the CG and weigh each part of the broom, which end would weigh more?

PIECE WITH BRUSH WEIGHS MORE

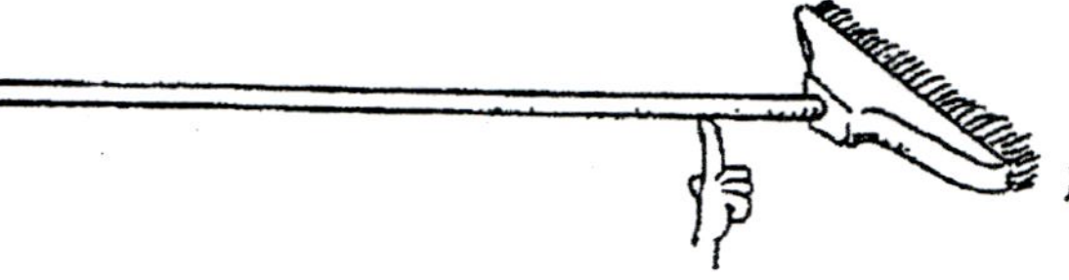

Explain why each end has or does not have the same weight?
(Hint: Compare this to one of the seesaw systems above.)

WEIGHT ON EITHER SIDE ISN'T THE SAME, BUT TORQUE IS! LIKE SEESAWS ABOVE, SHORTER LEVER ARM HAS MORE WEIGHT.

Name ______________________ Date __________

CONCEPTUAL Physics FUNDAMENTALS PRACTICE PAGE

Appendix B Linear and Rotational Motion

Torques and Rotation

1. Pull the string gently and the spool rolls. The direction of roll depends on the way the torque is applied.

   In (1) and (2) below, the force and lever arm are shown for the torque about the point where surface contact is made (shown by the triangular "fulcrum"). The lever arm is the heavy dashed line, which is different for each different pulling position.

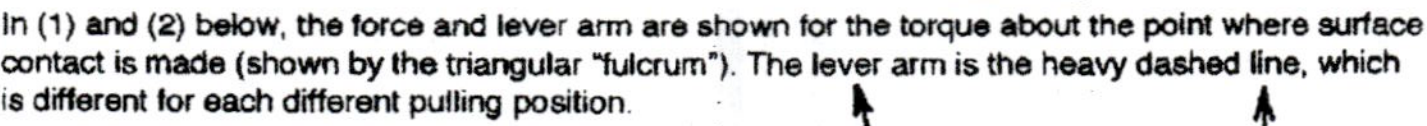

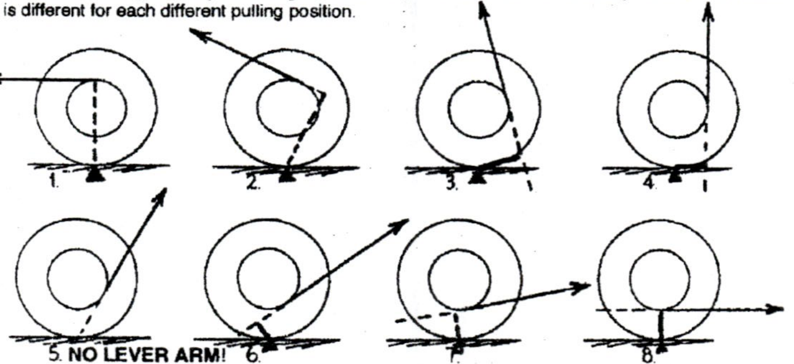

   a. Construct the lever arm for the other positions.

   b. Lever arm is longer when the string of the spool spindle is on the **[top]** [bottom].

   c. For a given pull, the torque is greater when the string is on the **[top]** [bottom].

   d. For the same pull, rotational acceleration is greater when the string is on the **[top]** [bottom] [makes no difference].

   e. At which position(s) does the spool roll to the left? 1,2,3,4

   f. At which position(s) does the spool roll to the right? 6,7,8

   g. At which position(s) does the spool not roll at all? 5

   h. Why does the spool slide rather than roll at this position?

   LINE OF ACTION EXTENDS TO FULCRUM; NO LEVER ARM, NO TORQUE.

   Be sure your right angle is between the force's *line of action* and the lever arm.

2. Relatively few people know that the reason a ball picks up rotational speed rolling down an incline is because of a torque. In sketch A, we see the ingredients of the torque acting on the ball—the force due to gravity and the lever arm to the point where surface contact is made.

   A B C

   a. Construct the lever arms for positions B and C.

   b. As the incline becomes steeper, the torque **[increases]** [decreases].

CONCEPTUAL Physics FUNDAMENTALS PRACTICE PAGE

Appendix B Linear and Rotational Motion

Acceleration and Circular Motion

Newton's 2nd law, $a = F/m$, tells us that net force and its corresponding acceleration are always in the same direction. But force and acceleration vectors are not always in the direction of velocity (another vector).

1. You're in a car at a traffic light. The light turns green and the driver "steps on the gas." The sketch shows the top view of the car. Note the direction of the velocity and acceleration vectors.

   a. Your body tends to lurch [forward] [not at all] **[backward]**.

   b. The car accelerates **[forward]** [not at all] [backward].

   c. The force on the car acts **[forward]** [not at all] [backward].

2. You're driving along and approach a stop sign. The driver steps on the brakes. The sketch shows the top view of the car. Draw vectors for velocity and acceleration.

   a. Your body tends to lurch **[forward]** [not at all] [backward].

   b. The car accelerates [forward] [not at all] **[backward]**.

3. You continue driving, and round a sharp curve to the left at constant speed.

   a. Your body tends to lean [inward] [not at all] **[outward]**.

   b. The direction of the car's acceleration is **[inward]** [not at all] [outward].

   c. The force on the car acts **[inward]** [not at all] [outward].

   Draw vectors for velocity and acceleration of the car.

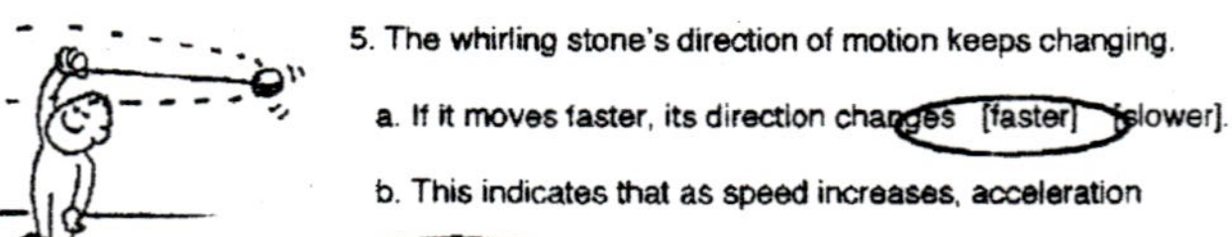

4. In general, the directions of lurch and acceleration, and therefore the directions of lurch and force are [the same] [not related] **[opposite]**.

5. The whirling stone's direction of motion keeps changing.

   a. If it moves faster, its direction changes **[faster]** [slower].

   b. This indicates that as speed increases, acceleration **[increases]** [decreases] [stays the same].

6. Like Question 5, consider whirling the stone on a shorter string—that is, of smaller radius.

   a. For a given speed, the rate that the stone changes direction is [less] **[more]** [the same].

   b. This indicates that as the radius decreases, acceleration **[increases]** [decreases] [stays the same].

thanx to Jim Harper

Name ______________________ Date ________

Appendix B Linear and Rotational Motion

The Flying Pig

The toy pig flies in a circle at constant speed. This arrangement is called a conical pendulum because the supporting string sweeps out a cone. Neglecting the action of its flapping wings, only two forces act on the pig—gravitational mg, and string tension T.

Vector Component Analysis:
Note that vector T can be resolved into two components—horizontal T_x, and vertical T_y. These vector components are dashed to distinguish them from the tension vector T.

Circle the correct answers:

1. If T were somehow replaced with T_x and T_y the pig

   [would] [would not] behave identically to being supported by T.

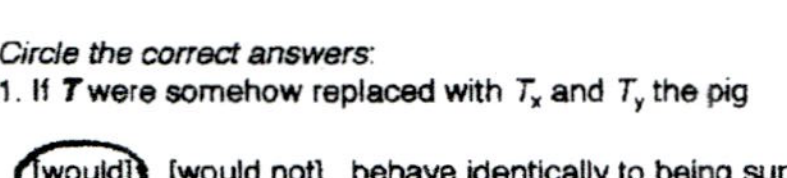

2. Since the pig doesn't accelerate vertically, compared with the magnitude of mg, component T_y,

   must be [greater] [less] [equal and opposite].

3. The velocity of the pig at any instant is [along the radius of] [tangent to] its circular path.

4. Since the pig continues in circular motion, component T_x must be a

   [centripetal] [centrifugal] [nonexistent] force, which equals [zero] [mv^2/r].

   Furthermore, T_x is [along the radius] [tangent to] the circle swept out.

Vector Resultant Analysis:

5. Rather than resolving T into horizontal and vertical components, use your pencil to sketch the resultant of mg and T using the *parallelogram rule*.

6. The resultant lies in a [horizontal] [vertical]

   direction, and [toward] [away from] the

   center of the circular path. The resultant of mg

   and T is a [centripetal] [centrifugal] force.

For straight-line motion with no acceleration, $\Sigma F = 0$.
But for uniform circular motion, $\Sigma F = mv^2/r$.

thanx to Pablo Robinson and Miss Piggy

Appendix B Linear and Rotational Motion

Banked Airplanes

An airplane banks as it turns along a horizontal circular path in the air. Except for the thrust of its engines and air resistance, the two significant forces on the plane are gravitational mg (vertical), and lift L (perpendicular to the wings).

Vector Component Analysis:
With a ruler and a pencil, resolve vector L into two perpendicular components, horizontal L_x, and vertical L_y. Make these vectors dashed to distinguish them from L.

Circle the correct answers:

1. The velocity of the airplane at any instant is

   [along the radius of] [tangent to] its circular path.

2. If L were somehow replaced with L_x and L_y,

   the airplane [would] [would not] behave the

   same as being supported by L.

3. Since the airplane doesn't accelerate vertically, component L_y must be

   [greater than] [less than] [equal and opposite to] mg.

4. Since the plane continues in circular motion, component L_x must equal [zero] [mv^2/r], and be a

   [centripetal] [centrifugal] [nonexistent] force. Furthermore, L_x is

   [along the radius of] [tangent to] the circular path.

Vector Resultant Analysis:

5. Rather than resolving L into horizontal and vertical components, use your pencil to sketch the resultant of mg and L using the *parallelogram rule*.

6. The resultant lies in a [horizontal] [vertical]

   direction, and [toward] [away from] the center

   of the circular path. The resultant of mg and L is a

   [centripetal] [centrifugal] force.

7. The resultant of mg and L is the same as [L_x] [L_y].

Challenge: Explain in your own words why the resultant of two vectors can be the same as a single component of one of them. **For any pair of vectors to be added, if $\Sigma V_y = 0$, and $\Sigma V_x \neq 0$, the resultant will be ΣV_x.**

CONCEPTUAL *Physics* FUNDAMENTALS PRACTICE PAGE

Appendix B Linear and Rotational Motion
Banked Track

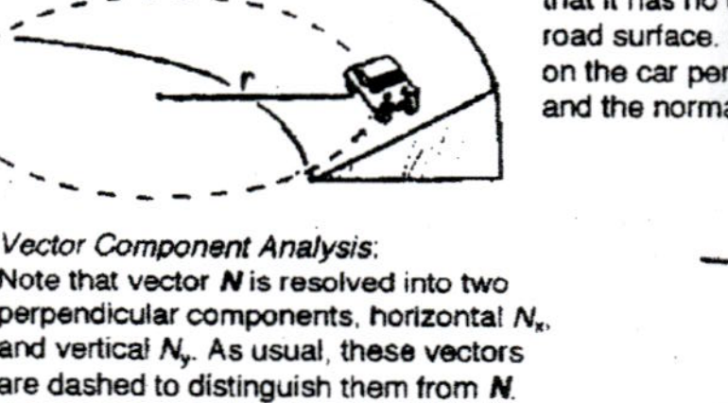
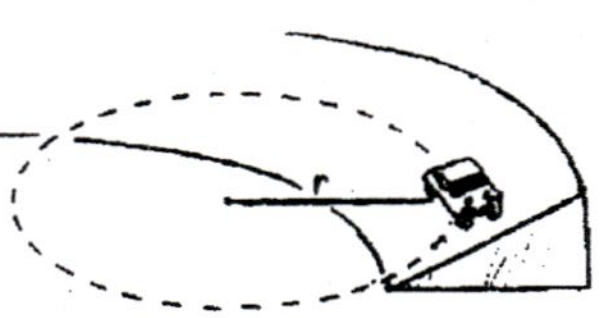

A car rounds a banked curve with just the right speed so that it has no tendency to slide down or up the banked road surface. Shown below are two main forces that act on the car perpendicular to its motion—gravitational mg and the normal force N (the support force of the surface).

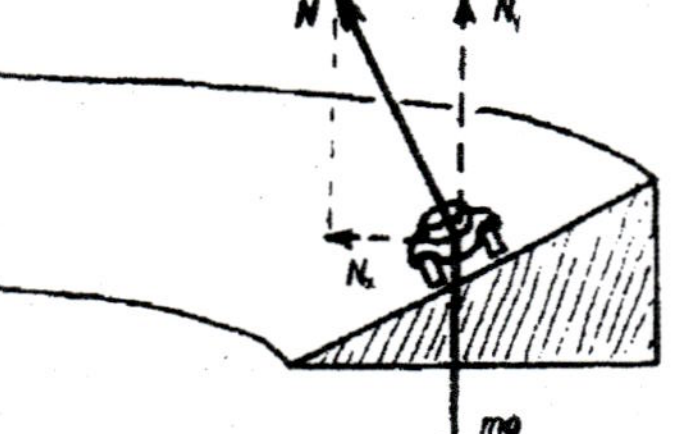

Vector Component Analysis:
Note that vector N is resolved into two perpendicular components, horizontal N_x, and vertical N_y. As usual, these vectors are dashed to distinguish them from N.

Circle the correct answers:

1. If N were somehow replaced with N_x and N_y, the car [**would**] [would not] behave identically to being supported by N.
2. Since the car doesn't accelerate vertically, component N_y must be [greater than] [**equal and opposite to**] [less than] mg.
3. The velocity of the car at any instant is [along the radius of] [**tangent to**] its circular path.
4. Since the car continues in uniform circular motion, component N_x must equal [zero] [**mv^2/r**] and be a [**centripetal**] [centrifugal] [nonexistent] force. Furthermore, N_x [**lies along the radius of**] [is tangent to] the circular path.

Vector Resultant Analysis:

5. Rather than resolving N into horizontal and vertical components, use your pencil to sketch the resultant of mg and N using the *parallelogram rule*.
6. The resultant lies in a [**horizontal**] [vertical] direction, and [**toward**] [away from] the center of the circular path. The resultant of mg and N is a [**centripetal**] [centrifugal] force.
7. The resultant of mg and N is the same as [**N_x**] [N_y], and provides the [**centripetal**] [centrifugal] force.

Notice that when a component of N makes up a centripetal force, $N > mg$.

CONCEPTUAL *Physics* FUNDAMENTALS PRACTICE PAGE

Appendix B Linear and Rotational Motion
Leaning On

When turning a corner on a bicycle, everyone knows that you've got to **lean "into** the curve." What is the physics of this leaning? It involves torque, friction, and **centripetal force** (mv^2/r).

First, consider the simple case of riding a bicycle along a straight-line path. Except for the force that propels the bike forward (friction of the road in the direction of motion) and air resistance (friction of air against the direction of motion), only two significant forces act: weight mg and the normal force N. (The vectors are drawn side-by-side, but actually lie along a single vertical line.)

Circle the correct answers:

1. Since there is no vertical acceleration, we can say that the magnitude of [$N > mg$] [$N < mg$] [**$N = mg$**] which means that in the vertical direction, [$\Sigma F_y > 0$] [$\Sigma F_y < 0$] [**$\Sigma F_y = 0$**]
2. Since the bike doesn't rotate or change in its rotational state, then the total torque is [**zero**] [not zero].

Now consider the same bike rounding a corner. In order to safely make the turn, the bicyclist leans in the direction of the turn. A force of friction pushes sideways on the tire toward the center of the curve.

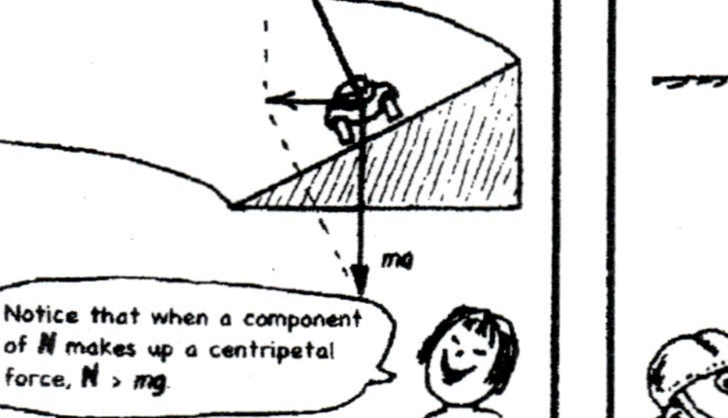

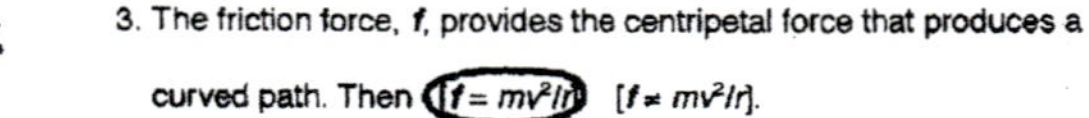

3. The friction force, f, provides the centripetal force that produces a curved path. Then [**$f = mv^2/r$**] [$f \neq mv^2/r$].
4. Consider the net torque about the center of mass (CM) of the bike-rider system. Gravity produces no torque about this point, but N and f do. The torque involving N tends to produce [**clockwise**] [counterclockwise] rotation, and the one involving f tends to produce [clockwise] [**counterclockwise**] rotation.

   These torques cancel each other when the resultant of vectors N and f pass through the CM.
5. With your pencil, use the parallelogram rule and sketch in the resultant of vectors N and f. Label your resultant R. Note the R passes through the center of mass of the bike-rider system. That means that R produces [a clockwise] [a counterclockwise] [**no**] torque about the CM. Therefore the bike-rider system [topples clockwise] [topples counterclockwise] [**doesn't topple**]

When learning how to turn on a bike, you lean so that the sum of the torques about your CM is zero. You may not be calculating torques, but your body learns to feel them.

thanx to Pablo Robinson

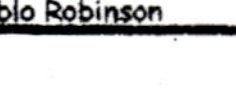

Name ______________________ Date ________

CONCEPTUAL Physics FUNDAMENTALS PRACTICE PAGE

Appendix B Linear and Rotational Motion
Simulated Gravity and Frames of Reference

Suzie Spacewalker and Bob Biker are in outer space. Bob experiences Earth-normal gravity in a rotating habitat, where centripetal force on his feet provides a normal support force that feels like weight. Suzie hovers outside in a weightless condition, motionless relative to the stars and the center of the habitat.

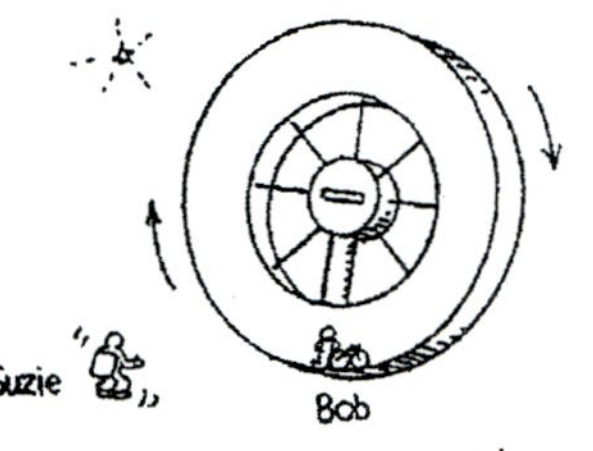

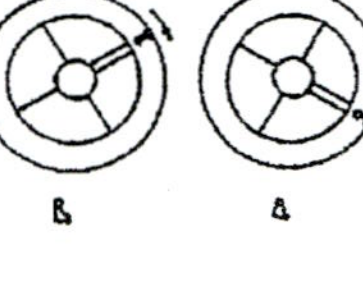
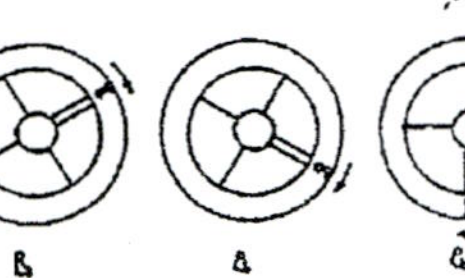
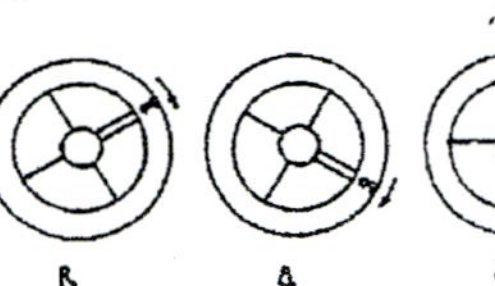

1. Suzie sees Bob rotating clockwise in a circular path at a linear speed of 30 km/h. Suzie and Bob are facing each other, and from Bob's point of view, he is at rest and he sees Suzie moving [clockwise] [counterclockwise].

Bob at rest on the floor
Suzie hovering in space

2. The rotating habitat seems like home to Bob—until he rides his bicycle. When he rides in the opposite direction as the habitat rotates, Suzie sees him moving [faster] [slower].

Bob rides counter-clockwise

3. As Bob's bicycle speedometer reading increases, his rotational speed [decreases] [remains unchanged] [increases] and the normal force that feels like weight [decreases] [remains unchanged] [increases]. So friction between the tires and the floor [decreases] [remains unchanged] [increases].

4. When Bob nevertheless gets his speed up to 30 km/h, as indicated on his bicycle speedometer, Suzie sees him [moving at 30 km/h] [motionless] [moving at 60 km/h].

thanx to Bob Becker

CONCEPTUAL Physics FUNDAMENTALS PRACTICE PAGE

Appendix B Linear and Rotational Motion
Simulated Gravity and Frames of Reference—continued

5. Bounding off the floor a bit while riding at 30 km/h, and neglecting wind effects, Bob [drifts toward the ceiling in midspace as the floor whizzes by him at 30 km/h] [falls as he would on Earth] [slams onto the floor with increased force] and finds himself [in the same frame of reference as Suzie] [as if he rode at 30 km/h on Earth's surface] [pressed harder against the bicyclist seat].

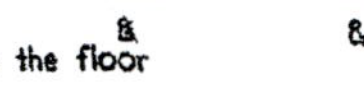
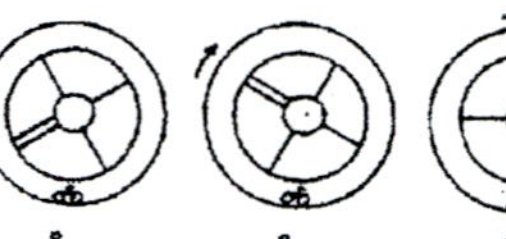
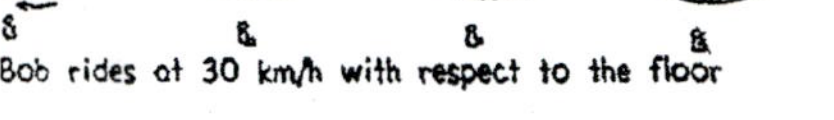
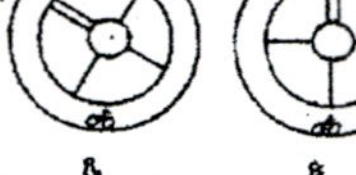
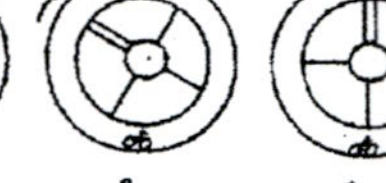

Bob rides at 30 km/h with respect to the floor

6. Bob maneuvers back to his initial condition, whirling at rest with the habitat, standing beside his bicycle. But not for long. Urged by Suzie, he rides in the opposite direction, clockwise with the rotation of the habitat.
Now Suzie sees him moving [faster] [slower].

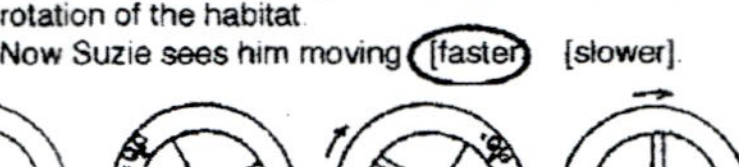

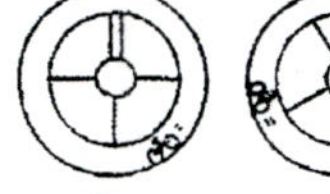

Bob rides clockwise

7. As Bob gains speed, the normal support force that feels like weight [decreases] [remains unchanged] [increases].

8. When Bob's speedometer reading gets up to 30 km/h, Suzie sees him moving [30 km/h] [not at all] [60 km/h] and Bob finds himself [weightless like Suzie] [just as if he rode at 30 km/h on Earth's surface] [pressed harder against the bicyclist seat].

Next, Bob goes bowling. You decide whether the game depends on which direction the ball is rolled!

Name ______________________ Date __________

CONCEPTUAL Physics FUNDAMENTALS PRACTICE PAGE

Appendix C Vectors

Vectors and Sailboats

(Please do not attempt this until you have studied Appendix D!)

1. The sketch shows a top view of a small railroad car pulled by a rope. The force ***F*** that the rope exerts on the car has one component along the track, and another component perpendicular to the track.

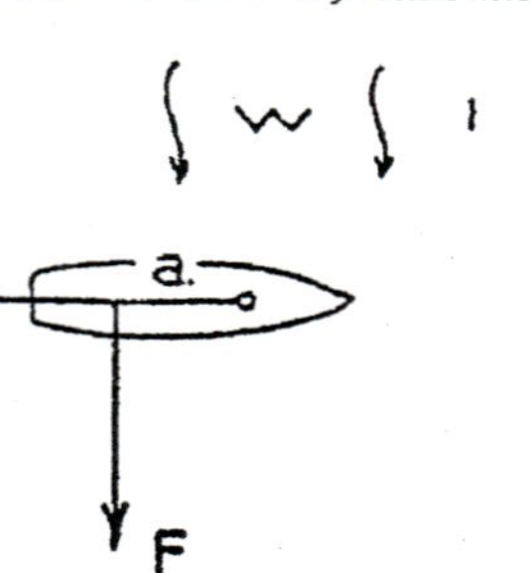

a. Draw these components on the sketch. Which component is larger?

PERPENDICULAR COMPONENT

b. Which component produces acceleration?

COMPONENT PARALLEL TO TRACK

c. What would be the effect of pulling on the rope if it were perpendicular to the track?

NO ACCELERATION

2. The sketches below represent simplified top views of sailboats in a cross-wind direction. The impact of the wind produces a FORCE vector on each as shown. (We do NOT consider *velocity* vectors here!)

W I N D

a. F

b. F

a. Why is the position of the sail above useless for propelling the boat along its forward direction? (Relate this to Question 1.c above where the train is constrained by tracks to move in one direction, and the boat is similarly constrained to move along one direction by its deep vertical fin—the *keel.*)

AS IN 1.c ABOVE, THERE'S NO COMPONENT PARALLEL TO DIRECTION OF MOTION.

b. Sketch the component of force parallel to the to the direction of the boat's motion (along its keel), and the component perpendicular to its motion. Will the boat move in a forward direction? (Relate this to Question 1.b above.)

YES, AS IN 1.b ABOVE, THERE IS A COMPONENT PARALLEL TO DIRECTION TO MOTION.

Hewitt Drew it!

Appendix C Vectors

***Vectors and Sailboats**—continued*

3. The boat to the right is oriented at an angle into the wind. Draw the force vector and its forward and perpendicular components.

   a. Will the boat move in a forward direction and tack into the wind? Why or why not?

   YES, BECAUSE THERE IS A COMPONENT OF FORCE PARALLEL TO THE DIRECTION OF MOTION.

W I N D

4. The sketch below is a top view of five identical sailboats. Where they exist, draw force vectors to represent wind impact on the sails. Then draw components parallel and perpendicular to the keels of each boat.

   a. Which boat will sail the fastest in a forward direction?

   BOAT 4 (WILL USUALLY EXCEED BOAT 1)

   b. Which will respond least to the wind?

   BOAT 2 (OR BOAT 3)*

   c. Which will move in a backward direction?

   BOAT 5

   d. Which will experience decreasing wind impact with increasing speed?

   BOAT 1 (NO IMPACT AT WIND SPEED)

***THE WIND MISSES THE SAIL OF BOAT 2, AND THERE'S NO COMPONENT PARALLEL TO THE KEEL FOR BOAT 3.**

W I N D

1 2 3 4 5

I may be a 6 and you may be an 8...

but together we're a perfect 10!

Hewitt Drew It!

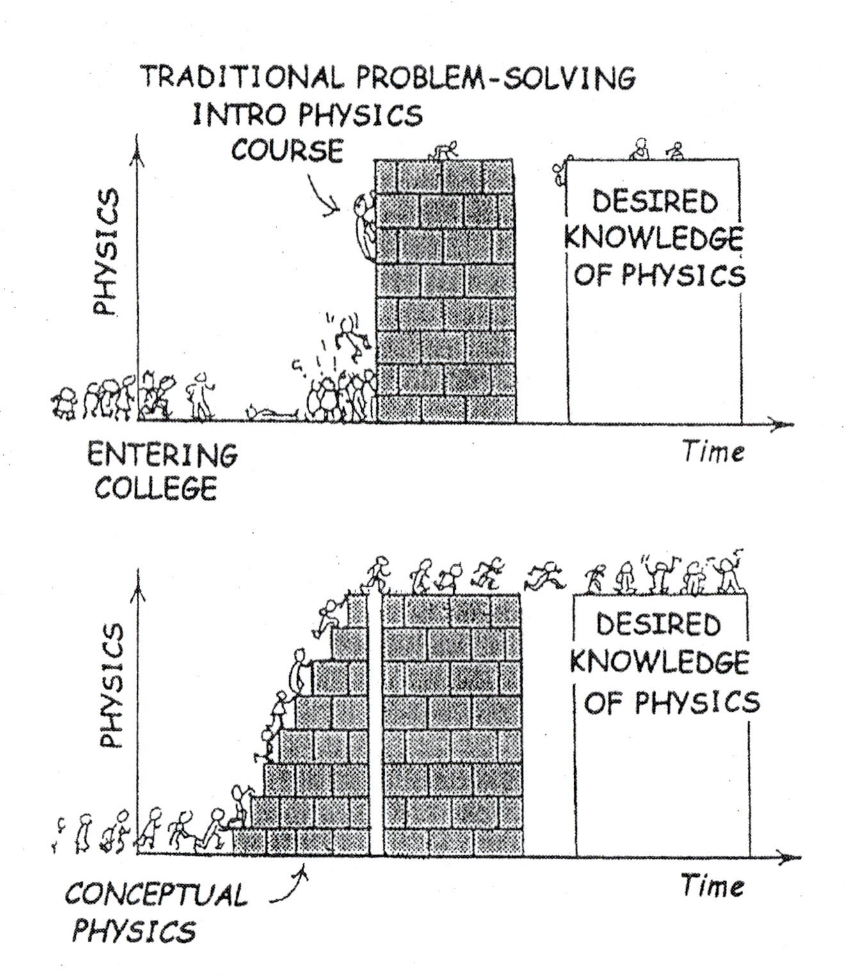
TRADITIONAL PROBLEM-SOLVING
INTRO PHYSICS
COURSE
PHYSICS
DESIRED
KNOWLEDGE
OF PHYSICS
ENTERING
COLLEGE
Time
PHYSICS
DESIRED
KNOWLEDGE
OF PHYSICS
CONCEPTUAL
PHYSICS
Time

SOLUTIONS TO THE ODD-NUMBERED EXERCISES AND PROBLEMS FROM CONCEPTUAL PHYSICS FUNDAMENTALS

SOLUTIONS TO CHAPTER 1 EXERCISES

1. The penalty for fraud is professional excommunication.

3. Aristotle's hypotheses was partially correct, for material that makes up the plant comes partly from the soil, but mainly from the air and water. An experiment would be to weigh a pot of soil with a small seedling, then weigh the potted plant later after it has grown. The fact that the grown plant will weigh more is evidence that the plant is composed of more material than the soil offers. By keeping a record of the weight of water used to water the plant, and covering the soil with plastic wrap to minimize evaporation losses, the weight of the grown plant can be compared with the weight of water it absorbs. How can the weight of air taken in by the plant be estimated?

5. The examples are endless. Knowledge of electricity, for example, has proven to be extremely useful. The number of people who have been harmed by electricity who understood it is far fewer than the number of people who are harmed by it who don't understand it. A fear of electricity is much more harmful than useful to one's general health and attitude.

7. What is likely being misunderstood is the distinction between theory and hypothesis. In common usage, "theory" may mean a guess or hypothesis, something that is tentative or speculative. But in science a theory is a synthesis of a large body of validated information (e.g., cell theory or quantum theory). The value of a theory is its usefulness (not its "truth").

SOLUTIONS TO CHAPTER 2 EXERCISES

1. One.

3. The average speed of molecules increases.

5. The cat leaves a trail of molecules and atoms on the grass. These in turn leave the grass and mix with the air, where they enter the dog's nose, activating its sense of smell.

7. The atoms that make up a newborn baby or anything else in this world originated in the explosions of ancient stars. (See Figure 2.8, my daughter Leslie.) The *molecules* that make up the baby, however, were formed from atoms ingested by the mother and transferred to her womb.

9. Of the substances listed, H_2, He, Na, and U are pure elements. H_2O and NaCl are compounds made of two elements, and three different elements contribute to H_2SO_4.

11. Brownian motion is the result of more atoms or molecules bumping against one side of a tiny particle than the other. This produces a net force on the particle, which it is set in motion. Such doesn't occur for larger particles because the numbers of bumps on opposite sides is more likely equal, producing no net force. The number of bumps on a baseball is practically the same on all sides, with no net force and no change in the baseball's motion.

13. Individual carbon atoms have less mass than individual oxygen atoms, so equal masses of each means more carbons than oxygens.

15. Nine.

17. The element is copper, atomic number 29. Any atom having 29 protons is by definition copper.

19. Lead.

21. An atom gains an electron to become a negative ion. Then it has more electrons than protons.

23. The capsule would be arsenic.

25. Germanium has properties most like silicon, as it is in the same column, Group XIV, as silicon in the periodic table.

27. Protons contribute more to an atom's mass, and electrons more to an atom's size.

29. Letting the formula KE = $= \frac{1}{2}mv^2$ guide your thinking, for the same speed the atom with greater mass has greater KE. Greater-mass carbon therefore has greater KE than hydrogen for the same speed.

31. You really are a part of every person around you in the sense that you are composed of atoms not only from every person around you, but from every person who ever lived on Earth! And the atoms that now compose you will make up the atomic pool that others will draw upon.

33. They assumed hydrogen and oxygen were single-atom molecules with water's formula being H_8O.

35. Open-ended.

SOLUTIONS TO CHAPTER 2 PROBLEMS

1. There are 16 grams of oxygen in 18 grams of water. We can see from the formula for water, H_2O, there are twice as many hydrogen atoms (each of atomic mass 1) as oxygen atoms (each of atomic mass 16). So the molecular mass of H_2O is 18, with 16 parts oxygen by mass.

3. The atomic mass of element A is $\frac{3}{2}$ the mass of element B. Why? Gas A has three times the mass of Gas B. If the equal number of molecules in A and B had equal numbers of atoms, then the atoms in Gas A would simply be three times as massive. But there are twice as many atoms in A, so the mass of each atom must be half of three times as much—$\frac{3}{2}$.

5. From the hint:

$$\frac{\text{number of molecules in thimble}}{\text{number of molecules in ocean}} = \frac{\text{number of molecules in question}}{\text{number of molecules in thimble}}$$

$$\frac{10^{23}}{10^{46}} = \frac{x}{10^{23}};\ x = \frac{10^{46}}{10^{46}} = 1$$

7. The total number of people who ever lived ($6 \times 10^9 \times 20 = 120 \times 10^9$. This is roughly 10^{11} people altogether) is enormously smaller than 10^{22}. How does 10^{22} compare to 10^{11}? 10^{22} is $(10^{11})^2$! Multiply the number of people who ever lived by the same number, and you'll get 10^{22}, the number of air molecules in a breath of air. Suppose each person on Earth journeyed to a different planet in the galaxy and every one of those planets contained as many people as the Earth now contains. The total number of people on all these planets would still be less than the number of molecules in a breath of air. Atoms are indeed small—and numerous!

SOLUTIONS TO CHAPTER 3 EXERCISES

1. Aristotle would likely say the ball slows to reach its natural state. Galileo would say the ball is encountering friction, an unbalanced force that slows it.

3. When rolling down it is going with gravity. Going up, against. (There are force components in the direction of motion, as we shall see later.) When moving horizontally, gravity is perpendicular, neither speeding or slowing the ball.

5. The piece of iron has more mass, but less volume. The answers are different because they address completely different concepts.

7. Like the massive ball that resists motion when pulled by the string, the massive anvil resists moving against Paul when hit with the hammer. Inertia in action.

9. The weight of a 10-kg object on the Earth is 98 N, and on the Moon $\frac{1}{6}$ of this, or 16.3 N. The mass would be 10 kg in any location.

11. From $\Sigma F = 0$, the upward forces are 400 N, and the downward forces are 250 N + weight of the staging. So the staging must weigh 150 N.

13. Each scale shows half her weight.

15. Yes, for it doesn't change its state of motion (accelerate). Strictly speaking, some friction does act so it is close to being in equilibrium.

17. The upward force, the support force, isn't the only force acting. Weight does also, producing a net force of zero.

19. Constant speed implies the net force on the cabinet is zero. So friction is 600 N in the opposite direction.

21. Constant velocity means constant direction, so your friend should say "at a constant *speed* of 100 km/h."

23. Not very, for his speed will be zero relative to the land.

25. 10 m/s.

27. The ball slows by 10 m/s each second, and gains 10 m/s when descending. The time up equals the time down if air resistance is nil.

29. Zero, for no change in velocity occurred. Misreading the question might mean missing the word steady (no change).

31. The ball on B finishes first, for its average speed along the lower part as well as the down and up slopes is greater than the average speed of the ball along track A.

SOLUTIONS TO CHAPTER 3 PROBLEMS

1. (a) 30 N + 20 N = 50 N. (b) 30 N − 20 N = 10 N.

3. From $\Sigma F = 0$, friction equals weight, *mg*, = (100 kg)(9.8 m/s^2) = 980 N.

5. $a = \dfrac{\text{change in velocity}}{\text{time interval}} = \dfrac{-100\text{ km/h}}{10\text{ s}} = -10\text{ km/h}\cdot\text{s}$. (The vehicle decelerates at 10 km/h·s.)

7. Since it starts going up at 30 m/s and loses 10 m/s each second, its time going up is 3 s. Its time returning is also 3 s, so it's in the air for a total of 6 s. Distance up (or down) is $\frac{1}{2}gt^2 = 5 \times 3^2 = 45$ m.

   Or from $d = vt$, where average velocity is $(30 + 0)/2 = 15$ m/s, and time is 3 s, $d = 15 \text{ m/s} \times 3 \text{ s} = 45$ m.

9. $d = v_{ave}t = \frac{v_f + v_0}{2} \times \frac{v_f - v_0}{a} = \frac{v_f^2 + v_f v_0 + v_f v_0 - v_0^2}{2a} = \frac{v_f^2 - v_0^2}{2a}.$

   $d = v_{ave}t = \left(\frac{v_f + v_0}{2}\right) \times \left(\frac{v_f - v_0}{a}\right) = \frac{v_f^2 - v_0^2}{2a}.$

SOLUTIONS TO CHAPTER 4 EXERCISES

1. Poke or kick the boxes. The one that more greatly resists a change in motion is the one with the greater mass—the one filled with sand.

3. The massive cleaver tends to keep moving when it encounters the vegetables, cutting them more effectively.

5. Newton's first law again—when the stone is released it is already moving as fast as the ship, and this horizontal motion continues as the stone falls. Much more about this in Chapter 6.

7. You exert a force to overcome the force of friction. This makes the net force zero, which is why the wagon moves without acceleration. If you pull harder, then net force will be greater than zero and acceleration will occur.

9. Let Newton's second law guide the answer to this; $a = F/m$. As m gets less (much the mass of the fuel), acceleration a increases for a constant force.

11. The sudden stop involves a large acceleration. So in accord with $a = F/m$, a large a means a large F. Ouch!

13. When air resistance affects motion, the ball thrown upward returns to its starting level with less speed than its initial speed; and also less speed than the ball tossed downward. So the downward thrown ball hits the ground below with a greater speed.

15. 100 N, the same reading it would have if one of the ends were tied to a wall instead of tied to the 100-N hanging weight. Although the net force on the system is zero, the tension in the rope within the system is 100 N, as shown on the scale reading.

17. (a) Two force pairs act; Earth's pull on the apple (action), and the apple's pull on the Earth (reaction). Hand pushes apple upward (action), and apple pushes hand downward (reaction). (b) If air resistance can be neglected, one force pair acts; Earth's pull on apple, and apple's pull on Earth. If air resistance counts, then air pushes upward on apple (action) and apple pushes downward on air (reaction).

19. Neither a stick of dynamite nor anything else "contains" force. We will see later that a stick of dynamite contains *energy,* which is capable of producing forces when an interaction of some kind occurs.

21. When the barbell is accelerated upward, the force exerted by the athlete is greater than the weight of the barbell (the barbell, simultaneously, pushes with greater force against the athlete). When acceleration is downward, the force supplied by the athlete is less.

23. 1000 N.

25. As in the preceding exercise, the force on each cart will be the same. But since the masses are different, the accelerations will differ. The twice-as-massive cart will undergo only half the acceleration of the less massive cart and will gain only half the speed.

27. In accord with Newton's 3rd law, the force on each will be of the same magnitude. But the effect of the force (acceleration) will be different for each because of the different mass. The more massive truck undergoes less change in motion than the motorcycle.

29. The person with twice the mass slides half as far as the twice-as-massive person. That means the lighter one slides 4 feet and the heavier one slides 8 feet (for a total of 12 feet).

31. In accord with Newton's third law, Steve and Gretchen are touching each other. One may initiate the touch, but the physical interaction can't occur without contact between both Steve and Gretchen. Indeed, you cannot touch without being touched!

33. The terminal speed attained by the falling cat is the same whether it falls from 50 stories or 20 stories. Once terminal speed is reached, falling extra distance does not affect the speed. (The low terminal velocities of small creatures enables them to fall without harm from heights that would kill larger creatures.)

35. Before reaching terminal velocity, weight is greater than air resistance. After reaching terminal velocity both weight and air resistance are of the same magnitude. Then the net force and acceleration are both zero.

37. Air resistance is not really negligible for so high a drop, so the heavier ball does strike the ground first. (This idea is shown in Figure 4.15.) But although a twice-as-heavy ball strikes first, it falls only a little faster, and not twice as fast, which is what followers of Aristotle believed. Galileo recognized that the small difference is due to friction and would not be present if there were no friction.

39. A hammock stretched tightly has more tension in the supporting ropes than one that sags. The tightly stretched ropes are more likely to break.

41. No, for the component of your velocity in a direction perpendicular to the water flow (directly across the river) does not depend on stream speed. The total distance you travel while swimming across, however, does depend on stream speed. For a swift current you'll be swept farther downstream, but the crossing time will remain the same.

43. (a) The other vector is upward as shown.

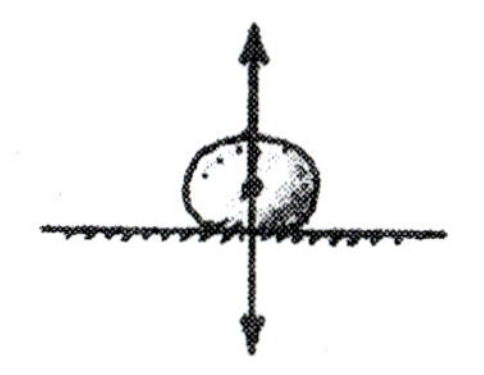

(b) It is called the normal force.

45. (a) As shown.

(b) Upward tension force is greater to result in upward net force.

47. The acceleration of the stone at the top of its path, or anywhere where the net force on the stone is mg, is g.

49. (a) As shown.

(b) Note the resultant of the normals is equal and opposite to the stone's weight.

SOLUTIONS TO CHAPTER 4 PROBLEMS

1. The given pair of forces produces a net force of 200 N forward, which accelerates the cart. To make the net force zero, a force of 200 N backward must be exerted on the cart.

3. Acceleration $a = F_{net}/m = (40\text{ N} - 24\text{ N})/4\text{ kg} = 16\text{ N}/4\text{ kg} = 4\text{ m/s}^2$.

5. Acceleration $a = F_{net}/m = (4 \times 250{,}000\text{ N})/330{,}000\text{ kg} = 3\text{ m/s}^2$.

7. $F_{net} = (mg - f) = (800\text{ N} - f) = ma = 80\text{ kg} \times 4\text{ m/s}^2 = 320\text{ N}$.
So $f = 800\text{ N} - 320\text{ N} = 480\text{ N}$.

9. (a) Force of air resistance will be equal to her weight, mg, or 500 N.

(b) She'll reach the same air resistance, but at a smaller speed, 500 N.

(c) The answers are the same, but for different speeds. In each case she attains equilibrium (no acceleration).

11. By the Pythagorean theorem, $V = \sqrt{[(3\text{ m/s})^2 + (4\text{ m/s})^2]} = 5\text{ m/s}$.

13. By the Pythagorean theorem, $V = \sqrt{[(120\text{ m/s})^2 + (90\text{ m/s})^2]} = 150\text{ m/s}$.

15. (a) The net force on the sled with only Phil on it is Ma, and remains the same when Zephram is added. Acceleration of sled = net force/total mass = $Ma/(M + m) = [M/(M + m)]a$.

(b) Acceleration = $[M/(M + m)]a = 70\text{ kg}/(70 + 45\text{ kg})3.6\text{ m/s}^2 = 2.2\text{ m/s}^2$.

SOLUTIONS TO CHAPTER 5 EXERCISES

1. Supertankers are so massive, that even at modest speeds their motional inertia, or *momenta*, are enormous. This means enormous impulses are needed for changing motion. How can large impulses be produced with modest forces? By applying modest forces over long periods of time. Hence the force of the water resistance over the time it takes to coast 25 kilometers sufficiently reduces the momentum.

3. The extra thickness extends the time during which momentum changes and reduces impact force.

5. Crumpling allows more time for reducing the momentum of the car, resulting in a smaller force of impact on the occupants.

7. Its momentum is the same (its weight might change, but not its mass).

9. The momentum of recoil of the world is 10 kg · m/s. Again, this is not apparent because the mass of the Earth is so enormous that its recoil velocity is imperceptible. (If the masses of Earth and person were equal, both would move at equal speeds in opposite directions.)

11. When a boxer hits his opponent, the opponent contributes to the impulse that changes the momentum of the punch. When punches miss, no impulse is supplied by the opponent—all effort that goes into reducing the momentum of the punches is supplied by the boxer himself. This tires the boxer. This is very evident to a boxer who can punch a heavy bag in the gym for hours and not tire, but who finds by contrast that a few minutes in the ring with an opponent is a tiring experience.

13. In jumping, you impart the same momentum to both you and the canoe. This means you jump from a canoe that is moving away from the dock, reducing your speed relative to the dock, so you don't jump as far as you expected to.

15. If no momentum is imparted to the ball, no oppositely directed momentum will be imparted to the thrower. Going through the motions of throwing has no net effect. If at the beginning of the throw you begin recoiling backward, at the end of the throw when you stop the motion of your arm and hold onto the ball, you stop moving too. Your position may change a little, but you end up at rest. No momentum given to the ball means no recoil momentum gained by you.

17. If the rocket and its exhaust gases are treated as a single system, the forces between rocket and exhaust gases are internal, and momentum in the rocket-gases system is conserved. So any momentum given to the gases is equal and opposite to momentum given to the rocket. A rocket attains momentum by giving momentum to the exhaust gases.

19. For the system comprised of ball + Earth, momentum is conserved for the impulses acting are internal impulses. The momentum of the falling apple is equal in magnitude to the momentum of the Earth toward the apple.

21. This exercise is similar to the previous one. If we consider Bronco to be the system, then a net force acts and momentum changes. In the system composed of Bronco alone, momentum is not conserved. If, however we consider the system to be Bronco and the world (including the air), then all the forces that act are internal forces and momentum is conserved. Momentum is conserved only in systems not subject to external forces.

23. If the air is brought to a halt by the sail, then the impulse against the sail will be equal and opposite to the impulse on the fan. There will be no net impulse and no change in momentum. The boat will remain motionless. Bouncing counts!

25. In terms of force: When the sand lands on the cart it is brought up to the cart's speed. This means a horizontal force provided by the cart acts on the sand. By action-reaction, the sand exerts a force on the cart in the opposite direction—which slows the cart. In terms of momen-

tum conservation: Since no external forces act in the horizontal direction, the momentum after the cart catches sand equals the momentum before. Since mass is added, velocity must decrease.

27. We assume the equal strengths of the astronauts means that each throws with the same speed. Since the masses are equal, when the first throws the second, both the first and second move away from each other at equal speeds. Say the thrown astronaut moves to the right with velocity V, and the first recoils with velocity $-V$. When the third makes the catch, both she and the second move to the right at velocity $V/2$ (twice the mass moving at half the speed, like the freight cars in Figure 5.12). When the third makes her throw, she recoils at velocity V (the same speed she imparts to the thrown astronaut) which is added to the $V/2$ she acquired in the catch. So her velocity is $V + V/2 = 3V/2$, to the right—too fast to stay in the game. Why? Because the velocity of the second astronaut is $V/2 - V = -V/2$, to the left—too slow to catch up with the first astronaut who is still moving at $-V$. The game is over. Both the first and the third got to throw the second astronaut only once!

29. Your friend does twice as much work ($4 \times \frac{1}{2} > 1 \times 1$).

31. Work done by each is the same, for they reach the same height. The one who climbs in 30 s uses more power because work is done in a shorter time.

33. Agree, because speed itself is relative to the frame of reference (Chapter 3). Hence $\frac{1}{2}mv^2$ is also relative to a frame of reference.

35. KE depends on the square of speed, so the faster one, the lighter golf ball, has the greater KE.

37. If the ball is given an initial KE, it will return to its starting position with that KE (moving in the other direction!) and hit the instructor. (The usual classroom procedure is to release the ball from the nose at rest. Then when it returns it will have no KE and will stop short of bumping the nose.)

39. The 100 J of potential energy that doesn't go into increasing her kinetic energy goes into thermal energy—heating her bottom and the slide.

41. You agree with your second classmate. The coaster could just as well encounter a low summit before or after a higher one, so long as the higher one is enough lower than the initial summit to compensate for energy dissipation by friction.

43. If KEs are the same but masses differ, then the ball with smaller mass has the greater speed. That is, $\frac{1}{2}MV^2 = \frac{1}{2}mv^2$. Likewise with molecules, where lighter ones move faster on the average than more massive ones. (We will see in Chapter 8 that temperature is a measure of average molecular KE—lighter molecules in a gas move faster than same-temperature heavier molecules.)

45. The information needed is the distance of rock penetration into the ground. The work that the rock does on the ground is equal to its PE before being dropped, mgh = 100 joules. Without knowing the distance the force acts upon the ground, the force of impact cannot be calculated. (If we knew the time during which the impulse occurs we could calculate the force from the impulse-momentum relationship—but not knowing the distance or time of the rock's penetration into the ground, we cannot calculate the force.)

47. The ball strikes the ground with the *same* speed, whether thrown upward or downward. The ball starts with the same energy at the same place, so they will have the same energy when they reach the ground. This means they will strike with the same speed. This is assuming negligible air resistance, for if air resistance is a factor, then the ball thrown upward will dissipate more energy in its longer path and strike with somewhat less speed. Both hit the ground at the same speed (but at different *times*).

49. The question can be restated; Is $(30^2 - 20^2)$ greater or less than $(20^2 - 10^2)$? We see that $(30^2 - 20^2) = (900 - 400) = 500$, which is considerably greater than $(20^2 - 10^2) = (400 - 100) = 300$. So KE changes more for a given Δv at the higher speed.

51. When the mass is doubled with no change in speed, both momentum and KE are doubled.

53. Both have the same momentum, but the 1-kg one, the faster one, has the greater KE.

55. Net momentum before the lumps collide is zero and is zero after collision. Momentum is indeed conserved. Kinetic energy after is zero, but was greater than zero before collision. The lumps are warmer after colliding because the initial kinetic energy of the lumps transforms into thermal energy. Momentum has only one form. There is no way to "transform" momentum from one form to another, so it is conserved. But energy comes in various forms and can easily be transformed. No single form of energy such as KE need be conserved.

57. An engine that is 100% efficient would not be warm to the touch, nor would its exhaust heat the air, nor would it make any noise, nor would it vibrate. This is because all these are transfers of energy, which cannot happen if all the energy given to the engine is transformed to useful work.

59. Your friend is correct, for changing KE requires work, which means more fuel consumption and decreased air quality.

SOLUTIONS TO CHAPTER 5 PROBLEMS

1. $\frac{F}{m} = \frac{\Delta v}{\Delta t}$. Cross multiply and get $F\Delta t = m\Delta v$. With m constant, $F\Delta t = \Delta(mv)$.

3. (a) $Ft = \Delta mv$; $F = \Delta mv/t = (8\text{ kg})(2\text{ m/s})/0.5\text{ s} = 32\text{ N}$.
   (b) 32 N, in accord with Newton's third law.

5. Momentum of the caught ball is $(0.15\text{ kg})(40\text{ m/s}) = 6.0\text{ kg}\cdot\text{m/s}$.
   (a) The impulse to produce this change of momentum has the same magnitude, $6.0\text{ N}\cdot\text{s}$.
   (b) From $Ft = \Delta mv$, $F = \Delta mv/t = [(0.15\text{ kg})(40\text{ m/s})]/0.03\text{ s} = 200\text{ N}$.

7. Let m be the mass of the freight car, and $4m$ the mass of the diesel engine, and v the speed after both have coupled together. Before collision, the total momentum is due only to the diesel engine, $4m(5\text{ km/h})$, because the momentum of the freight car is 0. After collision, the combined mass is $(4m + m)$, and combined momentum is $(4m + m)v$. By the conservation of momentum equation:
   $\text{Momentum}_{\text{before}} = \text{momentum}_{\text{after}}$
   $4m(5\text{ km/h}) + 0 = (4m + m)v$

   $v = \frac{(20m\cdot\text{km/h})}{5m} = 4\text{ km/h}$ (Note that you don't have to know m to solve the problem.)

9. By momentum conservation, asteroid mass $\times$ 800 m/s = Superman's mass $\times$ v.
   Since asteroid's mass is 1000 times Superman's,
   $(1000m)(800\text{ m/s}) = mv$
   $v = 800{,}000\text{ m/s}$. This is nearly 2 million miles per hour!

11. (a) PE + KE = Total E; KE = 10,000 J − 1,000 J = 9,000 J.

    (b) When Bernie's KE is 9,000 J, his PE is reduced to 1,000 J, which places him 0.1 the flagpole height above water when his KE is 9,000 J.

13. $(F \times d)_{\text{in}} = (F \times d)_{\text{out}}$
    $50\text{ N} \times 1.2\text{ m} = W \times 0.2\text{ m}$
    $W = [(50\text{ N})(1.2\text{ m})]/0.2\text{ m} = 300\text{ N}$.

15. $Fd = mad = ma(\frac{1}{2}at^2) = \frac{1}{2}maat^2 = \frac{1}{2}m(at)^2$. With $at = v$, we get $Fd = \frac{1}{2}mv^2$.

17. (a) The mass ends up at rest, so its momentum has changed by an amount mv. The magnitude of the impulse acting on the mass must also equal mv. From $Ft = \Delta(\text{momentum}) = mv \Rightarrow t = \frac{mv}{F}$.

(b) $t = \frac{mv}{F} = \frac{(20.0\,\text{kg})(3.0\frac{\text{m}}{\text{s}})}{(15.0\,\text{N})} = 4.0\frac{\text{kg}\cdot\frac{\text{m}}{\text{s}}}{\text{kg}\cdot\frac{\text{m}}{\text{s}^2}} = \mathbf{4.0\,s}$.

19. (a) Since no external forces act on the astronaut-hammer system the momentum of the system is conserved. The initial momentum of the system is zero. Afterward, the astronaut's momentum and the hammer's momentum are equal in magnitude but opposite in direction. If we use v to represent speeds we can write $(\text{momentum})_{\text{astronaut}} = (\text{momentum})_{\text{hammer}} \Rightarrow Mv_{\text{astronaut}} = mv_{\text{hammer}} \Rightarrow \text{speed}_{\text{astronaut}} = \frac{mv}{M}$.

(b) $\text{Speed}_{\text{astronaut}} = \frac{mv}{M} = \frac{(15\text{ kg})\left(4.5\frac{\text{m}}{\text{s}}\right)}{110\text{ kg}} = 0.6\,\frac{\text{m}}{\text{s}}$.

21. (a) Since speed $v = x/t$, momentum $= mv = \mathbf{mx/t}$.

(b) $\text{KE} = \frac{1}{2}mv^2 = \frac{1}{2}M\left(\frac{x}{t}\right)^2 = \mathbf{\frac{Mx^2}{2t^2}}$.

(c) For our units to work out we'll need to convert 250 km to m and 8.0 hours to seconds:

$250\,\text{km} \times \frac{1000\,\text{m}}{1\,\text{km}} = 250{,}000\text{ m};\ 8.0\,\text{h} \times \frac{3600\,\text{s}}{1\,\text{h}} = 28{,}800\text{ s}$.

So $\text{KE} = \frac{Mx^2}{2t^2} = \frac{(9.0 \times 10^7\,\text{kg})(250{,}000\,\text{m})^2}{2(28{,}800\text{ s})^2} = 1.4 \times 10^{10}\,\text{kg}\cdot\frac{\text{m}^2}{\text{s}^2} = \mathbf{3.4 \times 10^9 J}$.

23. (a) $d = ?$ From $W = Fd$ and $W = \Delta\text{KE} \Rightarrow Fd = \Delta\text{KE} \Rightarrow d = \frac{\frac{1}{2}mv^2}{F} = \mathbf{\frac{mv^2}{2F}}$.

(b) If both the distance and the force are doubled, four times as much work is done, which produces four times as much change in kinetic energy. Formally, $F_0d_0 = W_0 = (\Delta\text{KE})_0$ becomes $(2F_0)(2d_0) = 4(F_0d_0) = 4W_0 = \mathbf{4(\Delta KE)_0}$.

25. (a) Since the ice drops a vertical distance h its PE decreases by mgh. Its KE then increases by the same amount. $\text{KE}_{\text{at bottom}} = \frac{1}{2}mv^2$ and $\text{KE}_{\text{at bottom}} = mgh \Rightarrow \frac{1}{2}mv^2 = mgh \Rightarrow v = \mathbf{\sqrt{2gh}}$. Note that the mass doesn't enter into the solution. *Any* mass of ice sliding friction free will arrive at the bottom of the ramp with the same speed.

(b) $v = \sqrt{2gh} = \sqrt{2(9.8\frac{\text{m}}{\text{s}^2})(1.5\,\text{m})} = \mathbf{5.4\frac{m}{s}}$.

SOLUTIONS TO CHAPTER 6 EXERCISES

1. Nothing to be concerned about on this consumer label. It simply states the universal law of gravitation, which applies to *all* products. It looks like the manufacturer knows some physics and has a sense of humor.

3. The force of gravity is the same on each because the masses are the same, as Newton's equation for gravitational force verifies.

5. Astronauts are weightless because they lack a support force, but they are well in the grips of Earth gravity, which accounts for them circling the Earth rather than going off in a straight line in outer space.

7. Less, because an object there is farther from Earth's center.

9. Letting the equation for gravitation guide your thinking, twice the mass means twice the force, and twice the distance means one-quarter the force. Combined, the astronaut weighs half as much.

11. By the geometry of Figure 6.6, tripling the distance from the small source spreads the light over 9 times the area, or 9 m^2. Five times the distance spreads the light over 25 times the area or 25 m^2, and for 10 times as far, 100 m^2.

13. The gravitational force on a body, its weight, depends not only on mass but distance. On Jupiter, this is the distance between the body being weighed and Jupiter's center—the radius of Jupiter. If the radius of Jupiter were the same as that of the Earth, then a body would weigh 300 times as much because Jupiter is 300 times more massive than Earth. But Jupiter is also much bigger than the Earth, so the greater distance between its center and the CG of the body reduces the gravitational force. The radius is great enough to make the weight of a body only 3 times its Earth weight. The radius of Jupiter is in fact eleven times that of Earth.

15. In a car that drives off a cliff you "float" because the car no longer offers a support force. Both you and the car are in the same state of free fall. But gravity is still acting on you, as evidenced by your acceleration toward the ground. So, by definition, you would be weightless (until air resistance becomes important).

17. The pencil has the same state of motion that you have. The force of gravity on the pencil causes it to accelerate downward alongside of you. Although the pencil hovers relative to you, it and you are falling relative to the Earth.

19. First of all, it would be incorrect to say that the gravitational force of the distant Sun on you is too small to be measured. It's small, but not immeasurably small. If, for example, the Earth's axis were supported such that the Earth could continue turning but not otherwise move, an 85-kg person would see a gain of $\frac{1}{2}$ newton on a bathroom scale at midnight and a loss of $\frac{1}{2}$ newton at noon. The key idea is *support.* There is no "Sun support" because the Earth and all objects on the Earth—you, your bathroom scale, and everything else—are continually falling around the Sun. Just as you wouldn't be pulled against the seat of your car if it drives off a cliff, and just as a pencil is not pressed against the floor of an elevator in free fall, we are not pressed against or pulled from the Earth by our gravitational interaction with the Sun. That interaction keeps us and the Earth circling the Sun, but does not press us to the Earth's surface. Our interaction with the Earth does that.

21. The misunderstanding here is not distinguishing between a theory and a hypothesis or conjecture. A theory, such as the theory of universal gravitation, is a synthesis of a large body of information that encompasses well-tested and verified hypothesis about nature. Any doubts about the theory have to do with its applications to yet untested situations, not with the theory itself. One of the features of scientific theories is that they undergo refinement with new knowledge. (Einstein's general theory of relativity has taught us that in fact there are limits to the validity of Newton's theory of universal gravitation.)

23. The crate will not hit the Porsche, but will crash a distance beyond it determined by the height and speed of the plane.

25. Minimum speed occurs at the top, which is the same as the horizontal component of velocity anywhere along the path.

27. Both balls have the same range (see Figure 6.22). The ball with the initial projection angle of 30°, however, is in the air for a shorter time and hits the ground first.

29. Any vertically projected object has zero speed at the top of its trajectory. But if it is fired at an angle, only its vertical component of velocity is zero and the velocity of the projectile at the top is equal to its horizontal component of velocity. This would be 100 m/s when the 141-m/s projectile is fired at 45°.

31. The hang time will be the same, in accord with the answer to the preceding exercise. Hang time is related to the vertical height attained in a jump, not on horizontal distance moved across a level floor.

33. Neither the speed of a falling object (without air resistance) nor the speed of a satellite in orbit depends on its mass. In both cases, a greater mass (greater inertia) is balanced by a correspondingly greater gravitational force, so the acceleration remains the same ($a = F/m$, Newton's 2nd law).

35. Gravity changes the speed of a cannonball when the cannonball moves in the direction of Earth gravity. At low speeds, the cannonball curves downward and gains speed because there is a component of the force of gravity along its direction of motion. Fired fast enough, however, the curvature matches the curvature of the Earth so the cannonball moves at right angles to the force of gravity. With no component of force along its direction of motion, its speed remains constant.

37. Consider "Newton's cannon" fired from a tall mountain on Jupiter. To match the wider curvature of much larger Jupiter, and to contend with Jupiter's greater gravitational pull, the cannonball would have to be fired significantly faster. (Orbital speed about Jupiter is about 5 times that for Earth.)

39. Hawaii is closer to the equator, and therefore has a greater tangential speed about the polar axis. This speed could be added to the launch speed of a satellite and thereby save fuel.

41. When the velocity of a satellite is everywhere perpendicular to the force of gravity, the orbital path is a circle.

43. If a wrench or anything else is "dropped" from an orbiting space vehicle, it has the same tangential speed as the vehicle and remains in orbit. If a wrench is dropped from a high-flying jumbo jet, it too has the tangential speed of the jet. But this speed is insufficient for the wrench to fall around and around the Earth. Instead it soon falls into the Earth.

45. Communication satellites only appear motionless because their orbital period coincides with the daily rotation of the Earth.

47. The half brought to rest will fall vertically to Earth. The other half, in accord with the conservation of linear momentum will have twice the initial velocity, overshoot the circular orbit, and enter an elliptical orbit whose apogee (highest point) is farther from the Earth's center.

49. The satellite experiences the greatest gravitational force at A, where it is closest to the Earth; and the greatest speed and the greatest velocity at A, and by the same token the greatest momentum and greatest kinetic energy at A, and the greatest gravitational potential energy at the farthest point C. It would have the same total energy (KE + PE) at all parts of its orbit. It would have the greatest acceleration at A, where F/m is greatest.

SOLUTIONS TO CHAPTER 6 PROBLEMS

1. From $F = GmM/d^2$, $\frac{1}{5}$ of d squared is $\frac{1}{25}^{\text{th}}$ d^2, which means the force is 25 times greater.

3. $$\frac{F\text{ on neutron star}}{F\text{ on Earth}} = \frac{GmM_n/d_n^2}{GmM_E/d_E^2} = \frac{M_n d_E^2}{M_E d_n^2} = \frac{(3.0 \times 10^{30}\text{ kg})(6380\text{ km})^2}{(6 \times 10^{24}\text{ kg})(8\text{ km})^2} = 3.2 \times 10^{11}.$$

   This is about 300 billion times the force of gravity on Earth's surface.

5. (a) From $y = 5t^2 = 5(30)^2 = 4{,}500$ m, or 4.5 km high (4.4 km if we use $g = 9.8\text{ m/s}^2$).

   (b) In 30 seconds the falling engine travels horizontally 8400 m ($d = vt = 280\text{ m/s} \times 30\text{ s} = 8400\text{ m}$).

   (c) The engine is directly below the airplane. (In a more practical case, air resistance is overcome for the plane by its engines, but not for the falling engine, so the engine's speed is reduced by air drag and it covers less than 8400 horizontal meters, landing behind the plane.)

7. In accord with the work-energy theorem (Chapter 5) $W = \Delta KE$ the work done equals energy gained. The KE gain is 8 billion joules − 5 billion joules = 3 billion joules. The potential energy decreases by the same amount that the kinetic energy increases, 3 billion joules.

9. If we use the equation of the previous problem:

   $$v = \sqrt{\frac{GM}{d}} = \frac{(6.67 \times 10^{-11})(2 \times 10^{30})}{1.5 \times 10^{11}} = 3 \times 10^4\text{ m/s}.$$

   Another way is: $v = \textit{distance/time}$ where distance is the circumference of the Earth's orbit and time is 1 year. Then

   $$v = \frac{d}{t} = \frac{2\pi r}{1\text{ year}} = \frac{2\pi(1.5 \times 10^{11}\text{ m})}{365\text{ day} \times \frac{24\text{ h}}{\text{day}} \times \frac{3600\text{ s}}{\text{h}}} = 3 \times 10^4\text{ m/s} = 30\text{ km/s}.$$

11. (a) $a = \dfrac{F}{m} = G\dfrac{mM/d^2}{m} = G\dfrac{M}{d^2}$.

    (b) Note that mass of the accelerating object, m, cancels for acceleration. Only the mass of the object M pulling on m affects acceleration (not the object being pulled).

13. (a) Because of the independence of horizontal and vertical components of velocity, the falling time depends only on height y. Since there is no initial velocity in the vertical direction, the vertical distance is simply $y = \frac{1}{2}gt^2$ (as we learned in Chapter 3).

    (b) $y = \frac{1}{2}gt^2 = \frac{1}{2}(9.8\text{ m/s}^2)(2\text{ s})^2 = 19.6$ m (or 20 m for $g = 10\text{ m/s}^2$).

    (c) To solve this problem one needs to know that the horizontal and vertical components of velocity are independent of each other. That means falling distance is not affected by the horizontal component of velocity. Hence the bridge height is simply the height required for a freely falling rock to fall a vertical distance y. And that's $y = \frac{1}{2}gt^2$ (as we learned in Chapter 3).

15. (a) $d = v_x t = vt$.
    The time t is the same time for the penny to fall distance y.

    From $y = \dfrac{1}{2}gt^2 \Rightarrow t^2 = \dfrac{2y}{g} \Rightarrow t = \sqrt{\dfrac{2y}{g}}$. So $d = v\sqrt{\dfrac{2y}{g}}$.

    (b) $d = v\sqrt{\dfrac{2y}{g}} = 3.5\text{ m/s}\sqrt{\dfrac{2(0.4\text{ m})}{9.8\text{ m/s}^2}} = 1.0$ m.

SOLUTIONS TO CHAPTER 7 EXERCISES

1. The scale measures force, not pressure, and is calibrated to read your weight. That's why your weight on the scale is the same whether you stand on one foot or both.

3. Like the loaf of bread in Figure 7.1, its volume is decreased. Its mass stays the same so the density increases. A whale is denser when it swims deeper in the ocean.

5. A person lying on a waterbed experiences less body weight pressure because more of the body is in contact with the supporting surface. The greater area reduces the support pressure.

7. A woman with spike heels exerts considerably more pressure on the ground than an elephant! Example: A 500-N woman with 1-cm^2 spike heels puts half her weight on each foot, distributed (let's say) half on her heel and half on her sole. So the pressure exerted by each heel will be (125 N/1 cm^2) = 125 N/cm^2. A 20,000-N elephant with 1000 cm^2 feet exerting $\frac{1}{4}$ its weight on each foot produces (5000N/1000 cm^2) = 5N/cm^2; about 25 times less pressure. (So a woman with spike heels will make greater dents in a new linoleum floor than an elephant will.)

9. In deep water, you are buoyed up by the water displaced and as a result, you don't exert as much pressure against the stones on the bottom. When you are up to your neck in water, you hardly feel the bottom at all.

11. As per the Floating Mountains in the chapter, mountain ranges are very similar to icebergs: Both float in a denser medium, and extend farther down into that medium than they extend above it.

13. Heavy objects may or may not sink, depending on their densities (a heavy log floats while a small rock sinks, or a boat floats while a paper clip sinks, for example). People who say that heavy objects sink really mean that dense objects sink. Be careful to distinguish between how heavy an object is and how dense it is.

15. The water level will fall. This is because the iron will displace a greater amount of water while being supported than when submerged. A floating object displaces its weight of water, which is more than its own volume, while a submerged object displaces only its volume. (This may be illustrated in the kitchen sink with a dish floating in a dishpan full of water. Silverware in the dish takes the place of the scrap iron. Note the level of water at the side of the dishpan, and then throw the silverware overboard. The floating dish will float higher and the water level at the side of the dishpan will fall. Will the volume of the silverware displace enough water to bring the level to its starting point? No, not as long as it is denser than water.)

17. The balloon will sink to the bottom because its density increases with depth. The balloon is compressible, so the increase in water pressure beneath the surface compresses it and reduces its volume, thereby increasing its density. Density is further increased as it sinks to regions of greater pressure and compression. This sinking is understood also from a buoyant force point of view. As its volume is reduced by increasing pressure as it descends, the amount of water it displaces becomes less. The result is a decrease in the buoyant force that initially was sufficient to barely keep it afloat.

19. Since both preservers are the same size, they will displace the same amount of water when submerged and be buoyed up with equal forces. Effectiveness is another story. The amount of buoyant force exerted on the heavy lead-filled preserver is much less than its weight. If you wear it, you'll sink. The same amount of buoyant force exerted on the lighter Styrofoam preserver is greater than its weight and it will keep you afloat. The *amount* of the force and the *effectiveness* of the force are two different things.

21. When the ice cube melts the water level at the side of the glass is unchanged (neglecting temperature effects). To see this, suppose the ice cube to be a 5 gram cube; then while floating it will displace 5 grams of water. But when melted it becomes the same 5 grams of water. Hence the water level is unchanged. The same occurs when the ice cube with the air bubbles melts.

Whether the ice cube is hollow or solid, it will displace as much water floating as it will melted. If the ice cube contains grains of heavy sand, however, upon melting, the water level at the edge of the glass will drop. This is similar to the case of the scrap iron of Exercise 15.

23. Because of surface tension, which tends to minimize the surface of a blob of water, its shape without gravity and other distorting forces will be a *sphere*—the shape with the least surface area for a given volume.

25. Some of the molecules in the Earth's atmosphere *do* go off into outer space—those like helium with speeds greater than escape speed. But the average speeds of most molecules in the atmosphere are well below escape speed, so the atmosphere is held to Earth by Earth gravity.

27. The density of air in a deep mine is greater than at the surface. The air filling up the mine adds weight and pressure at the bottom of the mine, and according to Boyle's law, greater pressure in a gas means greater density.

29. If the item is sealed in an airtight package at sea level, then the pressure in the package is about 1 atmosphere. Cabin pressure is reduced somewhat for high altitude flying, so the pressure in the package is greater than the surrounding pressure and the package therefore puffs outwards.

31. The can collapses under the weight of the atmosphere. When water was boiling in the can, much of the air inside was driven out and replaced by steam. Then, with the cap tightly fastened, the steam inside cooled and condensed back to the liquid state, creating a partial vacuum in the can which could not withstand the crushing force of the atmosphere outside.

33. A vacuum cleaner wouldn't work on the Moon. A vacuum cleaner operates on Earth because the atmospheric pressure pushes dust into the machine's region of reduced pressure. On the Moon there is no atmospheric pressure to push the dust anywhere.

35. Drinking through a straw is slightly more difficult atop a mountain. This is because the reduced atmospheric pressure is less effective in pushing soda up into the straw.

37. One's lungs, like an inflated balloon, are compressed when submerged in water, and the air within is compressed. Air will not of itself flow from a region of low pressure into a region of higher pressure. The diaphragm in one's body reduces lung pressure to permit breathing, but this limit is strained when nearly 1 m below the water surface. The limit is exceeded at more than a 1-m depth.

39. An object rises in air only when buoyant force exceeds its weight. A steel tank of anything weighing more than the air it displaces, so won't rise. A helium-filled balloon weighs less than the air it displaces and rises.

41. The force of the atmosphere is on both sides of the window; the net force is zero, so windows don't normally break under the weight of the atmosphere. In a strong wind, however, pressure will be reduced on the windward side (Bernoulli's Principle) and the forces no longer cancel to zero. Many windows are blown outward in strong winds.

43. (a) Speed increases (so that the same quantity of gas can move through the pipe in the same time).

    (b) Pressure decreases (Bernoulli's principle).

    (c) The spacing between the streamlines decreases, because the same number of streamlines fit in a smaller area.

45. The air density and pressure are less at higher altitude, so the wings (and, with them, the whole airplane) are tilted to a greater angle to produce the needed pressure difference between the upper and lower surfaces of the wing. In terms of force and air deflection, the greater angle of attack is needed to deflect a greater volume of lower-density air downward to give the same upward force.

47. The troughs are partially shielded from the wind, so the air moves faster over the crests than in the troughs. Pressure is therefore lower at the top of the crests than down below in the troughs. The greater pressure in the troughs pushes the water into even higher crests.

SOLUTIONS TO CHAPTER 7 PROBLEMS

1. A 5-kg ball weighs 49 N, so the pressure is 49 $N/cm^2 \times (100\ cm/1\ m)^2 = 490{,}000\ N/m^2$ = 490 kPa.

3. Pressure = weight density × depth = 9800 $N/m^3 \times 406\ m$ = 3,978, 800 N/m^2 = 3978.8 kPa. Total pressure, add that due to atmospheric: 3,978.8 kPa + 101.3 = 4,080.1 kPa.

5. Now $A = 5\ m \times 2\ m = 10\ m^2$; to find the volume V of barge pushed into the water by the weight of the block, which equals the volume of water displaced, we know that density $= \frac{m}{V}$.

Or from this, $V = \frac{\text{mass}}{\text{density}} = \frac{400\ \text{kg}}{1000\ \text{kg/m}^3} = 0.4\ \text{m}^3$.

So $h = \frac{V}{A} = \frac{0.4\ \text{m}^3}{10\ \text{m}^2} = 0.04$ m, which is 4 cm deeper.

(b) If each block will push the barge 4 cm deeper, the question becomes: How many 4-cm increments will make 15 cm? 15/4 = 3.75, so 3 blocks can be carried without sinking. Four blocks of the same weight will sink the barge.

7. 10% of ice extends above water. So 10% of the 9-cm thick ice would float above the water line; 0.9 cm. So the ice pops up. Interestingly, when mountains erode they become lighter and similarly pop up! Hence it takes a long time for mountains to wear away.

9. According to Boyle's law, the product of pressure and volume is constant (at constant temperature), so one-tenth the volume means ten times the pressure.

11. Since weight = mg, the mass of the displaced air is $m = W/g$ = (20,000 N)/(10 m/s^2) = 2000 kg. Since density is mass/volume, the volume of the displaced air is volume = mass/density = (2000 kg)/(1.2 kg/m^3) = 1700 m^3 (same answer to two figures if $g = 9.8\ m/s^2$ is used).

13. Lift will equal the difference in force below and above the wing surface. The difference in force will equal the difference in air pressure × wing area.

Lift = 0.04 PA = (0.04)($10^5\ N/m^2$)(100 m^2) = 4×10^5 N. (That's about 44 tons.)

SOLUTIONS TO CHAPTER 8 EXERCISES

1. Gas molecules move haphazardly at random speeds. They continually run into one another, sometimes giving kinetic energy to neighbors, sometimes receiving kinetic energy. In this continual interaction, it would be statistically impossible for any large number of molecules to have the same speed. Temperature has to do with average speeds.

3. You cannot establish by your own touch whether or not you are running a fever because there would be no temperature difference between your hand and forehead. If your forehead is a couple of degrees higher in temperature than normal, your hand is also a couple of degrees higher.

5. The hot coffee has a higher temperature, but not a greater internal energy. Although the iceberg has less internal energy per mass, its enormously greater mass gives it a greater total energy than that in the small cup of coffee. (For a smaller volume of ice, the fewer number of more energetic molecules in the hot cup of coffee may constitute a greater total amount of internal energy—but not compared to an iceberg.)

7. No, for a difference of 273 in 10,000,000 is insignificant.

9. Work is done in compressing the air, which in accord with the first law of thermodynamics, increases its thermal energy. This is evident by its increased temperature.

11. You do work on the liquid when you vigorously shake it, which increases its thermal energy. The temperature change should be noticeable.

13. The tires heat up, which heats the air within. The molecules in the heated air move faster, which increases air pressure in the tires.

15. The brick will cool off too fast and you'll be cold in the middle of the night. Bring a jug of hot water with its higher specific heat to bed and you'll make it through the night.

17. Different substances have different thermal properties due to differences in the way energy is stored internally in the substances. When the same amount of heat produces different changes in temperatures in two substances of the same mass, we say they have different specific heat capacities. Each substance has its own characteristic specific heat capacity. Temperature measures the average kinetic energy of random motion, but not other kinds of energy.

19. The higher specific heat of the water results in water absorbing more heat than the metal pan.

21. In winter months when the water is warmer than the air, the air is warmed by the water to produce a seacoast climate warmer than inland. In summer months when the air is warmer than the water, the air is cooled by the water to produce a seacoast climate cooler than inland. This is why seacoast communities and especially islands do not experience the high and low temperature extremes that characterize inland locations.

23. Water is an exception. Below 4 degrees Celsius, it expands when cooled.

25. When the rivets cool they contract. This tightens the plates being attached.

27. Cool the inner glass and heat the outer glass. If it's done the other way around, the glasses will stick even tighter (if not break).

29. Every part of a metal ring expands when it is heated—not only the thickness, but the outer and inner circumference as well. Hence the ball that normally passes through the hole when the temperatures are equal will more easily pass through the expanded hole when the ring is heated. (Interestingly, the hole will expand as much as a disk of the same metal undergoing the same increase in temperature. Blacksmiths mounted metal rims in wooden wagon wheels by first heating the rims. Upon cooling, the contraction resulted in a snug fit.)

31. The gap in the ring will become wider when the ring is heated. Try this: draw a couple of lines on a ring where you pretend a gap to be. When you heat the ring, the lines will be farther apart—the same amount as if a real gap were there. Every part of the ring expands proportionally when heated uniformly—thickness, length, gap and all.

33. Water has the greatest density at 4°C; therefore, either cooling or heating at this temperature will result in an expansion of the water. A small rise in water level would be ambiguous and make a water thermometer impractical in this temperature region.

35. At 0°C it will contract when warmed a little; at 4°C it will expand, and at 6°C it will expand.

37. If cooling occurred at the bottom of a pond instead of at the surface, ice would still form at the surface, but it would take much longer for ponds to freeze. This is because all the water in the pond would have to be reduced to a temperature of 0°C rather than 4°C before the first ice would form. Ice that forms at the bottom where the cooling process is occurring would be less dense and would float to the surface (except for ice that may form on material anchored to the bottom of the pond).

SOLUTIONS TO CHAPTER 8 PROBLEMS

1. (a) The amount of heat absorbed by the water is $Q = cm\Delta T = (1.0\text{ cal/g°C})(50.0\text{ g})(50°\text{C} - 22°\text{C}) = 1400\text{ cal}$. At 40% efficiency only 0.4 the energy from the peanut raises the water temperature, so the calorie content of the peanut is 1400/0.4 = 3500 cal.

   (b) The food value of a peanut is 3500 cal/0.6 g = 5.8 kilocalories per gram.

3. Each kilogram requires 1 kilocalorie for each degree change, so 100 kg needs 100 kilocalories for each degree change. Twenty degrees means twenty times this, which is 2000 kcal.

   By formula, $Q = cm\Delta T = (1\text{ cal/g°C})(100{,}000\text{ g})(20°\text{C}) = 2000\text{ kcal}$. We can convert this to joules knowing that 4.18 J = 1 cal. In joules this quantity of heat is 8,360 kJ.

5. Heat gained by water = heat lost by nails
   $(cm\Delta T)_{water} = (cm\Delta T)_{nails}$
   $(1)(100)(T - 20) = (0.12)(100)(40 - T)$, giving $T = 22.1°\text{C}$

7. By formula: $\Delta L = L_0 a \Delta T = (1300\text{ m})(11 \times 10^{-6}/°\text{C})(15°\text{C}) = 0.21\text{ m}$.

SOLUTIONS TO CHAPTER 9 EXERCISES

1. No, the coat is not a source of heat, but merely keeps the thermal energy of the wearer from leaving rapidly.

3. Copper and aluminum are better conductors than stainless steel, and therefore more quickly transfer heat to the cookware's interior.

5. In touching the tongue to very cold metal, enough heat can be quickly conducted away from the tongue to bring the saliva to sub-zero temperature where it freezes, locking the tongue to the metal. In the case of relatively nonconducting wood, much less heat is conducted from the tongue and freezing does not take place fast enough for sudden sticking to occur.

7. Heat from the relatively warm ground is conducted by the gravestone to melt the snow in contact with the gravestone. Likewise for trees or any materials that are better conductors of heat than snow, and that extend into the ground.

9. The conductivity of wood is relatively low whatever the temperature—even in the stage of red-hot coals. You can safely walk barefoot across red-hot wooden coals if you step quickly (like removing the wooden-handled pan with bare hands quickly from the hot oven in the previous exercise) because very little heat is conducted to your feet. Because of the poor conductivity of the coals, energy from within the coals does not readily replace the energy that transfers to your feet. This is evident in the diminished redness of the coal after your foot has left it. Stepping on red-hot *iron* coals, however, is a different story. Because of the excellent conductivity of iron, very damaging amounts of heat would transfer to your feet. More than simply ouch!

11. Disagree, for having the same KE does not mean having the same speed, unless all gas molecules have equal masses.

13. The smoke, like hot air, is less dense than the surroundings and is buoyed upward. It cools with contact with the surrounding air and becomes more dense. When its density matches that of the surrounding air, its buoyancy and weight balance and rising ceases.

15. If they have the same temperature, then by definition, they have the same kinetic energies per molecule.

17. As in the explanation of the previous exercise, the molecules of gas with the lesser mass will have the higher average speeds. A look at the periodic table will show that argon (A = 18) has less massive atoms than krypton (A = 36). The faster atoms are those of argon. This is the case whether or not the gases are in separate containers.

19. When we warm a volume of air, we add energy to it. When we expand a volume of air, we normally take energy out of it (because the expanding air does work on its surroundings). So the conditions are quite different and the results will be different. Expanding a volume of air actually lowers its temperature.

21. The heat you received was from radiation.

23. A good reflector is a poor radiator of heat, and a poor reflector is a good radiator of heat.

25. Put the cream in right away for at least three reasons. Since black coffee radiates more heat than white coffee, make it whiter right away so it won't radiate and cool so quickly while you are waiting. Also, by Newton's law of cooling, the higher the temperature of the coffee above the surroundings, the greater will be the rate of cooling—so again add cream right away and lower the temperature to that of a reduced cooling rate, rather than allowing it to cool fast and then bring the temperature down still further by adding the cream later. Also—by adding the cream, you increase the total amount of liquid, which for the same surface area, cools more slowly.

27. Turn your heater off altogether and save fuel. When it is cold outside, your house is constantly losing heat. How much is lost depends on the insulation and the difference in inside and outside temperature (Newton's law of cooling). Keeping ΔT high consumes more fuel. To consume less fuel, keep ΔT low and turn your heater off altogether. Will more fuel be required to reheat the house when you return than would have been required to keep it warm while you were away? Not at all. When you return, you are replacing heat lost by the house at an average temperature below the normal setting, but if you had left the heater on, it would have supplied more heat, enough to make up for heat lost by the house at its normal, higher temperature setting. (Perhaps your instructor will demonstrate this with the analogy of leaking water buckets.)

29. When it is desirable to reduce the radiant energy coming into a greenhouse, whitewash is applied to the glass simply to reflect much of the incoming sunlight. Energy reflected is energy not absorbed by the greenhouse.

31. For maximum warmth, wear the plastic coat on the outside and utilize the greenhouse effect.

33. In this hypothetical case evaporation would not cool the remaining liquid because the energy of exiting molecules would be no different than the energy of molecules left behind. Although internal energy of the liquid would decrease with evaporation, energy per molecule would not change. No temperature change of the liquid would occur. (The surrounding air, on the other hand, would be cooled in this hypothetical case. Molecules flying away from the liquid surface would be slowed by the attractive force of the liquid acting on them.)

35. A bottle wrapped in wet cloth will cool by the evaporation of liquid from the cloth. As evaporation progresses, the average temperature of the liquid left behind in the cloth can easily drop below the temperature of the cool water that wet it in the first place. So to cool a bottle of beer, soda, or whatever at a picnic, wet a piece of cloth in a bucket of cool water. Wrap the wet cloth around the bottle to be cooled. As evaporation progresses, the temperature of the water in the cloth drops, and cools the bottle to a temperature below that of the bucket of water.

37. As the bubbles rise, less pressure is exerted on them.

39. When the jar reaches the boiling temperature, further heat does not enter it because it is in thermal equilibrium with the surrounding 100°C water. This is the principle of the "double boiler."

41. As in the answer to the previous exercise, high temperature and the resulting internal energy given to the food are responsible for cooking—if the water boils at a low temperature (presumably under reduced pressure), the food isn't hot enough to cook.

43. The lid on the pot traps heat, which quickens boiling; the lid also slightly increases pressure on the boiling water which raises its boiling temperature. The hotter water correspondingly cooks food in a shorter time, although the effect is not significant unless the lid is held down as on a pressure cooker.

45. The wood, because its greater specific heat capacity means it will release more energy in cooling.

47. The answer to this is similar to the previous answer, and also the fact that the coating of ice acts as an insulating blanket. Every gram of water that freezes releases 80 calories, much of it to the fruit; the thin layer of ice then acts as an insulating blanket against further loss of heat.

SOLUTIONS TO CHAPTER 9 PROBLEMS

1. From $-273°C$ "ice" to $0°C$ ice requires $(273)(0.5) = 140$ calories.
   From 0°C ice to 0°C water requires 80 calories.
   From 0°C water to 100°C water requires 100 calories.
   The total is 320 calories.
   Boiling this water at 100°C takes 540 calories, considerably more energy than it took to bring the water all the way from absolute zero to the boiling point! (In fact, at very low temperature, the specific heat capacity of ice is less than 0.5 cal/g°C, so the true difference is even greater than calculated here.)

3. $0.5mgh = cm\Delta T$
   $\Delta T = 0.5mgh/cm = 0.5gh/c = (0.5)(9.8 \text{ m/s}^2)(100 \text{ m})/450 \text{ J/kg} = 1.1°C$.
   Note that the mass cancels, so the same temperature would hold for any mass ball, assuming half the heat generated goes into warming the ball. Note that the units check since $1 \text{ J/kg} = 1 \text{ m}^2/\text{s}^2$.

5. The final temperature of the water will be the same as that of the ice, 0°C. The quantity of heat given to the ice by the water is
   $Q = cm\Delta T = (1 \text{ cal/g°C})(50 \text{ g})(80°C) = 4000 \text{ cal}$. This heat melts ice. How much?
   From $Q = mL$, $m = Q/L = (4000 \text{ cal})/(80 \text{ cal/g}) = 50$ grams. So water at 80°C will melt an equal mass of ice at 0°C.

7. Note that the heat of vaporization of ethyl alcohol (200 cal/g) is 2.5 times more than the heat of fusion of water (80 cal/g), so in a change of phase for both, 2.5 times as much ice will change phase; $2.5 \times 2 \text{ kg} = 5 \text{ kg}$.
   Or via formula, the refrigerant would draw away $Q = mL = (2000 \text{ g})(200 \text{ cal/g}) = 4 \times 10^5$ calories. The mass of ice formed is then $(4 \times 10^5 \text{ cal})/(80 \text{ cal/g}) = 5000$ g, or 5 kg.

SOLUTIONS TO CHAPTER 10 EXERCISES

1. There are no positives and negatives in gravitation—the interactions between masses are only attractive, whereas electrical interactions may be attractive as well as repulsive. The mass of one particle cannot "cancel" the mass of another, whereas the charge of one particle can cancel the effect of the opposite charge of another particle.

3. Excess electrons rubbed from your hair leave it with a positive charge; excess electrons on the comb give it a negative charge.

5. Cosmic rays produce ions in air, which offer a conducting path for the discharge of charged objects. Cosmic-ray particles streaming downward through the atmosphere are attenuated by radioactive decay and by absorption, so the radiation and the ionization are stronger at high altitude than at low altitude. Charged objects more quickly lose their charge at higher altitudes.

7. Electrons are easily dislodged from the outer regions of atoms, but protons are held tightly within the nucleus.

9. The electrons don't fly out of the penny because they are attracted to the five thousand billion billion positively charged protons in the atomic nuclei of atoms in the penny.

11. The inverse-square law is at play here. At half the distance the electric force field is four times as strong; at $\frac{1}{4}$ the distance, 16 times stronger. At four times the distance, one-sixteenth as strong.

13. For both electricity and heat, the conduction is via electrons, which in a metal are loosely bound, easy flowing, and easy to get moving. (Many fewer electrons in metals take part in heat conduction than in electric conduction, however.)

15. The forces on the electron and proton will be equal in magnitude, but opposite in direction. Because of the greater mass of the proton, its acceleration will be less than that of the electron, and be in the direction of the electric field. How much less? Since the mass of the proton is nearly 2000 times that of the electron, its acceleration will be about $\frac{1}{2000}$ that of the electron. The greater acceleration of the electron will be in the direction opposite to the electric field. The electron and proton accelerate in opposite directions.

17. Voltage $= \dfrac{0.5\text{ J}}{0.0001\text{ C}} = 5000\text{ V}.$

19. The cooling system of an automobile is a better analogy to an electric circuit because like an electric system it is a closed system, and it contains a pump, analogous to the battery or other voltage source in a circuit. The water hose does not re-circulate the water as the auto cooling system does.

21. Your tutor is wrong. An ampere measures current, and a volt measures electric potential (electric pressure). They are entirely different concepts; voltage produces amperes in a conductor.

23. Current flows *through* electrical devices, just as water flows through a plumbing circuit of pipes. If a water pump produces water pressure, water flows through both the pump and the circuit. Likewise with electric current in an electric circuit. For example, in a simple circuit consisting of a battery and a lamp, the electric current that flows in the lamp is the same electric current that flows through the wires that connect the lamp and the same electric current that flows through the battery. Current flows through these devices. (As a side point, it is common to speak of electric current flowing in a circuit, but strictly speaking, it is electric *charge* that flows in an electric circuit; the flow of charge is current. So if you want to be precisely correct grammatically, say that current is in a circuit and charge *flows* in a circuit.)

25. A lie detector circuit relies on the likelihood that the resistivity of your body changes when you tell a lie. Nervousness promotes perspiration, which lowers the body's electrical resistance, and

increases whatever current flows. If a person is able to lie with no emotional change and no change in perspiration, then such a lie detector will not be effective. (Better lying indicators focus on the eyes.)

27. The thick filament has less resistance and will draw (carry) more current than a thin wire connected across the same potential difference. (Important point: It is common to say that a certain resistor "draws" a certain current, but this may be misleading. A resistor doesn't "attract" or "draw" current, just as a pipe in a plumbing circuit doesn't "draw" water; it instead "allows" or "provides for" the passage of current when an electrical pressure is established across it.)

29. Current will be greater in the bulb connected to the 220-volt source. Twice the voltage would produce twice the current if the resistance of the filament remained the same. (In practice, the greater current produces a higher temperature and greater resistance in the lamp filament, so the current is greater than that produced by 110 volts, but appreciably less than twice as much for 220 volts. A bulb rated for 110 volts has a very short life when operated at 220 volts.)

31. In the first case the current passes through your chest; in the second case current passes only through your arm. You can cut off your arm and survive, but you cannot survive without your heart.

33. Auto headlights are wired in parallel. Then when one burns out, the other remains lit. If you've ever seen an automobile with one burned out headlight, you have evidence they're wired in parallel.

35. (a) volt, (b) ampere, (c) joule.

37. The equivalent resistance of resistors in parallel is less than the smaller resistance of the two. So connect a pair of resistors in parallel for less resistance.

39. Agree with your friend, for in series, more resistances add to the circuit resistance. But in parallel, the multiple paths provide less resistance (just as more lines at a checkout counter lessens resistance to flow).

41. Zero. Power companies do not sell electrons; they sell energy. Whatever number of electrons flow into a home, the same number flows out.

43. Bulbs will glow brighter when connected in parallel, for the voltage of the battery is impressed across each bulb. When two identical bulbs are connected in series, half the voltage of the battery is impressed across each bulb. The battery will run down faster when the bulbs are in parallel.

45. Bulb C is the brightest because the voltage across it equals that of the battery. Bulbs A and B share the voltage of the parallel branch of the circuit and have half the current of bulb C (assuming resistances are independent of voltages). If bulb A is unscrewed, the top branch is no longer part of the circuit and current ceases in both A and B. They no longer give light, while bulb C glows as before. If bulb C is instead unscrewed, then it goes out and bulbs A and B glow as before.

47. The three circuits are equivalent. Each branch is individually connected to the battery.

49. Current divides in a branch with more passing in the branch of lower resistance. But current in a branch never reduces to zero unless the resistance of the branch become infinite. In a non-infinite resistor, a voltage across it will produce current in accord with Ohm's law.

51. Some current flows in every branch of a parallel circuit.

SOLUTIONS TO CHAPTER 10 PROBLEMS

1. From Coulomb's law, $F = k\dfrac{q_1 q_2}{d^2} = (9 \times 10^9)\dfrac{(1.0 \times 10^{-6})^2}{(0.03)^2} = 10$ N.

   This is the same as the weight of a 1-kg mass.

3. From Coulomb's law, the force is given by $F = \dfrac{kq^2}{d^2}$, so the square of the charge is

   $$q^2 = \frac{Fd^2}{k} = \frac{(20 \text{ N})(0.06 \text{ m})^2}{9 \times 10^9 \text{ N}\cdot\text{m}^2/\text{C}^2} = 8.0 \times 10^{-12} \text{ C}^2.$$

   Taking the square root of this gives $q = 2.8 \times 10^{-6}$ C, or 2.8 microcoulombs.

5. a. $\Delta V = \dfrac{\text{energy}}{\text{charge}} = \dfrac{12 \text{ J}}{0.0001\text{C}} = 120{,}000$ volts.

   b. ΔV for twice the charge is $\dfrac{24 \text{ J}}{0.0002} =$ same 120 kV.

7. From $I_{\text{total}} = I_1 + I_2 + I_3 \ldots I_n$ a substitution of $I = \dfrac{V}{R}$ for each current gives $\dfrac{V}{R_{\text{eq}}} = \dfrac{V}{R_1} + \dfrac{V}{R_2} + \dfrac{V}{R_3} \ldots + \dfrac{V}{R_n}$. Dividing each term by V gives $\dfrac{1}{R_{\text{eq}}} = \dfrac{1}{R_1} + \dfrac{1}{R_2} + \dfrac{1}{R_3} \ldots + \dfrac{1}{R_n}$.

9. From current $= \dfrac{\text{voltage}}{\text{resistance}}$, resistance $= \dfrac{\text{voltage}}{\text{current}} = \dfrac{120\text{V}}{20\text{A}} = 6\ \Omega$.

11. Ohm's law can be stated $V = IR$. Then $P = IV = I(IR) = I^2R$.

13. First, 100 watts = 0.1 kilowatt. Second, there are 168 hours in one week (7 days × 24 h/day = 168 hours). So 168 hours × 0.1 kilowatt = 16.8 kilowatt-hours, which at 20 cents per kWh comes to $3.36.

15. Since current is charge per unit time, charge is current × time: $q = It = (9 \text{ A})(60 \text{ s}) = (9 \text{ C/s})(60 \text{ s}) = 540$ C. (Charges of this magnitude on the move are commonplace, but this quantity of charge accumulated in one place would be incredibly large.)

17. The resistance of the toaster is $R = V/I = (120 \text{ V})/(10 \text{ A}) = 12\ \Omega$. So when 108 V is applied, the current is $I = V/R = (108 \text{ V})/(12\ \Omega) = 9.0$ A and the power is $P = IV = (9.0 \text{ A})(108\text{V}) = 972$ W, only 81 percent of the normal power. (Can you see the reason for 81 percent? Current and voltage are both decreased by 10 percent, and $0.9 \times 0.9 = 0.81$.)

SOLUTIONS TO CHAPTER 11 EXERCISES

1. All iron materials are not magnetized because the tiny magnetic domains are most often oriented in random directions and cancel one another's effects.

3. Refrigerator magnets have narrow strips of alternating north and south poles. These magnets are strong enough to hold sheets of paper against a refrigerator door, but have a very short range because the north and south poles cancel a short distance from the magnetic surface.

5. An electron always experiences a force in an electric field because that force depends on nothing more than the field strength and the charge. But the force an electron experiences in a magnetic field depends on an added factor: velocity. If there is no motion of the electron through the magnetic field in which it is located, no magnetic force acts. Furthermore, if motion is along the magnetic field direction, and not at some angle to it, then no magnetic force acts also. Magnetic force, unlike electric force, depends on the velocity of the charge relative to the magnetic field.

7. Apply a small magnet to the door. If it sticks, your friend is wrong because aluminum is not magnetic. If it doesn't stick, your friend might be right (but not necessarily—there are lots of nonmagnetic materials).

9. The net force on a compass needle is zero because its north and south poles are pulled in opposite directions with equal forces in the Earth's magnetic field. When the needle is not aligned with the magnetic field of the Earth, then a pair of torques (relative to the center of the compass) is produced (Figure 11.4). This pair of equal torques, called a "couple," rotates the needle into alignment with the Earth's magnetic field.

11. Yes, for the compass aligns with the Earth's magnetic field, which extends from the magnetic pole in the Southern Hemisphere to the magnetic pole in the Northern Hemisphere.

13. Back to Newton's 3rd law! Both A and B are equally pulling on each other. If A pulls on B with 50 newtons, then B also pulls on A with 50 newtons. Period!

15. Newton's 3rd law again: Yes, the paper clip, as part of the interaction, certainly does exert a force on the magnet—just as much as the magnet pulls on it. The magnet and paper clip pull equally on each other to comprise the single interaction between them.

17. An electron has to be moving across lines of magnetic field in order to feel a magnetic force. So an electron at rest in a stationary magnetic field will feel no force to set it in motion. In an electric field, however, an electron will be accelerated whether or not it is already moving. (A combination of magnetic and electric fields is used in particle accelerators such as cyclotrons. The electric field accelerates the charged particle in its direction, and the magnetic field accelerates it perpendicular to its direction, causing it to follow a nearly circular path.)

19. Associated with every moving charged particle, electrons, protons, or whatever, is a magnetic field. Since a magnetic field is not unique to moving electrons, there is a magnetic field about moving protons as well. However, it differs in direction. The field lines about the proton beam circle in one direction whereas the field lines about an electron beam circle in the opposite direction. (Physicists use a "right-hand rule." If the right thumb points in the direction of motion of a positive particle, the curved fingers of that hand show the direction of the magnetic field. For negative particles, the left hand can be used.)

21. If the particles enter the field moving in the same direction and are deflected in opposite directions (say one left and one right), the charges must be of opposite sign.

23. Cosmic ray intensity at the Earth's surface would be greater when the Earth's magnetic field passed through a zero phase. Fossil evidence suggests the periods of no protective magnetic field may have been as effective in changing life forms as X-rays have been in the famous heredity studies of fruit flies.

25. Magnetic levitation will reduce surface friction to near zero. Then only air friction will remain. It can be made relatively small by aerodynamic design, but there is no way to eliminate it (short of sending vehicles through evacuated tunnels). Air friction gets rapidly larger as speed increases.

27. The two pulses are opposite in direction. When the wire enters the magnetic field between the poles of the magnet, a pulse of voltage is induced in the wire, which is indicated by movement of the galvanometer needle. When the wire is lifted a pulse in the opposite direction is induced, and the needle moves in the opposite direction.

29. A cyclist will coast farther if the lamp is disconnected from the generator. The energy that goes into lighting the lamp is taken from the bike's kinetic energy, so the bike slows down. The work saved by not lighting the lamp will be the extra "force ? distance" that lets the bike coast farther.

31. As in the previous answer, eddy currents induced in the metal change the magnetic field, which in turn changes the ac current in the coils and sets off an alarm.

33. Input and output are reversed for the two devices. When mechanical energy is put into the device and electricity is produced, we call it a generator. When electrical energy is put in and it spins and does mechanical work, we call it a motor. (While there are usually some practical differences in the designs of motors and generators, some devices are designed to operate either as motors or generators, depending only on whether the input is mechanical or electrical.)

35. In accord with electromagnetic induction, if the magnetic field alternates in the hole of the ring, an alternating voltage will be induced in the ring. Because the ring is metal, its relatively low resistance will result in a correspondingly high alternating current. This current is evident in the heating of the ring.

37. If the light bulb is connected to a wire loop that intercepts changing magnetic field lines from an electromagnet, voltage will be induced which can illuminate the bulb. Change is the key, so the electromagnet should be powered with ac.

39. The iron core increases the magnetic field of the primary coil. The greater field means a greater magnetic field change in the primary, and a greater voltage induced in the secondary. The iron core in the secondary further increases the changing magnetic field through the secondary and further increases the secondary voltage. Furthermore, the core guides more magnetic field lines from the primary to the secondary. The effect of an iron core in the coils is the induction of appreciably more voltage in the secondary.

41. When the secondary voltage is twice the primary voltage and the secondary acts as a source of voltage for a resistive "load," the secondary current is half the value of current in the primary. This is in accord with energy conservation, or since the time intervals for each are the same, "power conservation.." Power input = power output; or $(\text{current} \times \text{voltage})_{\text{primary}} = (\text{current} \times \text{voltage})_{\text{secondary}}$: with numerical values, $(1 \times \text{V})_{\text{primary}} = (\frac{1}{2} \times 2\text{V})_{\text{secondary}}$. (The simple rule power = current × voltage is strictly valid only for dc circuits and accircuits where current and voltage oscillate in phase. When voltage and current are out of phase, which can occur in a transformer, the net power is less than the product current × voltage. Voltage and current are then not "working together." When the secondary of a transformer is open, for example, connected to nothing, current and voltage in both the primary and the secondary are completely out of phase—that is, one is maximum when the other is zero—and no net power is delivered even though neither voltage nor current is zero.)

43. The voltage impressed across the lamp is 120 V and the current through it is 0.1 A. We see that the first transformer steps the voltage down to 12 V and the second one steps it back up to 120 V. The current in the secondary of the second transformer, which is the same as the current in the bulb, is one-tenth of the current in the primary, or 0.1 A.

45. By symmetry, the voltage and current for both primary and secondary are the same. So 12 V is impressed on the meter, with a current of 1 A ac.

47. The bar magnet induces current loops in the surrounding copper as it falls. The current loops produce magnetic fields that tend to repel the magnet as it approaches and attract it as it leaves, exerting a vertical upward force on it, opposite to gravity. The faster the magnet falls, the stronger is this upward force. At some speed, it will match gravity and the magnet will be at terminal speed. From an energy point of view, some of the gravitational potential energy is being transformed to heat in the copper pipe. The plastic pipe, on the other hand, is an insulator. So no current and therefore no magnetic field are induced to oppose the motion of the falling magnet.

49. Agree with your friend, for light is electromagnetic radiation having a frequency that matches the frequency to which our eyes are sensitive.

SOLUTIONS TO CHAPTER 11 PROBLEMS

1. (120 V)/(500 turns) = (12 V)/x, so x = 50 turns.

3. From the transformer relationship,

$$\frac{\text{primary voltage}}{\text{primary turns}} = \frac{\text{secondary voltage}}{\text{secondary turns}}, \frac{\text{primary voltage}}{\text{secondary voltage}} = \frac{\text{primary turns}}{\text{secondary turns}}; \frac{120\text{ V}}{24\text{ V}} = \frac{5}{1}.$$

So there are 5 times as many primary turns as secondary turns.

5. The transformer steps up voltage by a factor 36/9 = 4. Therefore a 12-V input will be stepped up to 4 × 12 V = 48 V.

7. From $\frac{\text{primary voltage}}{\text{primary turns}} = \frac{\text{secondary voltage}}{\text{secondary turns}}$, simple rearrangement gives

$$\frac{\text{primary voltage}}{\text{secondary voltage}} = \frac{\text{primary turns}}{\text{secondary turns}} = \frac{120\text{ V}}{12000\text{ V}} = \frac{1}{100}.$$

SOLUTIONS TO CHAPTER 12 EXERCISES

1. Something that vibrates.

3. As you dip your fingers more frequently into still water, the waves you produce will be of a higher frequency (we see the relationship between "how frequently" and "frequency"). The crests of the higher-frequency waves will be closer together—their wavelengths will be shorter.

5. Shake the garden hose to-and-fro in a direction perpendicular to the hose to produce a sine-like curve.

7. The fact that gas can be heard escaping from a gas tap before it is smelled indicates that the pulses of molecular collisions (the sound) travel more quickly than the molecules migrate. (There are three speeds to consider: (1) the average speed of the molecules themselves, as evidenced by temperature—quite fast, (2) the speed of the pulse produced as they collide—about $\frac{3}{4}$ the speed of the molecules themselves, and (3) the very much slower speed of molecular migration.)

9. The carrier frequency of electromagnetic waves emitted by the radio station is 101.1 MHz.

11. The wavelength of the electromagnetic wave will be much longer because of its greater speed. You can see this from the equation speed = wavelength × frequency, so for the same frequency greater speed means greater wavelength. Or you can think of the fact that in the time of one period—the same for both waves—each wave moves a distance equal to one wavelength, which will be greater for the faster wave.

13. The electronic starting gun does not rely on the speed of sound through air, which favors closer runners, but gets the starting signal to all runners simultaneously.

15. Because snow is a good absorber of sound, it reflects little sound—which is responsible for the quietness.

17. The Moon is described as a silent planet because it has no atmosphere to transmit sounds.

19. If the speed of sound were different for different frequencies, say, faster for higher frequencies, then the farther a listener is from the music source, the more jumbled the sound would be. In that case, higher-frequency notes would reach the ear of the listener first. The fact that this jumbling doesn't occur is evidence that sound of all frequencies travel at the same speed. (Be glad this is so, particularly if you sit far from the stage, or if you like outdoor concerts.)

21. Sound travels faster in warm air because the air molecules that compose warm air themselves travel faster and therefore don't take as long before they bump into each other. This lesser time for the molecules to bump against one another results in a faster speed of sound.

23. An echo is weaker than the original sound because sound spreads and is therefore less intense with distance. If you are at the source, the echo will sound as if it originated on the other side of the wall from which it reflects (just as your image in a mirror appears to come from behind the glass). Also contributing to its weakness is the wall, which likely is not a perfect reflector.

25. The rule is correct: This is because the speed of sound in air (340 m/s) can be rounded off to $\frac{1}{3}$ km/s. Then, from distance = speed × time, we have distance = ($\frac{1}{3}$) km/s × (number of seconds). Note that the time in seconds divided by 3 gives the same value.

27. Marchers at the end of a long parade will be out of step with marchers nearer the band because time is required for the sound of the band to reach the marchers at the end of a parade. They will step to the delayed beat they hear.

29. A harp produces relatively softer sounds than a piano because its sounding board is smaller and lighter.

31. The lower strings are resonating with the upper strings.

33. Waves of the same frequency can interfere destructively or constructively, depending on their relative phase, but to *alternate* between constructive and destructive interference, two waves have to have different frequencies. Beats arise from such alternation between constructive and destructive interference.

35. The piano tuner should loosen the piano string. When 3 beats per second is first heard, the tuner knows he was 3 hertz off the correct frequency. But this could be either 3 hertz above or 3 hertz below. When he tightened the string and increased its frequency, a lower beat frequency would have told him he was on the right track. But the greater beat frequency told him he should have been loosening the string. When there is no beat frequency, the frequencies match.

37. No, the effects of shortened waves and stretched waves would cancel one another.

39. The Doppler shifts show that one side approaches while the other side recedes, evidence that the Sun is spinning.

41. The conical angle of a shock wave becomes narrower with greater speeds. We see this in the sketches:

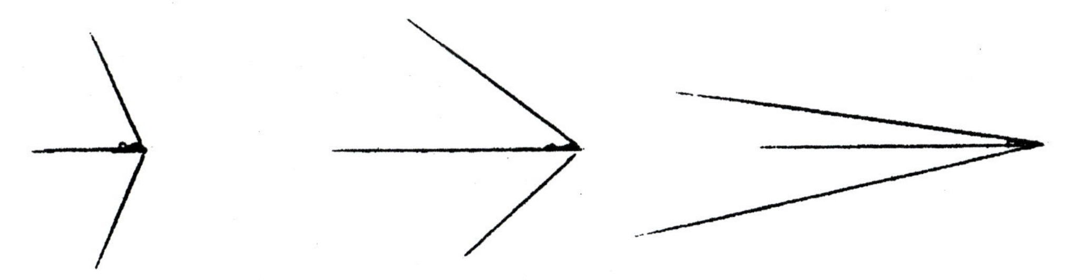

43. A shock wave and the resulting sonic boom are produced whenever an aircraft is supersonic, whether or not the aircraft has just become supersonic or has been supersonic for hours. It is a popular misconception that sonic booms are principally produced at the moment an aircraft becomes supersonic. This is akin to saying that a boat produces a bow wave at the moment it exceeds the wave-speed of water. It begins to produce a bow wave at this crucial moment, but if it moved no faster, the overlapping pattern of waves would not extend very far from the bow. Likewise with an aircraft. Both the boat and the aircraft must appreciably exceed wave speed to produce an ample bow and shock wave.

45. Resonance.

SOLUTIONS TO CHAPTER 12 PROBLEMS

1. f = (72 beats)/(60 s) = 1.2 Hz.; $T = 1/f = 1/(1.2\ \text{s}^{-1}) = 0.83$ s.

3. From $v = \lambda f$, $\lambda = v/f = (3.00 \times 10^8\ \text{m/s})/(2.45 \times 10^9\ \text{Hz}) = 0.122\ \text{m} = 12.2\ \text{cm}$

5. The ocean floor is 4590 meters down. The 6-second time delay means that the sound reached the bottom in 3 seconds. Distance = speed × time = 1530 m/s × 3 s = 4590 m.

7. Speed = distance traveled/time taken = (2 × 85 m)/0.5 s = 170 m/0.5 s = 340 m/s.

9. (a) Period = 1/frequency = 1/(256 Hz) = 0.00391 s, or 3.91 ms.

   (b) Speed = wavelength × frequency, so wavelength = speed/frequency = (340 m/s)/(256 Hz) = 1.33 m.

11. There are 3 possible beat frequencies, 2 Hz, 3 Hz, and 5 Hz. The beats consist of differences in fork frequencies: 261 − 259 = 2 Hz; 261 − 256 = 5 Hz; 259 − 256 = 3 Hz.

SOLUTIONS TO CHAPTER 13 EXERCISES

1. The fundamental source of electromagnetic radiation is oscillating electric charges, which emit oscillating electric and magnetic fields.

3. Ultraviolet has shorter waves than infrared. Correspondingly, ultraviolet also has the higher frequencies.

5. Sound requires a physical medium in which to travel. Light does not.

7. Radio waves and light are both electromagnetic, transverse, move at the speed of light, and are created and absorbed by oscillating charge. They differ in their frequency and wavelength and in the type of oscillating charge that creates and absorbs them.

9. The greater number of interactions per distance tends to slow the light and result is a smaller average speed.

11. Clouds are transparent to ultraviolet light, which is why clouds offer no protection from sunburn. Glass, however, is opaque to ultraviolet light, and will therefore shield you from sunburn.

13. The customer is being reasonable in requesting to see the colors in the daylight. Under fluorescent lighting, with its predominant higher frequencies, the bluer colors rather than the redder colors will be accented. Colors will appear quite different in sunlight.

15. We see not only yellow green, but also red and blue. All together, they mix to produce the white light we see.

17. If the yellow clothes of stage performers are illuminated with a complementary blue light, they will appear black.

19. Red and green produce yellow; red and blue produce magenta; red, blue, and green produce white.

21. The red shirt in the photo is seen as cyan in the negative, and the green shirt appears magenta—the complementary colors. When white light shines through the negative, red is transmitted where cyan is absorbed. Likewise, green is transmitted where magenta is absorbed.

23. Blue illumination produces black. A yellow banana reflects yellow and the adjacent colors, orange and green, so when illuminated with any of these colors it reflects that color and appears that color. A banana does not reflect blue, which is too far from yellow in the spectrum, so when illuminated with blue it appears black.

25. You see the complementary colors due to retina fatigue. The blue will appear yellow, the red cyan, and the white black. Try it and see!

27. At higher altitudes, there are fewer molecules above you and therefore less scattering of sunlight. This results in a darker sky. The extreme, no molecules at all, results in a black sky, as on the Moon.

29. Clouds are composed of atoms, molecules, and particles of a variety of sizes. So not only are high-frequency colors scattered from clouds, but middle and low frequencies as well. A combination of all the scattered colors produces white.

31. If we assume that Jupiter has an atmosphere which is similar to that of the Earth in terms of transparency, then the Sun would appear to be a deep reddish orange, just as it would when sunlight grazes 1000 kilometers of the Earth's atmosphere for a sunset from an elevated position. Interestingly enough, there is a thick cloud cover in Jupiter's atmosphere that blocks all sunlight from reaching its "surface." And it doesn't even have a solid surface! Your grandchildren may visit one of Jupiter's moons, but will not "land" on Jupiter itself—not intentionally, anyway. (Incidentally, there are only $3\frac{1}{3}$ planets with "solid" surfaces: Mercury, Venus, Mars, and $\frac{1}{3}$ of Earth! Dwarf planet Pluto also has a solid surface.)

33. The wavelengths of AM radio waves are hundreds of meters, much larger than the size of buildings, so they are easily diffracted around buildings. FM wavelengths are a few meters, borderline for diffraction around buildings. Light, with wavelengths a tiny fraction of a centimeter, show no appreciable diffraction around buildings.

35. Young's interference experiment produces a clearer fringe pattern with slits than with pinholes because the pattern is of parallel straight-line-shaped fringes rather than the fringes of overlapping circles. Circles overlap in relatively smaller segments than the broader overlap of parallel straight lines. Also, the slits allow more light to get through; the pattern with pinholes is dimmer.

37. There must be two reflecting surfaces for interference by reflection to occur. Light reflecting from the gasoline surface interferes with light reflected from the water surface beneath.

39. Blue, the complementary color. The blue is white minus the yellow light that is seen above.

SOLUTIONS TO CHAPTER 13 PROBLEMS

1. $f = c/\lambda = (3.00 \times 10^8 \text{ m/s})/(670 \times 10^{-9} \text{ m}) = 4.48 \times 10^{14} \text{ Hz}$.

3. $v_{Hydra}/c = (6.0 \times 10^7 \text{ m/s})/(3.0 \times 10^8 \text{ m/s}) = 0.2$, or 20% the speed of light.

5. As in the previous problem, $t = \dfrac{d}{v} = \dfrac{4.2 \times 10^{16} \text{ m}}{3 \times 10^8 \text{ m/s}} = 1.4 \times 10^8 \text{ s}$.

   Converting to years by dimensional analysis,

   $$1.4 \times 10^8 \text{ s} \times \frac{1 \text{ h}}{3600 \text{ s}} \times \frac{1 \text{ day}}{24 \text{ h}} \times \frac{1 \text{ yr}}{365 \text{ day}} = 4.4 \text{ yr}.$$

7. From $v = \lambda f$, $f = v/\lambda = (3.00 \times 10^8 \text{ m/s})/(360 \times 10^{-9} \text{ m}) = 8.33 \times 10^{14} \text{ Hz}$.

SOLUTIONS TO CHAPTER 14 EXERCISES

1. Only light from card number 2 reaches her eye.

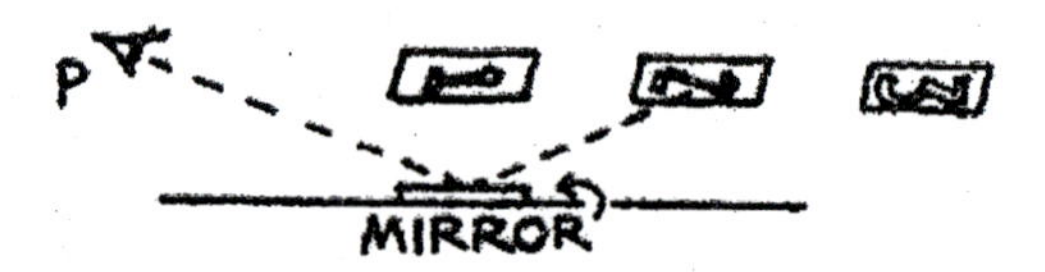

3. Light that takes a path from point A to point B will take the same reverse path in going from point B to point A, even if reflection or refraction is involved. So if you can't see the driver, the driver can't see you. (This independence of direction along light's path is the "principle of reciprocity.")

5. When you wave your right hand, image of the waving hand is still on your right, just as your head is still up and your feet still down. Neither left and right nor up and down are inverted by the mirror—but *front and back* are, as the author's sister Marjorie illustrates in Figure 14.3. (Consider three axes at right angles to each other, the standard coordinate system; horizontal x, vertical y, and perpendicular-to-the-mirror z. The only axis to be inverted is z, where the image is $-z$.)

7. When the source of glare is somewhat above the horizon, a vertical window will reflect it to people in front of the window. By tipping the window inward at the bottom, glare is reflected downward rather than into the eyes of passersby. (Note the similarity of this exercise and the previous one.)

9. The pebbly uneven surface is easier to see. Light reflected back from your headlights is what lets you see the road. The mirror-smooth surface might reflect more light, but it would reflect it forward, not backward, so it wouldn't help you see.

11. First of all, the reflected view of a scene is different than an inverted view of the scene, for the reflected view is seen from lower down. Just as a view of a bridge may not show its underside where the reflection does, so it is with the bird. The view reflected in water is the inverted view you would see if your eye were positioned as far beneath the water level as your eye is above it (and there were no refraction). Then your line of sight would intersect the water surface where reflection occurs. Put a mirror on the floor between you and a distant table. If you are standing, your view of the table is of the top. But the reflected view shows the table's bottom. Clearly, the two views are not simply inversions of each other. Take notice of this whenever you look at reflections (and of paintings of reflections—it's surprising how many artists are not aware of this).

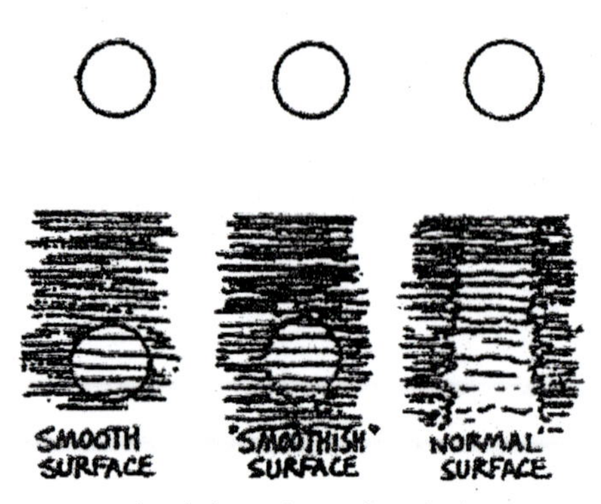

13. The half-height mirror works at any distance, as shown in the sketch. This is because if you move closer, your image moves closer as well. If you move farther away, your image does the same. Many people must actually try this before they believe it. The confusion arises because people know that they can see whole distant buildings

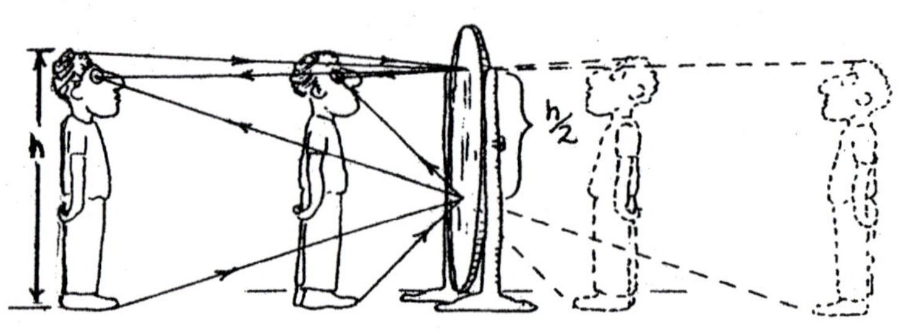

or even mountain ranges in a hand-held pocket mirror. Even then, the distance the object is from the mirror is the same as the distance of the virtual image on the other side of the mirror. You can see all of a distant person in your mirror, but the distant person cannot see all of herself in your mirror.

15. The wiped area will be half as tall as your face.

17. If the water were perfectly smooth, you would see a mirror image of the round Sun or Moon, an ellipse on the surface of the water. If the water were slightly rough, the image would be wavy. If the water were a bit more rough, little glimmers of portions of the Sun or Moon would be seen above and below the main image. This is because the water waves act like tiny parallel mirrors. For small waves only light near the main image reaches you. But as the water becomes choppier, there is a greater variety of mirror facets that are oriented to reflect sunlight or moonlight into your eye. The facets do not radically depart from an average flatness with the otherwise smooth water surface, so the reflected Sun or Moon is smeared into a long vertical streak. For still rougher water there are facets off to the side of the vertical streak that are tilted enough for sun or moon light to be reflected to you, and the vertical streak is wider.

19.

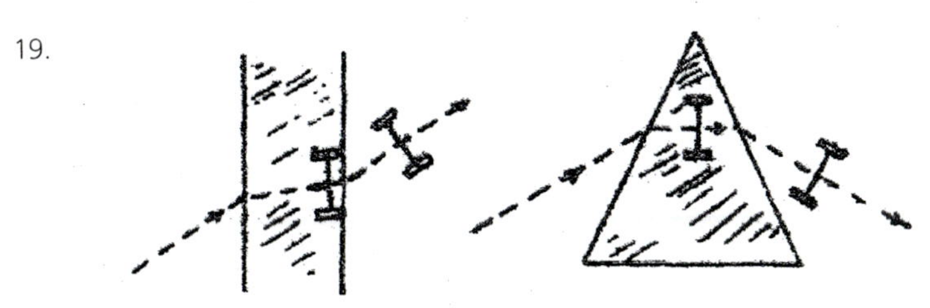

21. During a lunar eclipse the Moon is not totally dark, even though it is in the Earth's shadow. This is because the atmosphere of the Earth acts as a converging lens that refracts light into the Earth's shadow. It is the low frequencies that pass more easily through the long grazing path through the Earth's atmosphere to be refracted finally onto the Moon. Hence its reddish color—the refraction of the whole world's sunups and sunsets.

23. You would throw the spear below the apparent position of the fish, because the effect of refraction is to make the fish appear closer to the surface than it really is. But in zapping a fish with a laser, make no corrections and simply aim directly at the fish. This is because the light from the fish you see has been refracted in getting to you, and the laser light will refract along the same path in getting to the fish. A slight correction may be necessary, depending on the colors of the laser beam and the fish—see the next exercise.

25. A fish sees the sky (as well as some reflection from the bottom) when it looks upward at 45°, for the critical angle is 48° for water. If it looks at and beyond 48° it sees only a reflection of the bottom.

27. Total internal reflection occurs only for light rays that would gain speed in crossing the boundary they encounter. For light in air encountering a water surface, there is no total reflection. You can see this by sketching rays that go from water to air, and noting that light can travel in the other direction along all of these rays.

29. We cannot see a rainbow "off to the side," for a rainbow is not a tangible thing "out there." Colors are refracted in infinite directions and fill the sky. The only colors we see that aren't washed out by others are those that are along the conical angles between 40° and 42° to the Sun-antisun axis. To understand this, consider a paper-cone cup with a hole cut at the bottom. You can view the circular rim of the cone as an ellipse when you look at it from a near side view. But if you view the rim only with your eye at the apex of the cone, through the hole, you can see it only as a circle. That's the way we view a rainbow. Our eye is at the apex of a cone, the axis of which is the Sun-antisun axis, and the "rim" of which is the bow. From every vantage point, the bow forms part (or all) of a circle.

31. When the Sun is high in the sky and people on the airplane are looking down toward a cloud opposite to the direction of the Sun, they may see a rainbow that makes a complete circle. The shadow of the airplane will appear in the center of the circular bow. This is because the airplane is directly between the Sun and the drops or rain cloud producing the bow.

33. A projecting lens with chromatic aberration casts a rainbow-colored fringe around a spot of white light. The reason these colors don't appear inside the spot is because they overlap to form white. Only at the edges, which act as a circular prism, do they not overlap.

35. The average intensity of sunlight at the bottom is the same whether the water is moving or is still. Light that misses one part of the bottom of the pool reaches another part. Every dark region is balanced by a bright region—"conservation of light."

37. Normal sight depends on the amount of refraction that occurs for light traveling from air to the eye. The speed change ensures normal vision. But if the speed change is from water to eye, then light will be refracted less and an unclear image will result. A swimmer uses goggles to make sure that the light travels from air to eye, even if underwater.

39. If light had the same average speed in glass lenses that it has in air, no refraction of light would occur in lenses, and no magnification would occur. Magnification depends on refraction, which in turn depends on speed changes.

41. Your image is twice as far from the camera as the mirror frame. So although you can adjust the focus of your camera to clearly photograph your image in a mirror, and you can readjust the focus to clearly photograph the mirror frame, you cannot in the same photograph focus on both your image and the mirror frame. This is because they are at different distances from the camera.

43. Moon maps are upside-down views of the Moon to coincide with the upside-down image that Moon watchers see in a telescope.

45. The displays are polarized and depending on the angle of viewing, may not be seen at all. This is a serious liability of Polaroid sunglasses.

47. If the sheet is aligned with the polarization of the light, all the light gets through. If it is aligned perpendicular to the polarization of the light, none gets through. At any other angle, some of the light gets through because the polarized light can be "resolved" (like a vector) into components parallel and perpendicular to the alignment of the sheet.

49. You can determine that the sky is partially polarized by rotating a single sheet of Polaroid in front of your eye while viewing the sky. You'll notice the sky darken when the axis of the Polaroid is perpendicular to the polarization of the skylight.

SOLUTIONS TO CHAPTER 14 PROBLEMS

1. Set your focus for 6m, for your image will be as far in back of the mirror as you are in front.

3. 4 m/s. You and your image are both walking at 2 m/s.

5. When a mirror is rotated, its normal rotates also. See in the sketch how if the normal rotates 10°, the beam reflects at twice this, 20°. This is one reason that mirrors are used to detect delicate movements in instruments such as galvanometers. The more important reason is the amplification of displacement by having the beam arrive at a scale some distance away.

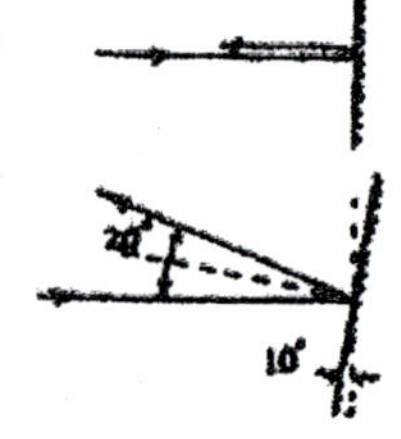

7. Use ratios: (1440 min)/(360 deg) = (unknown time)/(0.53 deg). So the unknown time is 0.53 × 1440/360 = 2.1 minutes. So the Sun moves a solar diameter across the sky every 2.1 minutes. At sunset, time is somewhat extended, depending on the extent of refraction. Then the disk of the setting Sun disappears over the horizon in a little longer than 2.1 minutes.

SOLUTIONS TO CHAPTER 15 EXERCISES

1. Higher-frequency ultraviolet light has more energy per photon.

3. Higher-frequency green light has more energy per photon.

5. Doubling the wavelength of light halves its frequency. Light of half frequency has half the energy per photon. Think in terms of the equation $c = f\lambda$. Since the speed of light c is constant, λ is inversely proportional to f.

7. The energy of red light is too low per photon to trigger the chemical reaction in the photographic crystals. Very bright light simply means more photons that are unable to trigger a reaction. Blue light, on the other hand, has sufficient energy per photon to trigger a reaction. Very dim blue light triggers fewer reactions only because there are fewer photons involved. It is safer to have bright red light than dim blue light.

9. The kinetic energy of ejected electrons depends on the frequency of the illuminating light. With sufficiently high frequency, the number of electrons ejected in the photoelectric effect is determined by the number of photons incident upon the metal. So whether or not ejection occurs depends on frequency, and how many electrons are ejected depends on the brightness of the sufficiently high-frequency light.

11. Ultraviolet photons are more energetic.

13. Particle nature.

15. The moving star will show a Doppler shift. Since the star is receding, it will be a redshift (to lower frequency and longer wavelength).

17. Atomic excitation occurs in solids, liquids, and gases. Because atoms in a solid are close packed, radiation from them (and liquids) is smeared into a broad distribution to produce a continuous spectrum, whereas radiation from widely-spaced atoms in a gas is in separate bunches that produce discrete "lines" when diffracted by a grating.

19. The same energy is needed both ways.

21. The acronym says it: *m*icrowave *a*mplification by *s*timulated *e*mission of *r*adiation.

23. Its energy is very concentrated in comparison with that of a lamp.

25. Your friend's assertion violates the law of energy conservation. A laser or any device cannot put out more energy than is put into it. Power, on the other hand, is another story, as is treated in the following exercise.

27. The metal is glowing at all temperatures, whether we can see the glow or not. As its temperature is increased, the glow reaches the visible part of the spectrum and is visible to human eyes. Light of the lowest energy per photon is red. So the heated metal passes from infra-red (which we can't see) to visible red. It is red hot.

29. Star's relative temperatures—lowish for reddish; middish for whitish; and hottish for bluish.

31. Six transitions are possible, as shown. The highest-frequency transition is from quantum level 4 to level 1. The lowest-frequency transition is from quantum level 4 to level 3.

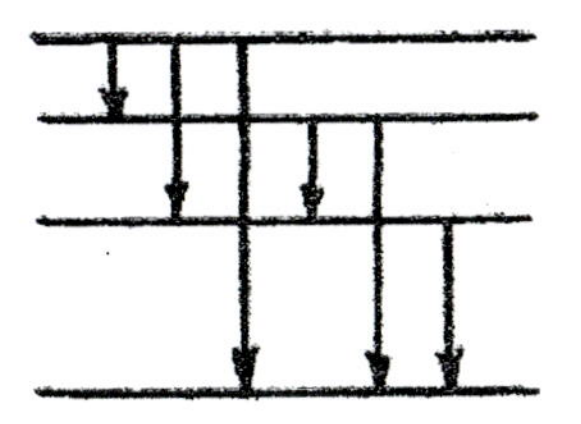

33. Photons behave like a wave when in transit, and like particles when they interact with a surface.

35. The electron microscope.

37. The more massive proton has more momentum, while the electron with its smaller momentum has the longer wavelength.

39. By de Broglie's formula, as velocity increases, momentum increases, so wavelength decreases.

41. The principal advantage of an electron microscope over an optical microscope is its ability to see things that are too small for viewing with an optical microscope. This is because an object cannot be discerned if it is smaller than the wavelength of the illuminating beam. An electron beam has a wavelength that is typically a thousand times shorter than the wavelength of visible light, which means it can discern particles a thousand times smaller than those barely seen with an optical microscope.

43. If somebody looks at an electron on the tip of your nose with an electron beam or a light beam, then its motion as well as that of surrounding electrons will be altered. We take the view here that passively looking at light after it has reflected from an object does not alter the electrons in the object. We distinguish between passive observation and probing. The uncertainty principle applies to probing, not to passive observation. (This view, however, is not held by some physicists who assert any measure, passive or probing, alters that being measured at the quantum level. These physicists argue that passive observation provides knowledge, and that without this knowledge, the electron might be doing something else or might be doing a mixture, a superposition, of other things.)

45. Heisenberg's uncertainty principle applies *only* to quantum phenomena. However, it serves as a popular metaphor for the macro domain. Just as the way we measure affects what's being measured, the way we phrase a question often influences the answer we get. So to various extents we alter that which we wish to measure in a public opinion survey. Although there are countless examples of altering circumstances by measuring them, the uncertainty principle has meaning only in the sub-microscopic world.

47. The question is absurd, with the implication that eradicating butterflies will prevent tornadoes. If a butterfly can, in principle, cause a tornado, so can a billion other things. Eradicating butterflies will leave the other 999,999,999 causes untouched. Besides, a butterfly is as likely to *prevent* a tornado as to cause one.

49. What waves is the probability amplitude.

51. Some of the behavior of light, such as interference and diffraction, can only be satisfactorily explained in terms of the wave model; while behavior such as the photoelectric effect can be satisfactorily explained only in terms of the particle model. Hence we say that light has both wave and particle properties.

53. Bohr's correspondence principle says that quantum mechanics must overlap and agree with classical mechanics in the domain where classical mechanics has been shown to be valid.

55. This is perhaps the extreme in altering that which is being measured by the process of measuring itself, as well as an extreme case of criminal stupidity and academic arrogance. The bristlecone pine, Old Methuselah, was the oldest known living thing in the world.

SOLUTIONS TO CHAPTER 15 PROBLEMS

1. (a) The B-to-A transition has twice the energy and twice the frequency of the C-to-B transition. Therefore it will have half the wavelength, or 300 nm. Reasoning: Since $c = f\lambda$, $\lambda = c/f$. Wavelength is inversely proportional to frequency. Twice the frequency means half the wavelength.
   (b) The C-to-A transition has three times the energy and three times the frequency of the C-to-B transition. Therefore it will have one-third the wavelength, or 200 nm.

3. De Broglie wavelength = Planck's constant/momentum, so we need the electron's momentum. It is $p = mv = (9.1 \times 10^{-31}\ \text{kg})(3.0 \times 10^{7}\ \text{m/s}) = 2.7 \times 10^{-23}\ \text{kg m/s}$. The de Broglie wavelength is then $\lambda = h/p = (6.6 \times 10^{-34})/(2.7 \times 10^{-23}) = 2.4 \times 10^{-11}\ \text{m}$.

SOLUTIONS TO CHAPTER 16 EXERCISES

1. Radioactivity is a part of nature, going back to the beginning of time.

3. A radioactive sample is always a little warmer than its surroundings because the radiating alpha or beta particles impart internal energy to the atoms of the sample. (Interestingly enough, the heat energy of the Earth originates with radioactive decay of the Earth's core and surrounding material.)

5. Alpha and beta rays are deflected in opposite directions in a magnetic field because they are oppositely charged—alpha are positive and beta negative. Gamma rays have no electric charge and are therefore undeflected.

7. The alpha particle has twice the charge, but almost 8000 times the inertia (since each of the four nucleons has nearly 2000 times the mass of an electron). Hence it bends very little compared to the much less massive beta particles (electrons). Gamma rays carry no electric charge and so are not affected by an electric field.

9. Gamma radiation produces not only the least change in mass and atomic numbers, but produces no change in mass number, atomic number, or electric charge. Both alpha and beta radiation do produce these changes.

11. The proton "bullets" need enough momentum to overcome the electric force of repulsion they experience once they get close to the atomic nucleus.

13. The strong nuclear force holds the nucleons of the nucleus together while the electric force pushes these nucleons apart.

15. No, it will not be entirely gone. Rather, after 1 day one-half of the sample will remain while after 2 days, one-fourth of the original sample will remain.

17. When radium (A = 88) emits an alpha particle, its atomic number reduces by 2 and becomes the new element radon (A = 86). The resulting atomic mass is reduced by 4. If the radium were of the most common isotope 226, then the radon iso¬tope would have atomic mass number 222.

19. Deuterium has 1 proton and 1 neutron; carbon has 6 protons and 6 neutrons; iron has 26 protons and 30 neutrons; gold has 79 protons and 118 neutrons; strontium has 38 protons and 52 neutrons; uranium has 92 protons and 146 neutrons.

21. The elements below uranium in atomic number with short half-lives exist as the product of the radioactive decay of uranium. As long as uranium is decaying, their existence is assured.

23. Although there is significantly more radioactivity in a nuclear power plant than in a coal-fired power plant, the absence of shielding for coal plants results in more radioactivity in the environment of a typical coal plant than in the environment of a typical nuclear plant. All nukes are shielded; coal plants are not.

25. Film badges monitor gamma radiation, which is very high-frequency X-rays. Like photographic film, the greater the exposure, the darker the film upon processing.

27. There are no fast-flying subatomic particles in gamma rays that might collide with the nuclei of the atoms within the food. Transformations within the nuclei of the atoms of the food, therefore, are not possible. Rather, the gamma rays are lethal to any living tissues within the food, such as those of pathogens. The gamma rays kill these pathogens, which helps to protect us from dangerous diseases such as botulism.

29. Stone tablets cannot be dated by the carbon dating technique. Nonliving stone does not ingest carbon and transform that carbon by radioactive decay. Carbon dating pertains to organic material.

31. A neutron makes a better "bullet" for penetrating atomic nuclei because it has no electric charge and is therefore not deflected from its path by electrical interactions, nor is it electrically repelled by an atomic nucleus.

33. Because plutonium triggers more reactions per atom, a smaller mass will produce the same neutron flux as a somewhat larger mass of uranium. So plutonium has a smaller critical mass than a similar shape of uranium.

35. Plutonium has a short half-life (24,360 years), so any plutonium initially in the Earth's crust has long since decayed. The same is true for any heavier elements with even shorter half-lives from which plutonium might originate. Trace amounts of plutonium can occur naturally in U-238 concentrations, however, as a result of neutron capture, where U-238 becomes U-239 and after beta emission becomes Np-239, and further beta emission to Pu-239. (There are elements in the Earth's crust with half-lives even shorter than plutonium's, but these are the products of uranium decay.)

37. A nucleus undergoes fission because the electric force of repulsion overcomes the strong nuclear force of attraction. This electric force of repulsion is of the very same nature as static electricity. So, in a way, your friend's claim that the explosive power of a nuclear bomb is due to static electricity is valid.

39. If the difference in mass for changes in the atomic nucleus increased tenfold (from 0.1% to 1.0%), the energy release from such reactions would increase tenfold as well.

41. Both convert mass to energy.

43. Energy would be released by the fissioning of gold and from the fusion of carbon, but by neither fission nor fusion for iron. Neither fission nor fusion will result in a decrease of mass for iron nucleons.

45. The radioactive decay of radioactive elements found under the Earth's surface warms the insides of the Earth and responsible for the molten lava that spews from volcanoes. The thermonuclear fusion of our Sun is responsible for warming everything on our planet's surface exposed to the Sun.

47. Such speculation could fill volumes. The energy and material abundance that is the expected outcome of a fusion age will likely prompt several fundamental changes. Obvious changes would occur in the fields of economics and commerce, which would be geared to relative abundance rather than scarcity. Already our present price system, which is geared to and in many ways dependent upon scarcity, often malfunctions in an environment of abundance. Hence we see instances where scarcity is created to keep the economic system functioning. Changes at the international level will likely be worldwide economic reform, and at the personal level a re-evaluation of the idea that scarcity ought to be the basis of value. A fusion age will likely see changes that will touch every facet of our way of life.

49. To create an abundant supply of molecular hydrogen will require an abundant source of energy, such as fusion power.

SOLUTIONS TO CHAPTER 16 PROBLEMS

1. At 2 meters the reading will be about 90 counts per minute. At 3 meters the reading will be about 40 counts per minute.

3. At 3:00 P.M. there are 0.125 grams left. At 6:00 P.M. there are 0.0156 grams left. At 10:00 P.M. there are 0.000977 grams left.

5. It will take four half-lives to decrease to one-sixteenth the original amount. Four half-lives of cesium-137 corresponds to 120 years.

7. Six percent corresponds to about one-sixteenth, which means that the carbon-14 has undergone about four half-lives. Four half-lives of carbon-14 equals 5730 years times 4 equals 22,920 years, about 23,000 years old.

9. $(2 \times 10^{10}\ \text{kcal}) = (1\ \text{kcal}/1°\text{C}\ 1\ \text{kg})(\text{mass})(50°\text{C})$

SOLUTIONS TO APPENDIX D QUESTIONS TO PONDER

1. The pond was half covered on the 29th day, and one-quarter covered on the 28th day.

3. At a steady inflation rate of 7%, the doubling time is 70/7% = 10 years; so every 10 years the prices of these items will double. This means the $20 theater ticket in 10 years will cost $40, in 20 years will cost $80, in 30 years will cost $160, in 40 years will cost $320, and in 50 years will cost $640. The $200 suit of clothes will similarly jump each decade to $400, $800, $1600, $3,200, and $6,400. For a $20,000 car the decade jumps will be $40,000, $80,000, $160,000, $320,000, and $640,000. For a $200,000 home, the decade jumps in price are $400,000, $800,000, $1,600,000, $3,200,000, and $6,400,000! Inflation often increases earnings more than prices, so we'll be able to pay for these things—and more.

5. All things being equal, doubling of food for twice the number of people simply means that twice as many people will be eating, and twice as many will be starving as are starving now!

7. On the 30th day your wages will be $5,368,709.12, which is one penny more than the $5,368,709.11 total from all the preceding days.